Workplace Strategies and Facilities Management

To Cliff Roberts

Workplace Strategies and Facilities Management

**Edited by Rick Best, Craig Langston
and Gerard de Valence**

OXFORD   AMSTERDAM   BOSTON   LONDON   NEW YORK   PARIS
SAN DIEGO   SAN FRANCISCO   SINGAPORE   SYDNEY   TOKYO

Butterworth-Heinemann
An imprint of Elsevier Science Limited
Linacre House, Jordan Hill, Oxford OX2 8DP
200 Wheeler Road, Burlington MA 01803

First published 2003

British Library Cataloguing in Publication Data
Workplace strategies and facilities management
  1. Facility management   2. Building – Cost control
  I. Best, Rick   II. Langston, Craig   III. De Valence, Gerard
  658.2

Library of Congress Cataloging in Publication Data
A catalog record for this book is available from the Library of Congress

ISBN 0 7506 51504

For information on all Butterworth-Heinemann publications
visit our website at www.bh.com

Composition by Genesis Typesetting Limited, Rochester, Kent
Printed and bound in Great Britain

Contents

Acknowledgements

We would like to thank some people who have helped in various ways in the production of this book.

Jackie Holding, who helped so much with the second volume and stayed with us through the long process of producing this volume, and Alex Hollingsworth, who believed us when we kept saying that it would, eventually, be finished.

Associate Professor Steve Harfield at the Faculty Design Architecture and Building at University of Technology Sydney for his assistance with the whole project.

Sally Beech who worked above and beyond the call of duty in preparing the many charts and diagrams.

All the contributors who accepted our criticism and suggestions with good humour, and kept their faith in our ability to complete the job.

List of Contributors

Editors

Rick Best – Senior Lecturer at the University of Technology Sydney (UTS). He has degrees in architecture and quantity surveying, and his research interests are related to low energy design, information technology in the AEC industry and energy supply systems. He recently completed a Masters degree for which he investigated the potential of cogeneration and district energy systems in Australia. He has begun research for a PhD that looks at comparative costs of construction in an international context.

Craig Langston – Professor of Construction Management at Deakin University, Geelong, Australia. Before becoming an academic, he worked for nine years in a professional quantity surveying office in Sydney. His PhD thesis was concerned with life-cost studies. He developed two cost-planning software packages (PROPHET and LIFECOST) that are sold throughout Australia and internationally, and is the author of two textbooks concerning sustainable practices in the construction industry and facility management.

Gerard de Valence – Senior Lecturer at UTS. He has an honours degree in Economics from the University of Sydney. He has worked in industry as an analyst and economist. His principle areas of research activity and interest include the measurement of project performance, the study of economic factors relevant to the construction industry, the analysis of the construction industry's role in the national and international economy, the study of interrelationships between construction project participants and the impact of emerging technologies.

Contributors

Deepak Bajaj – Course Co-ordinator of Project Management at UTS. He has a diverse technical, research and business background, with over 15 years of combined experience in contracting, consulting and academia, and has worked in the construction industry internationally with contractors on a range of projects and project types. He has an Engineering degree, a Masters in Construction Management. His PhD was in the area of strategic risk management, focusing on the development of a risk-averse business strategy in the procurement of constructed facilities.

Lydia Depuy – Project Manager with Fortis Real Estate Development, previously a researcher in the Department of Real Estate and Project Management at Delft University of Technology. From 1998 until 2000 she worked on a project dealing with real estate portfolio management originating within public organizations, part of a long-term co-operative effort between the Dutch Government Buildings Agency (GBA) and Delft University of Technology.

Bernard Devine – Managing Director of Butler and Devine Management Services in Sydney. He is a leader in asset management, property, support services and e-business solutions, with over 20 years' experience in the corporate real estate sector, during which he has advised some of Australasia's largest property owners on strategic and operational issues. He is a qualified accountant and economist with extensive experience in business case development, business process outsourcing, systems implementation, and property selection and management.

Geert Dewulf – Professor of Planning and Development and Chair of the Department of Construction Process Management in the Faculty of Technology and Management at the University of Twente, and an independent consultant on workplace strategies and corporate real estate.

Grace Ding – Lecturer in Construction Economics at UTS. She has a Diploma from Hong Kong Polytechnic, a Bachelors degree in quantity surveying from the University of Ulster and a Masters degree by thesis from the University of Salford. She has practised as a Quantity Surveyor in Hong Kong, England and Australia. Grace is currently completing a PhD study at UTS, and is involved in research and teaching in the area of environmental economics and general practice.

Robin Drogemuller – originally trained as an architect, he worked as architect and project manager in public and private practice. He gained additional qualifications in mathematics and computing, and taught at the Northern Territory University and James Cook University before moving to the Commonwealth Scientific and Industrial Research Organisation (CSIRO) where he leads a group developing software to support design and construction. He is heavily involved in the technical development of information exchange standards through the International Alliance for Interoperability and the STEP groups.

Thomas Froese – Associate Professor in the Department of Civil Engineering at the University of British Columbia (UBC), Vancouver, Canada. His interests are in construction management and computer applications, and he teaches undergraduate and graduate courses in construction, project management and computer applications for civil engineering. His research centres on computer applications and information technology to support construction management, particularly information models and standards of construction process data for computer-integrated construction. Thomas originally studied Civil Engineering at UBC before obtaining his PhD from Stanford University in 1992.

Virginia Gibson – Land Securities Trillium Fellow and Director of an innovative Post-Experience Masters programme in Corporate Real Estate and Facilities Management in the Department of Real Estate and Planning at the University of Reading. She has been involved in numerous research projects related to the way real estate is held, used and managed by both private and public sector organizations. She is also a member of the

RICS Research Advisory Board and Corporate Occupiers Management Group, a member of the Editorial Board for the *Journal of Corporate Real Estate* and Editor of *Management Digest*.

Mohammad A. Hassanain – Assistant Professor at the Department of Architectural Engineering, King Fahd University of Petroleum and Minerals, Saudi Arabia. His research interests are in information technology, facilities management, building systems and technologies. He earned his MSc and BSc in Architectural Engineering from King Fahd University of Petroleum and Minerals, Saudi Arabia. Mohammad obtained his PhD from the University of British Columbia, Vancouver, Canada, specializing in facilities engineering and management.

Constantine J. Katsanis – Associate Professor in the Faculty of Engineering and Applied Science at Ryerson University where he teaches project management and building economics. He received his academic training in civil and building engineering at Concordia University, and advanced training in engineering management and business management at The George Washington University and University of Montreal, where he completed his PhD. His current research focuses on network organizations, models of organizational strategy and structure for AEC firms, and the impact of technology and management practices on productivity.

Tom Kennie – a founding Director of the Ranmore Consulting Group, a visiting Professor within the Facilities Management Graduate Centre (FMGC) at Sheffield Hallam University and also an Adjunct Professor at the University of Technology Sydney. He is a Vice President of the International Federation of Surveyors. Ranmore specialize in supporting professional service organizations to enhance their business performance, using a range of interventions at board, team and individual levels.

Adrian Leaman – runs Building Use Studies, a UK-based consultancy that specializes in building monitoring and occupant feedback from the users' point of view. The aim is to improve buildings through better briefing for designers and managers, and more concern for occupants' needs. Details of the work and services may be found on www.usablebuildings.co.uk

David Leifer – is co-ordinator of the Masters Programme in Facilities Management at Sydney University, and a director of FM Solutions Pty in Brisbane, Australia. He is a registered architect and an incorporated engineer and has a PhD. He was formerly Queensland Chairman of both the Facility Management Association, and the Chartered Institution of Building Services Engineers. He previously lectured in facilities management at both the University of Brisbane and the University of Auckland.

Kirsty Máté – has a Bachelor of Architecture and Masters in Design. Her design background covers architecture, interior design and exhibition design, and she now runs her own consultancy in sustainable design, working with various companies on eco design. She was responsible for the first exhibition in Australia that combined design and environmental concerns under the one roof, and has had considerable influence in bringing environmental issues to the notice of corporate management.

Alison Muir – originally trained in interior design, she managed interior design groups in both the public and private sector in Sydney. Later, as First Assistant Secretary for the

Department of Administrative Services in Canberra, she developed her strategic facilities management skills, and now teaches Facility Planning at the University of NSW. She is a graduate of the Australian Institute of Company Directors, and a member of the Facility Management Association of Australia. Currently she is a partner in a private architectural and project management company in Sydney.

Ilfryn Price – originally a geologist, he spent 18 years in the oil industry, managing exploration, research and ultimately Business Process Review. He has been developing facilities related research at Sheffield Hallam since 1993 while also developing contributions to the theory and practice of complexity and memetics in organizations. He is co-author of the RICS guidelines on Practice Management, one book and numerous research papers.

Kaye Remington – after initial studies in structural engineering, she graduated in architecture from the University of Melbourne; now teaching in the Masters of Project Management programme at UTS. Since 1970 she has managed projects, programmes and portfolios in the fields of engineering, architecture, and, more recently, education, organizational change and development. Her post-graduate studies have been in the fields of psychology, social anthropology and organizational sociology.

Stuart Smith – holds a BSc, MSc and MMgt. He was formerly with the Physics Department at Macquarie University in Sydney before moving to a large facility management practice. He now runs his own intelligent building consultancy and has written extensively on the concept of intelligent buildings.

Danny Then – Associate Professor, Facility Management, Maintenance and Operations, in the Department of Building Services Engineering at Hong Kong Polytechnic University and current co-ordinator of CIB Working Commission W70 (Facilities Management and Maintenance). He has consulted and published widely in various areas of built assets management, strategic asset management and facilities management. He set up some of the first postgraduate courses in Facility Management in the UK and Australia before taking up his present post in Hong Kong.

Linda M. Thomas-Mobley – Assistant Professor, College of Architecture, Georgia Institute of Technology. She teaches in the areas of construction law, construction contracts, cost management, facility management, and safety and environmental issues. She holds a BS and MS in Civil Engineering as well as a Juris Doctor in Law and a PhD in Architecture. Her research interest includes indoor environment remediation, sick building syndrome and mould growth in buildings, and the application of artificial intelligence to construction and facility management problems.

Dana Vanier – Senior research officer with the Institute for Research in Construction, National Research Council Canada. He is involved in strategic and client research in the area of service life/asset management. He is internationally recognized in the fields of information technology and asset management in construction, and is a leading authority in a number of fields including computer-aided design, visualization and standards processing. He is also an Associate Editor of the *ITCON Journal*, the first scientific, peer-reviewed journal on information technologies on the World Wide Web, and an adjunct professor at the University of British Columbia.

Juriaan van Meel – Assistant Professor at the Department of Real Estate and Project Management of the Delft University of Technology in the Netherlands. He is also a partner of the Inter-Cultural Office Planners (ICOP) workplace consultants where he is involved in the design and implementation of alternative office solutions. He is a regular speaker at conferences and co-author of a number of publications on office design including *The Office, The Whole Office and Nothing but The Office'* (1999) and *The European Office* (2000).

Suzanne Wilkinson – a qualified and experienced civil engineer with a degree in Civil Engineering and a PhD in Construction Management, both from Oxford Brookes University, UK. She now works as a senior lecturer in engineering management at the University of Auckland where she is responsible for courses in project management, construction management, engineering administration and construction law. Her research interests are in the fields of project management, human resource management and law specifically as they apply to international construction industries.

Foreword

The tradition has been to measure economic growth of a country or a company by the level of capital. In today's knowledge economy, knowledge capital is more important. Some of the returns to this investment can be measured; others cannot, though they are no less important. The quantity and availability of information is growing and many businesses are facing information overload. The knowledge economy is growing in importance with people (and their knowledge) being the most valuable commodity. There will be at least 1 billion university graduates in 2020 compared with a few million in 1920. There will be several billion more sophisticated customers by 2020, who will be better informed and more demanding than ever before. The enlightened customer is driving change. We need to find ways of building and maintaining facilities more safely, to a lower unit cost, and with more certainty, which give value for money.

Globalization, connectivity, and information management are all phrases used to explain the changes that we see around us. The workplace is very different today from that of, say, three decades ago. At that time, how many had heard of electronic mail, the fax machine, or the photocopier?

There are fundamental changes in methods of working and the workplace itself. Whilst new 'managements' have appeared: facilities management, space management, information management, risk management, human resource management, asset management and so on, knowledge has remained vital, never more so than in today's knowledge economy.

Today's facilities managers gather, store and use data about assets, their operation and maintenance, increasingly relying on information technology for data capture, storage and retrieval. Yet the human factor remains a vital part of the process, a point reiterated many times throughout this book.

We know that people may choose to share or conceal knowledge, and that some individuals find it difficult to share knowledge. Their tacit knowledge is personal and context specific and so is difficult to formalize and communicate. It is deeply rooted in an individual's action and experience, ideals, values, or emotions. Therefore, the management of that knowledge involves the management of people and their work environments, and the exchange of knowledge rather than the collection. This makes workplace strategies very important.

Sustainability and whole life thinking demands knowledge of a facility from the cradle to the grave. Facility management integrated into the value chain is an important step, not only for the facilities management profession, but also for the sector and for a sustainable future.

This book is excellent. I wholly commend the contents to you; it gives knowledge, it deals with challenges for today and takes the thinking into tomorrow. It deals with theory and practice in a straightforward way, providing tools and techniques. Most importantly, the book gives knowledge on how we can change to serve our clients better.

Roger Flanagan
University of Reading

Preface

This is the third and final book in a series devoted to the concept of value in buildings and how those who are involved in building procurement may ultimately produce buildings that represent the best 'value-for-money' outcomes. Like its companion volumes, it is intended both for students in construction and property-related courses at tertiary level, and as a useful resource for industry professionals, particularly those who are working in the broad field of facility management.

The range of topics that have been included is based on the competencies that are generally recognized as relevant to facility management. As facility management embraces many activities, from routine building maintenance to strategic management of real property there is great diversity in the material addressed by the various authors. The aim of the book is, however, in line with those of the first two, i.e., to present, in a single volume, an introduction to many of the facets of value in buildings, specifically those that are the concern of the emerging discipline of facility management. Where the previous volumes considered value for money in the pre-design, and design and construction phases of buildings, the focus here is on value for money as it relates to buildings in use. The reference and bibliography lists at the end of each chapter give readers a starting point for further study of selected topics while most chapters are introduced by a short editorial piece that establishes the connection between the material discussed in individual chapters and the central theme of 'building in value'.

The book is broken into three parts: the first part sets the scene by discussing what facilities are and how the management of facilities has emerged as a professional discipline in recent years, the second outlines the various competencies that are required of modern facility managers, while the third looks at the factors that are driving change in the discipline including vital concerns such as sustainability, outsourcing and improving productivity in the workplace.

Contributions have come from many countries including Australia, New Zealand, Hong Kong, the USA, UK, Canada and The Netherlands – and the authors once again include academics and practitioners.

We believe that this book complements the others in the series and that together they provide a very useful introduction to the overall topic of value for money in buildings and how that may be achieved by the many individuals who contribute to the conception, design, construction and ongoing management of built facilities. We hope that this book and its companions will assist academics, students and professionals alike to better understand how value is embodied in buildings and how that value can be maximized for the good of building owners and occupants, and for the benefit of society at large.

Rick Best
Craig Langston
Gerard de Valence

1

Continuous improvement

Rick Best,* Craig Langston† and Gerard de Valence*

1.1 Introduction

> C-l-e-a-n, clean, verb active, to make bright, to scour. W-i-n, win, d-e-r, der, winder,
> a casement. When the boy knows this out of the book, he goes and does it.
> (Charles Dickens, *Nicholas Nickelby*, chapter 8)

The rather novel approach to education adopted by Mr Wackford Squeers embodies an
equally novel approach to the management of the facility known as Dotheboys Hall. It
also illustrates the two sides of facility management that are discussed and compared in
the rest of this book, the operational (the practical business of keeping the windows clean),
and the strategic and/or tactical (the integration of the occupants, their work environment
and the business functions of the organization). In fact, the system employed by Mr
Squeers fits quite snugly into the definition of facility management adopted by the Facility
Management Association of Australia (FMAA, 2002):

> Facility management is the practice of integrating the management of people and the
> business process of an organization with the physical infrastructure to enhance
> corporate performance.

Corporate performance was undoubtedly enhanced from Squeers' viewpoint and the
physical infrastructure integrated with the business process; however, the students (i.e.,
the 'customers'), who were expecting an education, were naturally less than pleased with
the system.

In Dickens' time the coining of the term 'facility management' (FM) was still more than
a century away but at least some of the functions of the facility manager were obviously
carried out by people, whether they had a job title or not – windows were cleaned,
equipment was serviced and repaired, roofs were re-thatched, supplies of candles and coal

* University of Technology Sydney, Australia
† Deakin University, Geelong, Australia

were ordered and stored, and so on. These were tasks of an operational nature, related to keeping a facility (probably a building) running and in reasonable repair. The poor conditions in which many people worked, in premises that were badly lit, largely unventilated, and with inadequate or non-existent plumbing and sanitation, did little more to promote the efficiency, productivity, comfort or good health of the occupants than the harsh regime of Dotheboys Hall did for the education of its students.

Today, FM is emerging as a discipline in its own right, and it embraces much more than the operational concerns of plumbing and lighting, and even more than the provision and maintenance of a productive and comfortable work environment. Increasingly the focus of FM is on the strategic management of facilities, with facility managers devoting their attention to a very broad range of concerns including human resource management, real estate portfolio management and quality management, as well as the more traditional operational concerns that relate largely to building maintenance.

Throughout the following chapters many authors argue that if FM is a true value-adding pursuit within a corporate framework then it must be primarily concerned with filling a strategic role, i.e., facility managers must be pro-active not reactive in their approach, and be able to forecast the needs of their organizations and make forward plans that will support the aims of the organization in the future. Clearly this is about more than window cleaning and plant maintenance schedules.

A number of common themes and catchphrases emerge: alignment with corporate goals, strategic planning, sustainability, change management, space management, value-adding, churn management, and so on – all these arise in the discussions provided by various authors as well as the more prosaic concerns of building maintenance, office design and financial management. Two main themes, however, become clear: there is considerable debate about what FM is, although most at least seem happy to agree that it is very broad in its coverage, and, if FM is to contribute as fully as it can to any organization and have its value recognized, then it must do more than keep the air-conditioning running and the carpets clean. These themes are explored in various ways by the contributors and the breadth of the topics that they cover gives some idea of the complex nature of FM.

1.2 Competencies and areas of expertise

Facility managers come from a range of backgrounds and given the diversity of concerns that the discipline covers, some specializations are inevitable – it is unlikely that many people will have qualifications and experience in, say, services engineering, human resources management and corporate real estate, yet these are only a few of the areas that are routinely brought together under the collective banner of FM.

An examination of the competencies required of those who wish to be certified by the International Facility Management Association (IFMA, 2002) reveals the breadth of knowledge and experience that a successful applicant must have if they are to become IFMA certified facility managers. There are eight broad competencies areas (e.g., Operations and Maintenance, Real Estate, and Human and Environmental Factors), 22 competencies (e.g., oversee acquisition, installation, operation, maintenance and disposition of grounds and exterior elements, Manage real estate assets, and Develop and manage emergency preparedness procedures) and 127 'performances' or work tasks relating to the competencies in detail. The scope is very broad and ranges from the very practical concerns of building repairs to the more abstract concerns of strategic facility planning such as evaluating the effects of economic change on real estate assets.

The FMAA has adopted a different approach to accreditation, having set up three levels of certification that reflect the varying emphasis on the operational versus the influential. The three levels (Parts 1, 2 and 3) are defined as follows:

- Part 1 – practicing FM (operational concerns outweigh strategic concerns)
- Part 2 – managing the practice of FM (operational and strategic roughly equal)
- Part 3 – leading the practice of FM (strategic outweighs operational).

The FMAA competencies are grouped somewhat differently to those of the IFMA but naturally cover much of the same ground. The broad categories are:

- use organizational understanding to manage facilities
- develop strategic facility response
- manage risk
- manage facility portfolio
- improve facility performance
- manage the delivery of services
- manage projects
- manage financial performance
- arrange and implement procurement/sourcing
- facilitate communication
- manage workplace relationships
- manage change.

The range of skills and knowledge required of facility managers, if they are to successfully carry out all of these functions, is quite alarming as it includes everything from computer networking and mechanical engineering to human resources management theory, occupational health and safety legislation, contract negotiation, future financial planning (e.g., budgeting, life costing, discounting), subcontract administration, construction management – the list is endless. The perfect facility manager may be someone who is a services engineer with majors in project management and law, with great charisma and interpersonal skills, coupled with certificates in accountancy, real estate and an MBA, not to mention a keen interest in the protection of the environment.

1.3 Key concerns

At the heart of modern FM is the concept of continuous improvement. Any organization should be striving to improve its operations, whether from the point of customer satisfaction, increased productivity, better quality of output, better environmental performance or any of a host of other performance indicators. The facility manager's role embraces all of the concerns to some degree as the physical facility (building, workplace, office, complex, space – 'facility' covers many alternatives) must accommodate and support the organization's activities in ways that allow the organization to service its customers in the best possible way. That concept of customer service (together with customer satisfaction) can be seen as the key driver of FM, given that the 'customer' may be an external entity (i.e., the classic customer who buys goods or services from an organization) or equally the employees who occupy the organization's space, or the organization itself. It is every bit as important that the facility serve the occupants as it is that it serve the customer or client who does business with the organization.

Within this framework the facility manager must seek to add value to the company's operation through a combination of strategic and operational activities covering all parts of the organization's business. These activities, apart from being divided into strategic and operational, may be grouped according to whether they are concerned with the physical aspects of the facility (e.g., maintenance planning, energy auditing, upgrades, refurbishment, retrofits), human concerns (e.g., recruitment, productivity, communications, change management, dispute resolution), business-related activity (e.g., corporate real estate management), operations management (e.g., outsourcing, security), and so on. Naturally there is a good deal of overlap between the various branches of FM, and also some blurring of the boundaries between the functions of FM and those of other departments or units within organizations, such as finance, human resources and IT.

1.3.1 Sustainability

It is now trite to say that concern for the natural environment has become of increasing importance in recent years – it is obvious that environmental awareness has become a vital concern for people in all walks of life in most parts of the world. This concern is changing the way that we do many things and FM is no exception. As we move slowly but inexorably towards to the goal of sustainability a combination of legislation, public pressure and corporate expediency is making it imperative for facility managers to look for more environmentally friendly materials, technologies and procedures to use in their work. Environmental assessment of buildings and their performance, and measurement of greenhouse emissions and energy usage are becoming more common and legislation in various parts of the world will doubtless make these procedures mandatory in most places in the near future.

1.3.2 Information and communications technology (ICT)

It is equally trite to point out that microchip and computer technology has changed forever the way that many things are done in the developed world. Apart from reducing the drudgery of many repetitive tasks it has enabled the globalization of business and given unprecedented access to knowledge and people for anyone with the wherewithal to purchase a personal computer and an Internet connection. The application of ICT to FM has given facility managers the tools to monitor, record and respond to events in areas under their control more quickly, and even to install systems that can monitor and respond remotely and automatically. It also provides a range of software tools that give the facility manager the power to collect, store and manage a great deal of useful data about their facilities, and to use that data to do a myriad of things such as tracking the location and condition of tangible assets (e.g., furniture or computer hardware), recording the maintenance history of plant and equipment or even spaces in buildings or whole buildings, keeping employee records, recording and analysing energy usages – hardly any part of the FM function does not have a computer or at least an embedded microchip involved at some point.

1.3.3 Outsourcing

Outsourcing refers to any situation where an organization contracts with another organization for the provision of a service that could equally be provided by a person, unit or department within the organization that requires the service. Typical examples are the provision of security services (including surveillance systems and access control as well as the more obvious provision of security personnel on site) and catering and laundry services in hospitals, where it is no longer common for the hospital to have laundry and kitchen facilities on site, and instead these services are taken care of by external organizations.

A large part of the functions of many facility managers is the management of outsourced services – monitoring the level of service, selecting providers, negotiating, managing and reviewing service contracts, and so on. Not all outsourcing has proved to be as successful as the promoters of the concept have claimed; while it is established and seems to work well in some areas, such as those mentioned above, in some other areas the same cannot be said. Provision of ICT functions by external providers is one example of a less than successful outcome of outsourcing and there are several reasons for this lack of success:

- slow response to problems – it is recognized that one measure of success in FM is how quickly problems are addressed and resolved, e.g., rectification of air-conditioning problems – when there is a problem with a computer system, users expect a rapid and effective response, and this often cannot be provided by an external provider as well as it can by a dedicated IT manager within the organization; logging a call with an outside company is not the same as calling someone you know in your organization and asking for assistance.
- lack of understanding of the organization's business – IT professionals are experts in IT, not in the business activities of the organization who hires them, and this can lead to frustration for both parties as one knows what they need from the point of view of their business operation but does not have the expertise to implement it, while the other has the expertise but not the experience to apply it to the specific situation that arises in a particular business setting.

1.3.4 The building/occupant relationship

In the period 1927–32, Elton Mayo, a professor at the Harvard Business School, conducted a long series of experiments at the Western Electric Hawthorne Works in Chicago, aimed at determining whether changes in the workplace promoted improved worker productivity (Accel-Team, 2001). These experiments followed an earlier study on the effect of lighting levels on productivity, which suggested that there was no detectable correlation between the two. The conclusion drawn by Mayo suggested that productivity improved as a result of social interaction within teams of workers and the positive reaction of workers when someone takes an interest in them. One long-term result of the study was that the idea that there was any connection between the physical characteristics of the workplace and the productivity of the people working there was discounted – productivity would be improved, the theory went, by organizing teams differently and promoting the emotional well-being of the workers through recognition, security and a sense of belonging.

This idea prevailed almost to the end of the twentieth century, but more recent research has shown that there is a clear connection between the physical workplace and productivity (Romm and Browning, 1994, 1995). This presents the facility manager with a great opportunity to add value to the organization – the provision and maintenance of a work environment that improves productivity, reduces absenteeism and allows the free flow of ideas, information and motivation must be a key goal for any facility manager, and every organization should not only demand that of its facility managers but support them in their pursuit of that goal.

1.3.5 Managing the intangibles

Contributing to office design, looking after building maintenance, auditing energy usage, and many other FM functions are basically practical concerns, but there are a number of less tangible concerns that also fall within the purview of the facility manager. These include risk management (including planning for unseen disasters, something that has been highlighted worldwide by the events at the World Trade Centre in 2001), conflict management and quality management.

Systematic risk management is becoming commonplace in many areas of business, mostly as a kind of forward planning with contingency plans already formulated and the consequences of the occurrence of identified risks considered and costed as a safeguard against potentially catastrophic consequences, should some possible events actually occur. Examples include mirroring computer networks at other locations so that in the event of a disaster at the prime location business can restart with minimum delay using the remote backup network. The Stock Exchange in Sydney is a good example – the whole operation is mirrored at another site in Sydney, well away from the main operation, and should something occur that destroys or cripples the main site the Stock Exchange can re-open within 24 hours and carry on trading as normal.

Conflict management is a key issue for FM, as the potential for conflict at many levels and with many degrees of seriousness is ever-present in any organization. Disputes range from the tiniest concerns, such as who left a mess of dirty cups in the tearoom, to large-scale disputes between organizations that involve years of expensive litigation. Many of these disputes will land in the facility manager's lap and will require some action before they are resolved. Often they will connect with other areas of FM, such as office planning where territorial disputes arise or work practices will change as a result of planning modifications.

The commonly held view of quality management is that it is about the quality of a physical product, i.e., the number of defective widgets per thousand, and how can that number be reduced. Quality management in FM is related to a physical product, inasmuch as the facility may be viewed as a product, but generally it focuses more on customer satisfaction and, as suggested earlier, the customer may be one person, many people or a whole organization.

1.4 Conclusion

The underlying premise is that space, ultimately, is not about real estate. It is about using all of the organization's scarce resources to their fullest potential to meet pressing business challenges.   (Becker, 2000)

Those scarce resources include finances, physical assets (from whole buildings down to pieces of furniture), information and, above all, people and their skills and knowledge. The facilities, which include buildings, computer networks, virtual workspaces and databases, provide the framework within which the organization operates, and optimization of this framework through a process of continuous improvement is the aim. User-friendly working environments, both real and virtual, must support the individual and the organization in their pursuit of improved performance in all segments of the triple bottom line, or now even the quadruple bottom line, as environmental performance assumes greater and greater importance and legitimacy.

Organizational change, whether described as re-engineering, downsizing, restructuring, re-positioning, integrating, harmonizing or whatever, can have profound effects on the business operation and the people involved in it. Managing these changes will inevitably involve FM whether the facility managers are leading the change, are part of the change management team or are simply in damage control as the decisions of others impact on the people and the business operations that they are engaged in. Flexibility is a key in a changing world, and that means not only organizational flexibility and some sort of flexibility in the physical workplace, but also flexibility in the attitudes of the people, from senior management to the operatives at the coalface, the frontline troops.

Clearly FM is a dynamic and growing area. Whether it has been clearly established as a discipline in its own right is, perhaps, still a subject for debate, as is the question of the operational versus the strategic and whether they are two parts of the same discipline or whether they should be seen as separate albeit related pursuits. What is undeniable is that the concerns discussed throughout this book are all vital parts of an overall picture of running a successful organization, particularly one faced with the increased pressures of globalization, the digital economy, an increasingly litigious environment and the demands by governments and society at large for better environmental performance. There is little doubt that the increased emphasis on FM in recent years will not abate and more likely the facility manager will aspire to, and fulfil, a much more dominant role in the running and planning of the operations of organizations and businesses throughout the world.

References and bibliography

Accel-Team (2001) Elton Mayo's Hawthorne Experiments. *Employee Motivation Theory and Practice.* www.accel-team.com/motivation/hawthorne_02.html

Becker, F. (2000) *Offices That Work: Balancing Cost, Flexibility, and Communication.* Cornell University International Workplace Studies Program (New York: IWSP).

FMAA (2002) Facility Management Association of Australia. www.fma.com.au/main.htm

IFMA (2002) *Certification.* International Facilities Management Association. www.ifma.org/certification/index.cfm?actionbig=8

Romm, J.J. and Browning, W.D. (1994) *Greening the Building and the Bottom Line [increasing productivity through energy efficient design]* (Snowmass, CO: Rocky Mountain Institute). www.rmi.org

Romm, J.J. and Browning, W.D. (1995) Energy efficient design can lead to productivity gains that far exceed energy savings. *The Construction Specifier*, June, 44–51.

PART 1

2

Defining facilities

Stuart Smith*

Facility management concerns *people* and *places*. People are generally the single biggest cost centre for any business or organization and its single biggest asset. Having good people means there is capacity, potential, creativity, responsiveness, continuity and a likelihood of success. Keeping people happy and enabling them to be productive in their daily activities is not only critical in gaining and retaining a strong workforce, but also in delivering overall business prosperity and growth.

Built facilities are typically the places where people work. Facilities are another major cost centre, in many cases the second largest expenditure category regardless of whether space is owned or leased. While it is important that facilities are well designed, efficiently managed and used to their best advantage, it is more important that they support core business goals by enabling people to be at their most productive. Improvements in worker productivity can lead to financial gains that outweigh facility operating costs such as energy, cleaning, maintenance and the like.

Technology is another major cost centre. It involves communication and information equipment and its support, software tools and data management. Once thought to be a separate area of expertise, it is now becoming so germane to the way in which people work that it not only affects facilities, but can substitute for them. A focus on performance, constant change and upgrade, and the need for timely support has led to a close connection between facilities and technology provision and an integrated approach to their management. People can now work remotely, be mobile, flexible, more in control and hence more efficient, often leading to increased job satisfaction as well as less reliance on dedicated workspace.

Notwithstanding technology needs, facilities are not necessarily confined to buildings. It is preferable to consider facilities as *infrastructure* that supports people, either

* KIBT Consultancy, Sydney, Australia

individually or collectively, to realize their goals. Examples of facilities include a cruise ship, theatre staging, a mining town, a hydro-electric generator, an orbiting space station, defence weapons installations, a sewage treatment plant, an airport transit lounge, a golf course, a waterfront container terminal, outdoor recreational areas, even a computer network. All require ongoing management if they are to remain aligned with their intended support function.

Facility management is therefore about empowering people through provision of infrastructure that adds value to the processes that they support. Facility managers are charged with the responsibility of ensuring that the infrastructure is available, operational, strategically aligned, safe and sustainable. Above all, however, facilities must encourage high productivity through a continual search for ways to improve quality, reduce cost and minimize risk.

2.1 Introduction

> A horse! A horse! My kingdom for a horse!
> (*King Richard III*, act 5, scene 4)

King Richard scans the battlefield and cannot see a successful outcome. He contemplates the future and will do anything to secure his life even if it means forfeiting the kingdom. Hopeless and forlorn, he attempts to continue in battle. His army is decimated and his strategy is unravelling at every turn.

The facility manager, consumed by a strategy driven by cost reduction, finds that the relationship between the accommodation strategy and (the long-term success of) the organization is unravelling (Becker and Joroff, 1995). The executive, alarmed by the chasm that is forming, is left with no alternative but to allocate scarce resources in an attempt to re-position the facility as a contributor to business success instead of a drain on capital and people, so that further problems do not arise and lead to a costly revision of the accommodation strategy.

In both cases, the determined strategy is not enough. King Richard's crown is lost; the organization is suffering due to the lack of congruence with the facility type.

Fate turns against Richard and the facility management team. The end is ruthless in its reckoning. The King is dead and the facility is cast adrift from its organizational umbilical.

The King is dead – long live the King. Can this be said of facilities? Yes.

2.2 Context

This chapter is about 'understanding' facilities. There is no neatly packaged dictionary definition for a facility in this context: a facility is a place for work and a place where social bonds are forged and broken, a facility is also a place for play and a place where people learn. To keep things simple, in this context, the facility to be defined, and understood, is the workplace. It is the most familiar. This is not a retrospective of 'what is a workplace?' – as with Shakespeare, there is more to the story. What follows is a forward-looking view of some key elements that underscore the importance of the facility as part of the corporate infrastructure. The elements are the characteristics that give the workplace relevance.

2.3 The workplace is dead, long live the workplace

At the most fundamental level the rise of the collaborative workplace has seen the demise of the traditional, hierarchically based, workplace (Levine *et al.*, 1997). In an era when hierarchical organizational structure defined an organization, the office was the pinnacle of corporate success. Facilities were defined using the office as the fundamental tenet – its *raison d'être*. Everyone had an office or desired one – and the number of square metres occupied was directly proportional to the position. This was the 'old economy'.

In a 'newish economy', the mantra of 'open plan' has meant the divine right to an office has come to an end – a combination of technology and economic rationalism has ensured that (Duffy, 1995). The jury is still out on whether this is space planning or a misguided collectivist solution for the white-collar factory (Brill and Weidemann, 2001). No longer is the office a symbol of power, no longer is the office a place of privacy where workers can concentrate on the task at hand, retreat from the demands of subordinates or slip into the occasional snooze over the keyboard – glass walls have all but eliminated those possibilities. The traditional means to an end no longer exists as there are now spaces specially designed for each activity or work setting (Cook, 1993).

Project rooms, quiet rooms, meeting rooms, break-out rooms, task rooms, conference rooms, hot-desks and hotelling spaces are all available to optimize productivity (Brill *et al.*, 1985). Make a booking with the concierge and check the time. Your workspace and your personal possessions will be laid out in anticipation of your arrival. Figure 2.1 illustrates a sample of the workplace or work setting possibilities. Each provides the worker with an appropriate environment to optimize the desired work outcome.

Think of a void, a true void without walls or restrictions. This void is three-dimensional. Work surfaces are at any angle from zero degrees and up. There is no workstation, no storage unit and no personal locker neatly located underneath the workstation in a beautiful shade of brushed metal (Myerson, 1999). There is no chair. There is an extensive collection of building blocks from which the team can choose and so construct individual work settings to create a workplace in their personal and functional image. And don't worry about the wires because there aren't any. This is the future (Grimshaw and Cairns, 2000).

(1)   (2)   (3)

Figure 2.1 (1) A home office, (2) Microsoft, Redman, Seattle, WA and (3) Sun Microsystems, Palo Alto, CA (Duffy, 1997).

2.4 The sustainable workplace

To set the scene, sustainable development can be defined (WCED, 1987) as 'development that meets the needs of the present without compromising the ability of future generations to meet their own needs'. Popular understanding focuses on an ecological view of sustainable development with greenhouse gas reduction as the key to a sustainable future. It is important to recognize that reducing greenhouse gases is an output of a much more deliberate strategy to change the attitudes, behaviours and beliefs of individuals and corporations in relation to the production of goods and services. This also applies to facilities.

Organizational sustainability has strong links with ecological sustainability (Dunphy *et al.*, 2000). In a new era where there is an understanding of the importance of the interdependence between ecological sustainability, facility design and management, organization and business, facilities will add sustainable value to organizations, and hence create value-driven organizations (Dunphy, 1998).

Focusing on an ecological future is only part of the story; sustainability now includes organizational development, strategy and function. The 'sustainable organization' described by Dunphy *et al.* (2000) extends the traditional organizational definition (Mintzberg, 1973) by taking many of the sustainable elements and philosophical guidance from its ecological cousin, and applying them to the survival and success of organizations in a radically changeable economic environment (Pralahad and Hamel, 1994). Sustainability is not only about weathering the storm of social, economic and organizational uncertainty, it is about recognizing the relationship that all of the factors have to the organization and acquiring the ability to integrate processes and structure so that the organization can adapt to each new challenge (Hinterhuber and Levin, 1994).

A well-worn cliché, that is appropriate to sustainability, is 'the value of the whole is greater than the sum of its parts'. Think triple bottom line, greenhouse gas reduction, waste stream minimization, whole of life economics, intergenerational equity, social value, organizational renewal and even office ergonomics, and sustainability starts to become a little clearer. At the same time it becomes a little daunting.

Achieving a balance between sustainable ecological, physical and emotionally relevant environments creates value (Nadine, 1999). Applying these ideas to facilities is the great challenge of our time. To achieve the necessary conditions in which organizations, practising sustainability, are accommodated in sustainable facilities requires benchmarks to be established for speculative design. In this scenario, change is a continuum, in which managing risk in all its forms, achieving financial outcomes that benefit the short- and long-term financial viability of the organization, and understanding the enabling role of technology are the key elements (Pralahad and Hamel, 1997).

2.5 Defining facilities – an organizational view

> All the world's a stage,
> And all the men and women merely players.
> They have their exits and their entrances,
> And one man in his time plays many parts,
> His acts being seven ages.
> (*As You Like It*, act 2, scene 7)

Jaques's view of the world proposes that there are many influences and changes. The world is not a simple place. It cannot be defined from one view only. The organizational context, and the business that an organization engages in at a particular time, shape facilities. This includes organizational culture, strategy, systems and people. Hence the facility is no longer defined only as the physical structure of the building alone (Smith, 1999a) – it now includes:

- the work space
- the client space
- the interrelationship between functional units and the organization's activities
- the technology supporting the transfer of knowledge internally and externally
- the optimization of financial strategies that deliver short-term cost advantage and long-term profit projections, without loss of quality
- the technology underlying the operation of the organization
- the overriding strategy that co-ordinates business objectives in an ever-changing business environment
- the interrelationship with society through corporate citizenship.

Additionally, given the access to 'virtual networking', where an employee can connect with the organization's IT resources from anywhere in the world, the facility is not bound to a single location (Amabile, 1997). Therefore the additional consideration of customs and culture (vernacular input) is required. Tightly woven within the fabric of the 'new facility' is the transformation of organizational culture to embrace a new workplace – that spans space and time – held together in an integrated technological environment. Integrated with this is a transition of management style and philosophy that can be mapped to the organization's activity or work setting (James, 1998). Table 2.1 illustrates the transition from a Taylorist view of the organization (Taylor, 1947) to a view espoused by current best practice management.

Embracing an organizational view is difficult given the deep-seated belief that the physical work environment is a neutral factor in improving productivity. Challenging the neutrality of facilities is not helped by a reliance on a rational or mechanistic (old) model of work and facilities performance measurement. This represents a major impediment to

Table 2.1 The transition of organizational context (Machan, 1994; Duffy, 1997; Worthington, 1997)

Traditional	Current (new)
attrition	sustainability
formal	informal
hierarchical	flat
autonomous	teamed
functional	synergy
individualistic	shared vision
operational focus	goal focus
routine	creative, knowledge
leadership	leadership/learning
conglomerate	network
physical	virtual
training	coaching

Table 2.2 Understanding the old and new behavioural view of organizations (Smith, 2000)

	Old	New
structure	hierarchical	self-managed, flat
culture	intimidation, paternalism	egalitarian, there is no one best way
political	bureaucracy, ideology, covert action, coalitions, dysfunctional power is centrally controlled	'adhocracy', overt action, shared values, power is distributed
human resource	workers as units of production	workers as valued members of the community

defining facilities to reflect the dynamics of organizational change to a transactional (new) environment. Table 2.2 illustrates the relationship of facilities to organizations and how this is changing.

A few examples of this relationship:

- a just-in-time production line is the facility type that best fits the needs of a car manufacturing business as it competes in a global market against the mutually exclusive foes cost and quality, as well as the particular demands of international trade
- a logical technology-based facility, existing primarily in cyberspace, is the facility type that best matches the needs of a knowledge centred global consultancy as it competes across time and space
- a small shopfront in the high street is the facility type that best fits the needs of a financial service provider in small communities, as it tries to find ways to improve its shareholders' return on investment and still maintain some sense of social connection with the community.

Using the framework of Table 2.2 it is possible to overlay corresponding workplace designs. These are shown in Table 2.3.

The translations identified in Tables 2.2 and 2.3 drive the creation of 'value' for the organization. These are productivity, life-cycle-focused accommodation and organizational sustainability.

Table 2.3 Mapping organizational change to facilities design (Smith, 2000)

	Old	New
structure	closed offices	personal retreats
culture	status driven, managers have workstations with views	non-territorial, workers have the workstations with views
political	strict protocols of communications that foster covert actions	informal gatherings anywhere, foster overt action
human resource	workers are subservient to the environment, they are punished	workers have control over their environment, they are nurtured

2.6 Defining facilities – a vehicle of change

> Yet Edmund was beloved:
> And one the other poison'd for my sake
> And after slew herself.
>
> *(King Lear,* act 5, scene 3)

Edmund utters these words as he dies. He has just ordered the death of Cordelia and Lear as the bodies of Lear's two wicked daughters are brought before him. One has killed the other and then herself. In a moment of awakening, Edmund embraces change.

The ability to embrace change, in all its forms, as a constant and not something best done once and forgotten, is of fundamental importance in life and business. We see change every day in our facilities: a wall up here and a wall down there, new furniture, new technology or a new manager that reorganizes everyone in the hope that the new adjacencies will motivate people. Every day we see this approach to change failing.

Watch closely the restaurants in any neighbourhood – they are constantly changing, refurbishing to create a fresh image, in the hope of re-igniting the interest of old customers and capturing the interest of new ones.

Defining facilities as a vehicle of change acknowledges the influence physical surroundings, virtual connectivity and technological systems integration have on shifts in thinking about how we work. The mantra of today is 'interaction and creativity'. Extrapolating from this, the mantra of tomorrow is a combination of interaction and creativity: interactivity. Creating space that allows organizational processes and wider thinking that borders on the edge of chaos to co-exist: this is the seedbed of new ideas (Turner, 1998).

What stops us from degenerating into chaos is the ability of the workplace to be both the agent of change and the stalwart of stability. In this rarefied environment ideas flourish. Competitive advantage is nurtured and competencies are challenged or enhanced.

In the final throes of the twentieth century we challenged the strongly held Victorian work ethic (Annunziato, 1999). Businesses – and workers – have been merged, acquired, globalized, consumed, outsourced and contracted. No longer is the 'organization man' the backbone of success; instead we are 'free agents' with our time, our values and our choice of work environment. Knowledge is the raw material of the business of the future (Buffini, 2001). The only constant through all of this is change and the unpredictability of achieving cultural cohesiveness.

A facility, as an agent of change, can be an agent of meshing cultures. There are three basic types. They are:

- the culture that exists within an organization that is renewing its workplace
- the cultures that collide when two or more organizations merge
- the integration of the 'customer' as an influence on the defined product value.

The facility is the one element each of these types has in common and it is the tangible element of change that people can grow with. As the new culture takes shape, the facility adapts with it (Jones and Goffee, 1996). The way that banking has changed provides a good working example. Technology has made it possible for many of activities that require the transfer of money, whether from account to account or for payment to suppliers of goods and services, to be done through the use of a virtual banking environment. Enter a bank and the bank teller does not just process our money, but also sells associated financial products, and counter services now take place in 'customer studios' with a 'client account manager'. Customers and banking staff have been required to adapt to a different experience. The physical environment has changed to reflect both the image of 'accessibility' and new 'customer service relationships'. The bank branch looks more like an airport lounge than the bastion of finance from a previous century.

There are other examples where facilities act as agents of change. Some are well known. They include:

● British Airways – England (Duffy, 1999)
● Campus MLC – Australia (LaBarre, 1999)
● Chiat/Day – America (Berger, 1999).

The physical change is more dynamic. The subtleties implied by the change are as complex as the banking example.

Facilities become a resource to be optimized and integrated with the organizational psyche. As the organization changes shape and focus, the facility must adapt with it (Hamilton et al., 1996). To do so they must be adaptable and responsive to physical, technological and emotional change. In the case of virtual organizations, a logical connection to the physical workplace is essential if the space and time opportunities are to be optimized as well (Kimball and Eunice, 1999), however, rationalization, not optimization, of space is still the order of the day. As a meal, it satisfies the urge to eat but fails to deliver any real sustenance.

2.7 Defining the facility – a product view

> What's in a name? That which we call a rose
> By any other word would smell as sweet.
> (*Romeo and Juliet*, act 2, scene 2)

These two short lines capture the central struggle and tragedy of the play: Romeo denounces his heritage in order to satisfy the need to be 'Juliet's lover' at any cost.

A product is defined as 'anything that can be offered to satisfy a need or want' and has two main attributes: features and benefits (Kotler, 1997). Features include functionality, ease of use, 'upgrade-ability', adaptability, aesthetics and useful life. These are the same characteristics that can be used to describe a facility. They are demonstrable and describable. Benefits are less tangible and therefore harder to pinpoint. They focus on questions such as: 'What's in it for me? Will it improve the way we do business? Will it increase productivity?' In essence a product is more than just its physical form, rather it is the physical embodiment of the service it offers or the solution it provides (Venkatraman, 1998). We closely associate products with organizations, e.g., we associate:

● a particular brand of cola with the shape of its bottle
● a passenger aeroplane with the shape of its wings (Figure 2.2)
● a restaurant chain with the shape of its emblem.

The Oval Office   Concorde

Figure 2.2 The image or brand power of products and facilities.

In the same way it is possible to associate facilities with organizations through recognition of their physical form (Becker, 2000). A few examples are:

- the Oval Office as the seat of democratic government in the United States of America
- a restaurant chain whose facilities are replicated in every location across the world.

It is not surprising, given the strength of the branding relationship of the product to the facility, that the restaurant chain is as successful as it is. The benefits or capabilities of the facility are essential to the functioning of the organization. Table 2.4 illustrates this concept using a television set for comparison.

This is a simple example. In the facility examples cited, success is founded on providing a product that dominates the market, unaffected by the wavering loyalties of customers. Most organizations do not dominate their market in the same way as the US government or the Catholic Church. Most organizations do not possess facilities that are instantly recognizable. Most organizations are continually matching their product to the needs of their customers. In cases where information is the raw material and knowledge the product, the mechanisms that defined facilities in the past are no longer relevant.

As more organizations create knowledge rather than manufacture products in the traditional sense, defining facilities will become more difficult. In more subtle cases, such

Table 2.4 Product characteristics (Smith, 1999a)

	Attribute	A new television set	A new office
feature	ease of use	remote control	open plan
	adaptability	camera input	plug and play furniture system
	aesthetics	chassis style	finishes
	useful life	until new viewing requirements emerge	until new working requirements emerge
	functionality	range of controls	diverse work spaces
benefit	what's in it for me?	better viewing conditions	better working conditions
	what improvement will it bring?	less strain on the eyes, enhanced viewing pleasure	productivity, lower absenteeism, cultural refresh, happier workplace

as the knowledge organization, it is possible to understand the relationship between facility type and organizational type using the mapping process outlined in Tables 2.2 and 2.3.

Table 2.5 lists two examples of where there the relationship between facility type and organizational type is identified.

The metaphor of 'facility as product' can be taken a step further. The development of mass customization enabled manufacturers to deliver a unique product offering to a unique customer (Gilmour and Hunt, 1993). Facilities, in turn, can be mass 'customerized' to offer employees the choice of the mode of working that best suits their requirements as well as those of the organization (Ahuja, 1999).

It is not uncommon for employees to be provided with the technology needed to complete a task. It is uncommon, however, that they are provided with facilities that enable the best use of the technology. For example, workers are provided with laptop computers that allow for location independent working; however, the workplace to support this work setting is not provided. A lingering reliance on 'command and control' management practices makes this a difficult change to make. An appropriate workspace solution is to allow employees to determine the type of work setting appropriate to their task (Becker, 2000). This is not the same as work anywhere, anytime, anyhow, nor is it the same as providing a suite of different work settings within a facility. Instead it is a conscious decision relating to allocation of resources – financial, technology and physical space as required in a mass 'customerized' manner (Gilmour and Hunt, 1993). Even more difficult is extending the concept to the logical representation of the workplace (Stone and Luchetti, 1985). This is not the virtual workplace but the extension of the organization into the hearts and minds of its customers through a ubiquitous web presence, whether it is business-to-business (B2B) or business-to-customer (B2C) (Ogilvie, 2001). Customers of the physical facility are expecting the same experience when navigating the web site and vice versa. The question then is: 'Is the physical workplace pre-eminent by comparison with the logical or virtual workplace?' Or will the logical workplace, given the current obsession with the Internet, become the dominant view of what the workplace should look like?

The connection between physical space and logical space is still in its infancy. The relationship between facilities and organizations, and how they work to maintain and improve organizational performance – while understood – is at best tenuous and at worst

Table 2.5 Congruence of effectiveness and perception from an organizational and workspace view

	Organization	Workspace
effectiveness	managers can interpret things from varying standpoints and so get a comprehensive picture of what is going on	being able to interpret design and its impact on the people who will occupy the space, facility planners get a comprehensive picture of the effect of different designs on the behaviour and productivity of the people who will work in it
perception	managers become more attuned to the people around them and so become more able to learn from them	by understanding the messages that different designs communicate, managers are better attuned to the design options that are available to them

non-existent. The relationship between people and their facility is even more tenuous. For sustainable organizations, trading up, down, opting out of a facility or forcing employees to work in inappropriate facilities will not be the most appropriate strategies for success. Sustainable organizations will define the facility as having a strong relationship with the culture of the organization. In this sense, emotion, symbol, ritual and story define the facility. The maxim 'the way we do business around here' will have as much to do with the shape, feel – and thought – of the work environment as it has to do with the processes and systems that are used by the organization to do its work.

Challenging the traditional facilities design paradigm, through the widespread application of strategies intended to produce convergent views of the virtual facility and its physical counterpart, has radical implications for the way that facilities will be defined in the future (Huang, 2001).

2.8 Defining facilities – a stakeholder view

Oft expectation fails, and most oft there
Where most it promises; and oft it hits
Where hope is coldest, and despair most fits.
(*All's Well That Ends Well*, act 2, scene 1)

The king is suffering from a terminal illness, but Helena claims to have a cure, and seeks to treat him. The king has doubts. Helena prevails through the skilful use of rhetoric and the king allows her to treat his illness, which she does successfully. In this scene the prospect of co-operation is jeopardized by experience and mistrust. There are compelling reasons for allowing Helena to treat the king but there are also other objectives that must be managed.

The stakeholders in the facility are primarily the client or building owner, on one side, and the tenant on the other. Money, time and function are the principal drivers. When evaluating facilities the stakeholders are often driven by conflicting agendas; the client is driven by a desire for short-term construction and the tenant by long-term occupation. One thing they can agree on is the desire to build/occupy a functionally purposeful building at the lowest cost premium.

They have different views that make it difficult to agree on a comprehensive definition of facilities. This is not unlike King Richard who underestimated his adversary in battle or Edmund who underestimated his ability to change.

Unfortunately, the many agendas often work against each other. To support the stakeholder view (and in turn the sustainable view) four requirements must be considered:

- to tilt the cost–benefit balance in favour of the benefits afforded by consultative design (life cycle costing and integrated facilities)
- to redefine the cost motivation as a profit motivation (obsolescence and adaptability)
- to make tangible those attributes of intelligent buildings that are currently intangible (building performance and its impact on business performance)
- to realign the office/facility as an organizational competence.

Table 2.6 illustrates the differing views taken by different stakeholders when defining facilities.

Table 2.6 Comparison of different stakeholder groups (Smith, 1999a)

Stakeholder group	Needs/benefits
client/owner	adequate return on investment costs based on sound value management strategies saleable/leased building excited customers
tenant	long-term cost management ability to align the workplace with work processes responsible facilities management
secondary	limited impact on the environment, use of non-toxic components and processes efficient processes that do not undermine quality or quality practices a building that contributes to the community

To counter the different views of the stakeholders the same emotional (social), organizational and financial conditions that apply to creating sustainable value need to be considered. There are four. They are:

- promote life cycle costing and integrated facilities management as the underlying driver for physical sustainability of the facility (Huston, 1999)
- redefine cost motivation as a profit motivation and so minimize obsolescence and enhance adaptability (Romm and Browning, 1994)
- make tangible those attributes of the facility that are currently intangible, such as building performance and its impact on business performance (Smith, 2000)
- include the facility – as an enabler of improved productivity – as an organizational competence (Aronoff and Kaplan, 1995).

Figure 2.3 illustrates the cost–benefit approach to defining facilities. Note that only the top right-hand quadrant is sustainable in the long term.

Essential requirements for the establishment of integrated facility environments that will define facilities for the foreseeable future are:

- flexibility – the mapping of current space needs to future space needs
- diversity – the type and range of work spaces

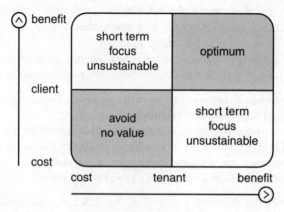

Figure 2.3 The cost–benefit matrix (Smith, 1999a).

- convergence – the fusing of technological systems
- interoperability – connectivity across different technology platforms
- strategic facilities management – aligning the design and management of the facility with the business strategy
- integration of building systems – to provide total technical integration of all systems that affect the performance of the organization
- a focus on life cycle costing and value management.

Together they present a holistic understanding of the facility.

2.9 Defining facilities – evaluating the risk

The fault dear Brutus, is not in the stars
But in ourselves, that we are underlings
(*Julius Caesar*, act 1, scene 2)

Fate is not what drives men to their decisions and actions, but rather the human condition. Fundamental to the human condition is living with risk. That we awake each day and survive the challenges placed before us indicates that as humans we have learned to evaluate and manage risk, and overcome unpredictability. Risk associated with facilities comes in six flavours. These are listed in Table 2.7.

Each of the risk types is interrelated and there is a different mix for different types of facilities, e.g., office buildings, manufacturing facilities, sporting arenas or shopping complexes.

New facilities have an innovative/change risk that applies when physical and emotional boundaries are challenged by new organizational space. Procurement risk, based on

Table 2.7 Types of risk

Risk type	Description
business risk	threat or act by the organization that negatively affects the financial performance, e.g., failure to select the best relocation option that results in a lower net present value (NPV) to capital investment (I) (Haresign, 1999)
public/social risk	threat or action that directly or indirectly affects members of the public that are not involved with the organization or facility, e.g., oil spills or inadequate demolition practices (Kirsh, 1996)
occupational health and safety risk	poor safety/health/ergonomic conditions that lead to worker injury or stoppages in work processes; a simple example is not complying with the provision of disabled access (Ross, 1999)
security risk	threat or action that may cause physical or psychological damage, or theft of physical or intellectual property from the facility (Watkins-Miller, 1998)
procurement risk	failure to assess the most appropriate procurement method; in simple terms whether to lease or own the facility (Becker and Sims, 2000)
innovation/change risk	failure to adopt new accommodation standards or environmental controls that improve organizational performance (Drucker, 1998)

understanding and integrating a well-defined accommodation strategy, is most important when acquiring or changing workspace type or location (Becker and Sims, 2000).

Of the six types of risk, the first three are well understood. Most important are business and social risk. These are the precursors to all risks that follow. They have a financial and emotional element, whether it relates to brand damage, personal and physical damage or environmental damage. Examples are BHP (Kirsh, 1996) and Bhopal (bhopal.org, 2002) where both financial and environmental damage occurred. The final three are less well understood. The influence that the procurement process has on the development and the relationship to innovation on the facilities options available can be overwhelming in the long term (Drucker, 1998), e.g., a standard design-construct procurement process tends to favour speculative design.

Table 2.8 highlights a number of issues that must be considered before choosing the most appropriate development strategy. This is an example of the development of a new facility for an organization that is relocating. Aspects to be understood, evaluated and managed – in terms of lost opportunities – are listed.

All of the risks mentioned in Table 2.8 can be reduced. Some of the strategies used include:

- effective information management founded on well developed systems and processes
- sound legal and contractual arrangements that enable all involved to understand the level of risk to be managed
- specification of performance and competency standards
- comprehensive due diligence that integrates the responsibilities of supplier and client and minimizes the opportunity for vicarious liability.

A procurement process that includes a facility planning process is more likely to consider the influence on the organization as an important factor on the design of the facility. This is because risk is not abated as the facility takes shape; during planning, design brief development, construction and operation, the risk focus changes but not the total risk

Table 2.8 Risk evaluation issues

Issue	Likelihood	Impact	Containment
funding not available to complete fit-out	L	L	the organization has selected the lowest cost option by way of leasing new accommodation
lack of office space in selected location	L	H	sites are available in the selected area for development, the organization will need to agree to a long-term lease to secure accommodation
construction quality risk	M	M	contract with the developer or owner is to clearly specify in detail the quality and standard of all items required
value for money result	M	H	a two-stage tender process is preferable to ensure value-for-money is achieved and appropriate terms are negotiated
staff concerns	H	H	there is strong concern from the staff about relocation, ongoing consultation and information sharing will be necessary to ensure that all staff are aware of and understand the options

H, high; M, medium; L, low.

does not. It is important then to understand the nature of risk and how it changes over time – not only during the development phase but also throughout the life of the facility. This is not a case of controlling the transfer of risk, or mapping risk from one phase to another, but of managing the translation of risk so that the total operational risk is minimized.

2.10 Defining facilities – a technological view

> We are such stuff
> As dreams are made on and our little life
> Is rounded with a sleep. . .
> *(The Tempest*, act 4, scene 1)

These words of Prospero celebrate the unique ability of humans to blend mind and matter.

In 1965 Gordon Moore proposed what is known as 'Moore's Law' (Moore, 1965). He observed that each new computer chip had roughly twice as much capacity as its predecessor and each chip was released 18–24 months after the previous chip. If this trend continued, he reasoned, computing power would rise exponentially over relatively brief periods of time. Only after 20 years has the limit of this law come into sight. In those 20 years, technology on the desk has gone from non-existent to a PC per desk integrated within a network environment that delivers instant global communication and access to information on a scale inconceivable even five years ago. This has been complemented and complicated by the complete metamorphosis of organizational shape. If Moore's Law has a diminishing return then it is not being demonstrated by organizations as they stretch, contract, and seek new and clean competitive space in which to operate. Throughout all of this chaotic activity, technology has marched onward toward a seamless blend of mind and matter.

Technology is defined as encompassing building systems, architectural structure's office automation, information technology, 'plug 'n' play' furniture systems, management practices and processes (Smith, 1995). This is more comprehensive than is generally understood. Technology is an 'enabling' tool – and not unlike paper was to parchment – it has 'enabled' the way we work, and play, and has taken on radically new forms that shape how we communicate and transfer information. Rather than the question being, 'What do I do with it?', the question now is, 'How can it help me do what I do better, faster and cheaper?' (Jensen, 2000). This applies not only to individuals, but to teams and even whole organizations.

As technology has been developed to support smarter working, so it has become more complex and at the same time access to it has become simpler. There is now a wide gap between those who use technology and those who understand it. At the fringes, it is those people or organizations that can cross the gap that are delivering new innovations, feeding the implementation of technology and furthering the innovation process. Organizations may be bordering on the chaotic in order to generate new ideas but there is limited integration of the technology that could turn those ideas into real products or services. The hardware necessary to allow communication with a friend using two tin cans could be built in ten minutes. Communicating with a friend is now only a few button pushes (or clicks of the mouse) away; however, the infrastructure that supports the communication process is far more complex. Technology in a can is still as relevant today as it was one hundred years ago – only the expectation has changed. We marvelled at the fact that we could talk and hear each other at the end of the string, accepting the poor quality due to the simplicity of the system. Yet we cannot accept it when the quality of a telephone call is poor even though the technology that supports the success of the call is far more complex.

Technology has increased the depth and breadth of information that is flowing to and through organizations. At the same time, due to market and innovation pressure the completion speed of every task has increased rapidly. Management has responded to this by relying less on functional independence and more on cross-functional interdependence in an effort to create 'zero time' communication – in keeping with the perceived speed of technology – to support interaction and integration (Lipnack and Stamps, 1997). The consolidation of information management and the development of knowledge management are furthering the need for facilities to provide a fertile place where the seeds of knowledge are nurtured and can flourish (Myers, 1996).

The workplace has been a driver of the technological revolution. 'All your technology in a can' symbolizes the need for instant communication and that all the ingredients are mixed together in a simple-to-prepare formula.

In the future, space will be more closely entwined with technology (Smith, 1999b). This will occur as the look and feel of the office more closely integrates with the view of the organization from cyberspace. This is not about 'clicks and mortar'; it is about providing the same experience in physical space as in cyberspace and not the other way around (Huang, 2001). Organizations are now networked, interconnected and organic. Organizations interact in an elastic and highly developed way in order to create the technological complexity needed to provide the marketplace with simple solutions. Corporations are taking on more than the complexity of biological systems (Harrison, 1997). Causal links between specific actions and specific organizational outcomes over the long term disappear in the complexity of the interaction between people in an organization, and between them and people in other organizations.

Large organizations will redefine themselves as 'federations' of autonomous business units in an attempt to compete with heterogeneous, smaller, strategically aligned networks of organizations (Hinterhuber and Levin, 1994), supported by a greater focus on logistics (Hinterhuber and Levin, 1994) and sophisticated technology. Alternatively, organizations will use technology to 'virtualize' themselves (Alexander, 1997). Similarly, organizations will learn to manage the discontinuities in the market by managing the market segment boundaries (Alexander, 1997).

The adage 'form follows function' will be further tested in the new technological environment and presents new challenges in defining facilities.

2.11 Conclusion

Through all of this the facility, as a gathering point for sharing information and knowledge and engaging in human contact, will remain steadfast. The need for social connectivity will become the major driver of facilities in the future. Facilities will not be workplaces, they will be 'centres of experience'. Here people will interact, share information, create knowledge and, at the end of the day, package it as a product or service. Facilities will exist as a series of optimally configured spaces designed for the required work setting. Every space will be a meeting room, a project room or a room for quiet contemplation. Design will take its cue from manufacturing and product mass 'customerization'. Facilities will be a mixture of what is current best practice today and a form not yet imagined.

In this time organizations will become more organic, even kaleidoscopic, in their form. What was relevant today will be reset and reconfigured for tomorrow. Management will not be a driver of new organizational forms – instead the capabilities of facilities will drive the organization. Management's role will be the optimization of the patterns of involvement that create the required productivity.

One thing that will not change is the basic elements of the facility: organizational sustainability, risk, technology, stakeholder relationships, innovation and change. In the words of Polonius:

> Be not too tame neither, but let your own
> Discretion be your tutor; suit the action
> to the word, the word to the action.
>
> (*Hamlet*, Act 4, scene 2)

References and bibliography

Ahuja, A. (1999) Design–build from the inside out. *Consulting–Specifying Engineer*, **25** (7) June, 36–8.

Alexander, M. (1997) Getting to Grips with the Virtual Organisation. *Long Range Planning* **30** (1), 122–4.

Amabile, T.M. (1997) Motivating creativity in organisations [on doing what you love and loving what you do]. *California Management Review*, **40** (1) Fall, 39–57.

Annunziato, L. (1999) The humanistic workplace of the future. *Facilities Design & Management*, December. www.fmlink.com/ProfResources/Magazines/article.cgi?Facilities:fac1299b.htm

Aronoff, S. and Kaplan, A. (1995) *Total Workplace Performance – Rethinking the Office Environment* (Ottawa: WDL Publications).

Baird, G., Gray, J., Isaacs, N., Kernohan, D. and McIndoe, G. (1995) *Building Evaluation Techniques* (New York: McGraw-Hill).

Baldry, C. (1997) The social construction of office space. *International Labour Review*, **136** (3), 365–79.

Becker, F. (2000) Integrated portfolio strategies for dynamic organisations. *Facilities*, **18** (10/11/12), 411–20.

Becker, F. and Joroff, M. (1995) *Reinventing the Workplace*. International Workplace Studies Program, Cornell University/Massachusetts Institute of Technology.

Becker, F. and Sims, W. (2000) *Managing Uncertainty – Integrated Portfolio Strategies for Dynamic Organisations*. The International Workplace Studies Program College of Human Ecology, Cornell University.

Berger, W. (1999) Lost in space. *Wired*. www.wired.com/wired/archive/7.02/chiat.html

Bhopal.org (2002) The Bhopal Medical Appeal. www.bhopal.org

Brill, M. and Weidemann, S. (2001) *Disproving Widespread Myths about Workplace Design* (New York: Kimball International for Bosti Associates).

Brill, M., Margulis, S. and Koner, E. (1985) Using office design to increase productivity. In: *Workplace Design and Productivity* (Buffalo, NY: Buffalo Organization for Social and Technological Innovation – BOSTI).

Buffini, F. (2001) The free agent nation. *The Financial Review – Boss Magazine*, June, 39–41.

Cook, R. (1993) New strategies and design methodologies for the virtual workplace. *Site Selection & Industrial Development*, December.

Drucker, P.F. (1998) The discipline of innovation. *Harvard Business Review*, November/December, 149–57.

Duffy, F., Laing, A. and Crisp, V. (1993) *The Responsible Workplace* (Oxford: Butterworth Architecture).

Duffy, F. (1995) Visions of the New Office. *Journal of the Professional Issues in Engineering Education and Practice*, **121** (4), 233–5.

Duffy, F. (1996) *Hot, Open, Chaotic – Workplace of the Future* (London: DEGW).

Duffy, F. (1997) *The New Office* (London: Conran Octopus).

Duffy, F. (1999) Office politics. *Perspective Magazine*, Winter, 30–6.

Duffy, F. (2000) Design and facilities management in a time of change. *Facilities*, **18** (10/11/12), 371–5.

Dunphy, D. (1998) *The Sustainable Corporation* (Sydney: Allen & Unwin).

Dunphy, D., Benveniste, J., Griffiths, A. and Sutton, P. (2000) *Sustainability – The Corporate Challenge of the 21st Century* (Sydney: Unwin & Allen).

Gilmour, P. and Hunt, R. (1993) *The Management of Technology* (Melbourne: Longman Cheshire).

Grimshaw, B. and Cairns, G. (2000) Chasing the miracle: managing facilities in a virtual world. *Facilities*, **18** (10/11/12), 392–401.

Hamilton, J., Baker, S. and Vlasic, B. (1996) The new workplace. *Business Week*, April 29. www.businessweek.com/1996/18/b34731.htm

Haresign, D. (1999) Is a corporate campus right for your business? *Site Selection*, December/January, 1118–21.

Harrison, A. (1997) Converging technologies for virtual organisations. In: Worthington, J. (ed.) *Reinventing the Workplace* (Oxford: Architectural Press).

Hinterhuber, H.H. and Levin, B.M. (1994) Strategic networks – the organisation of the future. *Long Range Planning*, **27** (3), 43–53.

Huang, J. (2001) Future space: a new blueprint for business architecture. *Harvard Business Review*, April, 149–57.

Huston, J. (1999) Building integration: mastering the facility. *Buildings*, **93** (12), 51–4.

James, D. (1998) Why the human factor will still be vital in the growing cyber-economy. *Business Review Weekly*, September 7, 72–3.

Jensen, B. (2000) *Simplicity – The New Corporate Advantage* (Cambridge, MA: Perseus Books).

Jones, G. and Goffee, R. (1996) What holds the modern company together? *Harvard Business Review*, November/December, 133–48.

Kimball, L. and Eunice, A. (1999) The virtual team: strategies to optimize performance. *Health Forum Journal*, May 1. www.caucus.com/pw-thevirtualteam.html

Kirsh, S. (1996) Cleaning up OK Tedi: settlement favours Yonggom people. *Journal of the International Institute*, **4** (1), Fall. www.umich.edu/~iinet/journal/vol4no1/oktedi.html

Kotler, P. (1997) *Marketing Management – Analysis, Planning, Implementation and Control* (Englewood Cliffs, NJ: Prentice-Hall).

LaBarre, P. (1999) The office of the future. *Fast Company*, **27**, 176.

Levine, N., Pitt, D., Beech, N. and Isaac, R. (1997) Organisations for the new millennium: facilities management and the new organisational environment. *International Journal of Facilities Management*, **1** (1), 11–20.

Lipnack, J. and Stamps, J. (1997) *Virtual Teams: Reaching across Space, Time and Organisations with Technology* (New York: John Wiley).

Mintzberg, H. (1973) Strategy-making in three modes. *California Management Review*, **16** (2), Winter, 44–53.

Machan, D. (1994) We're not Authoritarian Goons. *Forbes*, **154** (10), 246–8.

Mohamed, S. and Salzmann, A. (1999) Re-engineering traditional risk analysis approaches investment justification in large construction projects. In: *Proceedings 2nd International Conference on Construction Process Re-engineering*, Sydney, 13–20.

Moore, G. (1965) Cramming more components onto integrated circuits. *Electronics*, **38** (8), April, 114–17.

Moran, M. and Taylor, J. (1998) Outsourcing: managing risk. *Executive Briefs*, April, 1–4 (Cambridge, MA: International Society of Facilities Executives).

Myers, P. (1996) *Knowledge Management and Organisational Design: An Introduction* (Oxford: Butterworth-Heinemann).

Myerson, J. (1999) Britain's most creative offices. *Management Today*, April, 62–67.

Nadine, M. (1999) Sustainability can mean bigger profits, claim 'green backers'. *ENR*, **243** (19), 14.

Ogilvie, R.A. (2001) B2B vs B2C how they differ. *e-Access Magazine*, June, 38–43.

Plenty, T.C., Chen, S.E. and McGeorge, W.D. (1999) Re-engineering the construction process to facilitate early risk analysis and allocation. In: *Proceedings 2nd International Conference on Construction Process Re-engineering*, Sydney, 3–12.

Pralahad, C. and Hamel, G. (1994) *Competing for the Future* (Boston, MA: Harvard Business School Press).

Pralahad, C. and Hamel, G. (1997) Rethinking the Future. In: Gibson, R. (ed.) *Business, Principles, Competition, Control Leadership, Markets and the World* (London: Nicholas Brealey).

Romm, J.J. and Browning, W.D. (1994) *Greening the Building and the Bottom Line [increasing productivity through energy efficient design]* (Snowmass, CO: Rocky Mountain Institute). www.rmi.org

Ross, P. (1999) Unwired. *The Safety and Health Practitioner*, **17** (5), 26–8.

Smith, S. (1995) Integration of Building Services Communications Networks – The Use of ISDN as a Platform to Integrate Building Services with a Digital Telecommunications. *New Methods and Technologies in Planning and Construction of Intelligent Buildings*, Proceedings of the IB/IC Intelligent Building Congress, Tel Aviv Israel, pp. 146–55.

Smith, S. (1999a) The use of quality functional deployment in the design of the office of the future. In: *Proceedings 2nd International Conference on Construction Process Re-engineering*, Sydney, 35–46.

Smith, S. (1999b) The impact of communications technology on the office of the future. *FM Magazine*, April, 31.

Smith, S. (2000) Reframing approaches to strategic facility planning. In: *Proceedings of 'IdeaAction 2000'*, National Conference of the Facility Management Association of Australia.

Stone, P. and Luchetti, R. (1985) Your office is where you are. *Harvard Business Review*, March/ April, 102–17.

Taylor, F. (1947) *Scientific Management* (New York: Harper Bros).

Turner, G. (1998) Organisational culture change and office environments. *Journal of Management Learning*, **29** (2).

Venkatraman, N. (1998) Real strategies for virtual organisations. *Sloan Management Review*, Fall, 33–48.

Watkins-Miller, E. (1998) Secure places. *Buildings*, May, 64–6.

WCED (1987) *Our Common Future*. World Commission on Environment and Development (Oxford: Oxford University Press).

Worthington, J. (ed.) (1997) *Re-inventing the Workplace* (Oxford: Architectural Press).

3

Facility management as an emerging discipline

Ilfryn Price*

Editorial comment

Facility management (FM) is an emerging discipline. Its roots lie in the custodial role of a building superintendent/caretaker largely concerned with operational issues of maintenance, cleaning and tenant security. The growth in the complexity of buildings and the cost significance of their operation has led to a need to introduce both tactical and strategic management functions, thus raising the profile of the discipline alongside other support functions such as the management of human resources and information technology. As time progresses these three key areas will be further integrated into the ultimate pursuit of making people more productive.

FM is both *global* and *generic*. It is global, certainly in a business context, in that the marketplace is not limited by regional boundaries, and products and services are available increasingly to a diverse consumer base. It is generic in the sense that the processes that manage quality, cost and risk are not confined to conventional buildings but apply to all sectors and all types of infrastructure. Coupled with that, FM embraces more than just operational issues as it must consider opportunities for new acquisition, including development and project management processes.

In recent years FM has demonstrated significant growth as a profession in its own right due to a need for specialist people who can add value to businesses and organizations that control infrastructure. Much like project management, FM requires a broad knowledge base and skills that cross traditional discipline boundaries, and it shares a common strategic focus. It is a blend of technical expertise, business administration and entrepreneurialism which draws people from a range of backgrounds including architecture, design, engineering, business management, property and construction.

There is a clear deficiency at present in the identification of a body of knowledge that is distinct from other professions, as well as an absence of educational pathways at all

* Sheffield Hallam University, UK

levels of career development. Yet in the light of a diverse range of contemporary issues such as environmental impacts, internationalization, terrorism and corporate downsizing, facility managers are being elevated to higher levels of influence and responsibility that can only increase their ability to identify and deliver value to society.

Much of the history of the discipline presented in this chapter is drawn from personal research carried out by the author, much of it unpublished. Some of it is based on conversations with individuals, some drawn from speeches and conference meetings, and some from early magazine articles that predate the coining of the term 'facility management'. There is a diversity of information and opinion that makes it a fascinating insight into how this new professional discipline has emerged and gives a basis for speculation on where the discipline may go in the future.

3.1 Why a history?

History is bunk (attributed to Henry Ford)

Those who cannot remember the past are condemned to repeat it
(George Santayana).[1]

Why write a history of facility management (FM) in what is supposed to be a definitive global treatise on the subject as it is currently known and practised? Should we not, like Henry F, get on and establish the best way of doing it then put it into practice? Perhaps, but even the industry that Ford founded is currently struggling to escape its history.[2]

Facility, or facilities, management (the supposed difference is discussed below but we can otherwise just say FM) is struggling to define precisely what it is and where it is going. Some understanding of how it got where it is may help those involved appreciate where it is, or is not, going.

By some criteria FM is blooming. In the English speaking world it has three significant professional associations (International Facility Management Association (IFMA), British Institute of Facilities Management (BIFM) and Facilities Management Association of Australia (FMAA)) plus a growing presence in South Africa and, in the non-English speaking world, varying degrees of penetration and interpretation, but at least a level of recognition. In CIB 70 it has long-term support as a community of research and practice into 'Facilities Management and Maintenance'. As an industry, providing FM services is a global market, the size of which no-one agrees on, but the totality of which is measured in at least $US100 billions. Add to this the spend by organizations on their 'facilities' and the figures may reach the trillions. As a topic for research and further management education, FM is a lucrative niche market that many universities have found profitable to pursue, as an adjunct to architectural faculties, to construction centres, to property economic ones and to business schools. Yet there is no agreement on what it is!

Those within FM still debate its status as a profession or a market (Green and Price, 2000), its current status as a field of academic inquiry (Cairns and Beech, 1999; Grimshaw, 1999) and future direction (Nutt, 1999, 2000). It is beset by paradoxes, among them an aspiration to the status of a strategic discipline when most practitioners operate at an operational level in their respective organizations (Grimshaw, 1999), a low base of rigorous research which pays scant regard to wider developments in social and organizational science (Grimshaw, 1999; Cairns and Beech, 1999) and an uneasy balance of professional traditions (Nutt, 1999, 2000). In the UK there is an, albeit small, body of evidence correlating board level FM in local authorities and NHS Trusts with improved 'core business' performance (Rees and Clark, 2000), but there is no still no general body of public domain evidence or conceptual framework to link property initiatives to business performance (see Haynes *et al.*, 2001). FM failed to meet Alexander's (1994) prediction of becoming 'part of the language of business', or at least of any business except FM provision, by the turn of the century, yet it continues to command apparently growing attention. Perhaps a history can help us understand some of these paradoxes.

3.2 Research methods

The normal methods of formal history research, such as the study of primary sources in archives, were not available for this research; however, thanks largely to databases available via Proquest, there is a body of secondary source material, principally articles in professional and business journals. Academic work is more limited but, as part of the ongoing occupier.org initiative (www.occupier.org), a number of academics were recently commissioned to access and evaluate the literature (both academic and professional) concerning 'the impact of property on occupiers' business performance'. Some 400 pieces have been accessed and analysed (Haynes *et al.*, 2001; Heavisides, 2001; Price, ongoing). The opportunity for quite such a comprehensive review does not actually present itself very frequently; more usually we update our reading in smaller modules. This history of FM is, then, underpinned by a comprehensive search for the themes and evidence that has informed the growth of FM. Such evidence is, however, in short supply.

In the UK, some evidence of FM's spread can also be gathered by tracking the development and content of the journal *Facilities,* as it developed from an in-house publication of Frank Duffy's firm DEGW, into an academic journal. It has also been easier to tap into people's recollections and to current business and research activity as is shown in the next chapter.

The growth of online databases has facilitated another approach to the study of management fashions: as Abrahamson (1996) demonstrated with reference to quality circles, the frequency of mentions in professional literature offers a guide to the spread, and in the case of quality circles, subsequent decline of a particular managerial approach. Scarborough and Swan (1999) apply the same approach to 'knowledge management' and 'learning organizations' and make, incidentally, a case for the former displacing the latter. FM is more robust (Price, 2002 and below).

As FM has spread geographically and developed several nuances of meaning it has been helped by, and has enabled, the creation of various enterprises (professional associations, academic journals and industries) all of which depend on the continued propagation of some meaning of the term in managerial and business conversations. That meaning does not of course have to be consistent. In this respect it is no different to many other successful management fashions (Price and Shaw, 1996, 1998) but FM can lay some claim to having outlasted most. Another pointer to the global spread of FM can thus be found in the dates of foundation and the subsequent spread of professional associations. As part of this research the English speaking bodies were contacted and asked for details of their date of foundation and subsequent growth. The home pages of the national associations affiliated to EuroFM were accessed for similar statistical information.

Finally, in an attempt to gain a global perspective, a short questionnaire was designed concerning the professional background of individuals, the timing and nature of their involvement in FM, their views on its current definition and business impact, and their views on its future. The definitions and impact statements were those provided by the participants in a 1999 Delphi study (Green and Price, 2000) while the future options were the four trails identified by Nutt (2000) and explored in more detail by contributors to Nutt and McLennan (2000). The questionnaire was distributed by e-mail to authors contributing to this volume who were asked to pass it on. The response was poor, with only 13 replies received, but the responses do serve to demonstrate, once again, the divergent views that people have of FM and its future.

3.3 The 1970s: first signs

FM has a very distinct 'origin story' provided by the IFMA (2002):

> In the early 1970s, two significant, simultaneous events occurred that helped set the evolutionary course of facility management. The use of independent, freestanding screens in the office environment – popularized in the 1960s – were gradually replaced with today's increasingly sophisticated systems furniture. The introduction of the computer terminal into the workstation challenged facility managers to solve computer, wiring, lighting, acoustic and territory problems. The office scene was becoming more complex and the facility manager needed guidance.
>
> At this time, many facility professionals were members of other international organizations, but those groups could not supply the information needed to manage the offices of the future. The first step toward the formation of a more specialized organization occurred in December 1978 when Herman Miller Research Corp. hosted a conference, 'Facility Influence on Productivity,' in Ann Arbor, Michigan, USA.
>
> This conference was the meeting place for the three founders of IFMA.[3] George Graves of Texas Eastern Transmission Corp., Charles Hitch of Manufacturer's Bank in Detroit, Michigan, USA, and David Armstrong of Michigan State University voiced a need for an organization comprised of facility professionals from private industry.
>
> In May 1980, Graves hosted a meeting in Houston to establish a formal organizational base for a facility management association. By the end of the meeting, a new organization known as the National Facility Management Association (NFMA) had a constitution and bylaws, temporary officers and plans to expand nationally.

In the survey mentioned earlier participants were asked whether they knew of an alternative origin, and universally no-one did, despite one respondent having first heard the term in 1968, and one having heard of it and become involved in 1974. The literature search extends the origin some 10 years further back: when searching for 'Facility Management' the oldest record revealed (Scott, 1971) describes 'Facilities Management' as the practice of banks outsourcing their data-processing operations. Only the abstract survives on-line but it states:

The credit card opens a new set of problems requiring intelligent communication between credit card processing centres on a nationwide basis. With this requirement comes a need for large terminal networks to allow merchants to obtain immediate credit information and also to record and process all information pertinent to a sale.

Banking and then more general data processing outsourcing dominates the early professional literature and the first surviving definition (Anonymous, 1972) offers this definition:

Facilities management is the complete takeover and operation of a clients data processing by a service firm.

Only in 1977 (Anonymous) is an alternative outsourcing reported:

Allied Chemical Corporation in Morristown, New Jersey, and Exxon Corporation's mathematics, systems, and computer department in neighbouring Florham Park have contracted with a supplier of temporary services to manage and operate their mail handling and distribution.

Offices and workplaces do not appear until 1978 (Anonymous):

Office lighting, one of the most vital yet least understood of all the 'tools of administration', can affect a variety of operational factors critical to success and facilities management. These include: (1) the human factor; (2) equipment and installation costs; and (3) operating costs.

And then again three years later (Magnus, 1981):

Due to increased awareness about work environment, the development of office automation, and the desire for increased productivity, the facility management/space planning field is developing rapidly.

3.4 The 1980s: a decade of dispersal

3.4.1 Literature (on Proquest)

By 1980 two distinct terms are in existence: Facilities Management describing outsourcing (primarily of computing but also of other office services) and Facility Management focusing on the work place as the shadow of office automation looms. It seems this is what brought the two together. In 1980, Proquest lists seven hits on FM, all concerned with data processing. In 1981 there are eight, with seven of those concerned with data processing. In 1982, however, there are 11 – six concerned with DP and seven with offices!

By November, it is time to *Say Hello to the Facility Manager* (Mills and Davidson, 1982):

Due to spiralling office-lease costs and to changes created by office automation, a vital and emerging role in the corporate structure is that of facility manager. As part of a long-term survival program, a facility manager maximizes the use of total office space while minimizing the costs of maintaining and retaining it. For example, facility management designs administrative facilities with organization needs in mind, to avoid getting locked into increasingly costly space. Besides being linked to the Consumer Price Index, one lease escalation clause that is increasingly common and expensive entails operating costs. As more and more office space leases are for shorter periods of time in order to bargain for rent increases, facility management techniques become even more vital. Office redesign and decentralization are techniques that help businesses combat rising rents. Companies without facility manager know-how can turn to consultants for help in holding down office costs.

In 1983 the balance is even at six titles each, and the first UK articles (Spooner, 1983; Hennessy, 1983) preserved on Proquest[4] appears. By 1984, with 17 titles, the analysis breaks down: CAD, Local Area Networks and *The Information Revolution for the Facility Manager* (Trayer, 1984) appear in the office-orientated literature. Outsourcing of data processing has all but disappeared, or been subsumed into the wider issue of office management. Issues of office design and automation dominate the 25 references for 1985 and Kaufman (1985) can ask, *Who Should be Running the Office, You or the Experts*:

> Expert help is often necessary for the manager who must choose among the alternatives in a rapidly expanding technological field. There are three basic alternatives in seeking outside help. An office consulting firm helps identify needs and problems and comes up with a plan describing what to do and how to do it. However, such consultants may have vendor preferences that do not best serve the client's needs. A service bureau could handle peak loads and so reduce administrative costs by lowering capital investments in equipment, space, and payroll. The key to sending work out successfully is to check references with existing customers and to visit the production facilities to be sure work is in capable hands. *Facilities management* [sic] involves an outside company managing one or more of a firm's in-house needs.

FM then is up and running in its new guise of expert workplace management, with or without IT, but the outsourcing theme remains and 'Facilities Management' as outsourcing continues to be distinct from 'Facility Management' as a body of knowledge and practice. There are also signs, however, of FM becoming intertwined with (some would say subsumed into[5]) the corporate real estate agenda. Yee (1986) reports the first surviving survey of US corporate practice:

> A survey, conducted by *Corporate Design & Realty*, of corporate real estate executives at randomly selected Fortune 1000 companies yielded these results: (1) The respondents manage a median of 131 facilities each, employing an average staff of 66 to control space with a median of 5 million sq. ft. per company; (2) A wide variety of labels are used to describe the real estate function; (3) Included in real estate staffs are such professions as asset managers (42%), leasing agents (40%), and engineers (37%); (4) The average length of service for respondents with their employers in real estate is 16 years (5) While 90% of the respondents report to vice-presidents, 24% report directly to the chief executive officer; (6) The internal generation of corporate real estate financing is reported by 59% of respondents, while 70% report use of straight line depreciation; and (7) Respondents ranked their priorities as space leasing, asset deployment, and facility management in descending order.

Meanwhile, in London, there is an echo from the past: Clarke (1986) questions, for management accountants, whether the time has come to outsource expensive main-frame computers. With hindsight, and the subsequent rush to networks, one might see this as a diversion; however, the office emphasis continues with the beginnings of a call for a greater understanding of what really makes the white-collar office productive (Wilson, 1987):

> Facilities management has the responsibility of transforming an organization's building or buildings into a company resource. A survey of 22 companies, published by Building Use Studies, classified companies' different views of their buildings as: (1) containers; (2) prestige symbols; (3) vehicles for industrial relations; (4) instruments of efficiency; and (5) inspirational forces. Such a variety of views exists due to the shortage of systematic research on the relation between white-collar productivity and the office environment. However, the execution of numerous office surveys and consultation exercises with staff clearly indicate that office workers perceive a clear connection between their output level and the quality of the work environment. White-collar productivity might be enhanced if office design could move from the individual and the task to the group and communication, from comfort and efficiency to reinforcing value systems and stimulating creativity.

At the same time Becker was developing concepts of organizational ecology (Becker, 1990) and calling for FM to understand the patterns of an organization, what Schein (1990) described as 'shared patterns of thought, belief, feelings and values that result from shared experience and common learning'.

Unfortunately FM was, by and large, not going to do so. Indeed a major conclusion from our occupier.org review (Haynes *et al.*, 2001; Price 2002) was the failure of both FM and the property and real estate arena to take on board the messages of pioneers such as Becker. Perhaps the seeds were being sown by the late 1980s when it is also possible to see FM moving away from the workplace *per se* and into general issues of building management. In the USA, early FM education seems to have emphasized design (King, 1988):

> A decade ago, there were no university facilities management (FM) degree programs in existence. Today, five schools offer bachelor's degrees and two offer master's degrees, Cornell University's FM program began in 1980 and has strong concentrations in human and environment relations, environmental psychology, ergonomics, human factors, real estate development, and city planning. Michigan State University offers a Master of Arts in Interior Design & Human Environment, with a specialization in Facilities Design & Management. Students select courses concentrated in social sciences, business administration, finance, and technology.

Whereas more traditional property and business issues were featuring more strongly in the professional agenda (Anonymous, 1989):[6]

> Educators can now receive the International Facility Management Association's (IFMA) model curriculum for bachelor degree programs in facilities management (FM). Delineated levels in 10 fundamental curriculum topics reflect basic familiarity, deeper levels of comprehension of concepts, and intensive exposure to topics and materials. The 10 topics are: (1) facility planning and design; (2) facility

operation and maintenance; (3) human/environmental factors; (4) organizational management; (5) financial theory and practice; (6) real estate planning and development practices; (7) integrative and problem-solving skills; (8) communication skills; (9) research and analytical methods; and (10) professional practice.

IT outsourcing reappears strongly in the 1988–89 literature, when industrial outsourcing also appears. Meanwhile there is evidence of expansion, e.g., FM reached Japan (Graham, 1988) where the reaction (Makoto, 1990) was thorough:

> Facility management (FM) is a concept from the US that was introduced into Japan around 1985. Research on FM has been conducted by the Facility Management Research Committee. From its inception, the committee conducted research on various subjects related to the diffusion and advancement of facility management in Japan under three newly formed subcommittees.

The sub-committees covered 'the analysis of facilities as objects of FM from the standpoints of the form of ownership, the type of building, and the composition'; 'guidelines as reference material of facility managers with the co-operation of scholars and specialists' and 'the tasks of the facility manager'. Attempts to ascertain the subsequent outcomes have not been successful.

Meanwhile systems outsourcing still dominates that part of the literature concerned with FM as a business. The recession, re-engineering, transformation, the virtual economy and major public sector outsourcing are all ahead, but only one article (Mandell, 1990) heralds what is to come. FM has established itself, and re-established a concern with systems outsourcing but the real decade of growth still lies ahead.

3.4.2 Facilities: the UK story

More than anyone else, Francis Duffy is perceived as having brought American thinking on new workplaces and FM as workplace management back to the UK (Hannay, 1993). Thomson (1988) confirms this and cites Duffy's address to a 1987 IAM-FMG conference:

> Facilities management has been around a long time. Now it has become a profession. It is strongest and most active in Britain in two main sectors, *electronics and financial services* [emphasis added]. This already gives us the clue as to why it is becoming so prominent today and why it is certain to grow in importance in the future: because the organizations which need facilities management most are those which are changing most rapidly, particularly through the adaptation of that great catalyst of yet more change, information technology.
>
> Why should I, an architect, be so deeply involved in the professionalization of facilities management? The answer is that facilities management is an essential part of the development of a new vision of design, a new kind of architecture. It is this coincidence of interests that has given me the opportunity, a rare and cherished privilege, to be involved in a small way in the development of a new profession. Let me try to explain.
>
> The traditional architect's view of his responsibilities has been both timeless and total – a great building is erected on a new site which is complete in every detail,

which transcends contamination by users whose role is to admire, conserve, perhaps to adorn, but never to change. This is the fantasy. The reality users know as inconvenience and architects as the inevitable deterioration of everything they make, especially interiors.

In New York in the Sixties, I first came across serious designers (who called themselves space planners and who were then despised as hacks by conventional architects) who understood something of the changing relationship between complex organizations and their stock of space. The unit of analysis was no longer the individual building, it had become the space held by organizations over time. The emphasis was no longer on the exterior shell but on the office interior.

Duffy's concern with interior architecture is clear; however, the tension between operating facilities managers and architects remained. Thomson also quotes Tickle (1986), quoting in turn the then president and founder of the UK's Association of Facilities Managers:

For too long the area of facilities management has suffered from the '*anyone can do it*' syndrome. Well they can't. But because of the situation too many designers have been getting away with too much. They haven't had enough discipline imposed on them. There's too much attention to aesthetics, not enough to function.

Butcher believes the AFM will do something about that; 'Our aim is to improve the relationship with designers, which is not too good at the moment. We want them to talk to facilities managers first. If we can talk about their designs we can get the aesthetics and function right', he said.

Facilities began publication in 1983 as a DEGW[7] subscription publication on workplace issues. From the beginning it was promoting FM as a movement, hence in June:

The beginnings of a movement

Speaking in April to the Manchester Society of Architects, Sir Monty Finniston[8] had this to say:

I should like the architect to become involved in what I believe the Americans call facilities management, giving constant attention to improving the efficiency of the building ... enhancing the environment inside ... and keeping it in good repair (providing the after sales service which is often lacking).

And in October, on facility managers:

What we have heard time and time again is that they exist and have already evolved a wide variety of informal ways of keeping in touch.

Other snippets from *Facilities* give a picture of FM in the UK through the decade,[9] though most articles are firmly operational.

April 1984 – Trying to overcome early resistance and professional rivalry?

Space planning is an American term dating back to the 1950s and 1960s introduced into the UK in the early 1970s and it still arouses derision. How does it differ from interior design or architecture? And what has it to offer the British Facilities Manager?

April 1987 – Education takes notice

The Open University, 'despite being an idea of the hallucinatory and unfortunate sixties [sic]' is considering 'bolting on' a course specifically related to FM to an existing course on 'the effective manager'. Meanwhile Strathclyde, having completed a successful series of seminars on facility management is designing a course as part of a new MSc in Architecture and Building Science. Is FM a responsibility to be assumed by Design or Business Studies? Ultimately it should be self-regulating like 'Law, Medicine or Architecture'.

November 1988 – The shadow of outsourcing looms

'Will the FM Manager be contracted out of a job?'

May 1989 – More signs of schism[10]

A group of 'dissidents' from MIT reportedly 'affront' the IFMA by trying to pull 'the management rug from underneath their hard grafting feet'. It is suggested that 'perhaps there is a place in FM for special interest groups but not for quarrels between disaffected opportunists'.

November 1989 – Early moves into Europe

The IFMA and NEFMA are mentioned, as is news of an October 1988 FM course at Eindhoven; cultural differences are seen with The Netherlands having 'a customer oriented building culture finely tuned to the intimate psychology of the workplace'.

May 1990 – Portents of the 1990s

EuroFM is launched and holds its first international conference in Glasgow. Duffy attacks facility managers for 'the decade of lost opportunities' and The Centre for Facilities Management argues that, for FM, 'data is power'. The Japanese claim to be systematically measuring changes in staff response, though no further details emerge. The IFM and AFM are reported to be 'quarrelling', and schools contracting out FM become news.[11]

August 1990 – An attempt to get back to the origins

Tony Thomson (then recently moved from Hewlett Packard to DEGW) argues that 'Real Facilities Management' is to be found in facilities planning and says:

> I see the failure of facilities management in the UK as missing this most central of issues and immersing itself in long and complicated explanations, together with producing lists of activities that seemingly encourage more and more diverse skills into the fold and add to the overall confusion.

1991

Facilities transfers to MCB Press with Keith Alexander of the Centre for Facilities [sic] Management (CFM) as consulting editor. It begins to develop a reviewed section.

Summary

By the end of the decade then FM in the UK has established itself, but is also split into competing factions, professional and academic. All the signs of future problems are there.

3.5 Interlude: professional bodies

The history of the various professional bodies involved in the story of FM fails to fall, even approximately, into the story by decades. For completeness it is taken as here as a separate topic.

3.5.1 International Facility Management Association (IFMA)

Founded in 1980 as the National Facility Management Association, but formally International since 1982 when a first Canadian Chapter was established in Toronto, the IFMA is clearly the largest body in the game. Membership growth was rapid, with 2,363 members joining in the first six years. Growth since then is shown in Figure 3.1. Double figure percentages were sustained through the 1980s and averaged over 6% p.a. from 1991 to 2000.

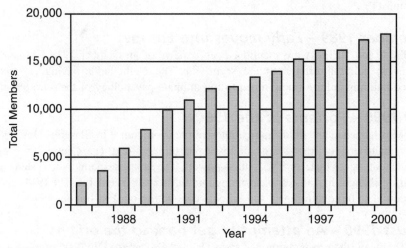

Figure 3.1 Membership growth by year: IFMA.

Outside North America, chapters were established in Hong Kong (1992), Sweden (1994), Belgium, Italy and Finland (1995), Germany (1996), Luxembourg (1997), Austria, France and Nigeria (1998), Argentina and Holland (1999), and the UK, the Czech Republic and India (2000). There is some correlation with later (or even no) establishment of chapters in countries where larger national associations were in place (Holland, UK and Australia).

3.5.2 British Institute of Facilities Management (BIFM)

The BIFM was formally born in 1994 out of the 1993 merger of two competing groups, IFM and AFM. There are two versions of the early history.

The former, and founding, chairman of the Association of Facilities Managers (AFM) reports the foundation of the AFM in 1986, following the efforts of a group of volunteers who first expressed a willingness to participate at a London Conference in 1984. Cost reduction was on the agenda from the beginning (an early slogan was 'Managers who can save millions'). The Association distinguished itself from IFMA from the outset, and saw its members as 'real facilities managers', i.e., people doing it for large corporations.

The Institute of Facilities Managers was formally founded in 1990 (sponsored by the Institution of Administrative Managers) but had been active as a discussion group (FMG) since around 1985, and had in fact registered the title before coming to use it.

Lingering memories of the differences survive to this day. One influential member from the AFM recollects 'we were *FMS* [facility managers, emphasis added] – they were simply office manager types' while one from the IFM suggests 'we were more strategic – it was a blow to the AFM founders when the membership voted for institution in the title of the new organization'.

Both groups claim that Frank Duffy was involved in their foundation and in a 1987 speech he himself stressed his involvement in both (Thomson, 1988). By the time of the merger the AFM was apparently the larger group and according to the AFM view, it was their administrative structure that transferred to BIFM. The IFM, however, claims that the precursors to professionalism in standards and educational requirements came from them. Whether either of these claims is entirely correct is unclear, but these views are further evidence of the conflicting operational and strategic design traditions in FM.

Membership growth since 1995 (Figure 3.2), while not spectacular, has, in percentage terms, exceeded that of the IFMA, and the institute now has a national membership of over 6000 and a seemingly secure niche.

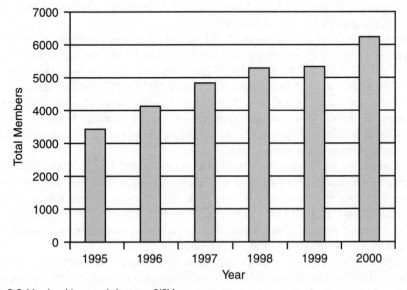

Figure 3.2 Membership growth by year: BIFM.

3.5.3 Diverging definitions

The preceding historical background gives some insight into how the two large FM institutes have arrived at different definitions. The IFMA remain true to what they term the classic definition of facility management as 'the practice of co-ordinating the physical workplace with the people and work of the organization. It integrates the principles of business administration, architecture and the behavioural and engineering sciences'.

For the BIFM facilities management is 'the integration of multidisciplinary activities within the built environment and the management of their impact upon people and the workplace'.

Note the difference: in the US definition, the workplace is treated as a potential lever that impacts on the organization; in the UK it is something that FM impacts. Those who were managing facilities, under whatever title, in effect adopted the new name but resisted some of the implications or widened the definition to incorporate what they did, i.e., procured and, sometimes, operated and maintained buildings.

3.5.4 The EuroFM Network

Variants on both definitions have developed in Europe; the websites of members of EuroFM (all visited in January 2001) provide a snapshot. 'Facility' rather than 'facilities' dominates (except in Denmark and Ireland) but the sense in which the word is used ranges from the formal IFMA focus on people, process and place through to a drive for cost reduction through services outsourcing.

The network was launched at the 1991 Glasgow conference mentioned above, facilitated by the Centre for Facilities Management (CFM, 2002) in collaboration with the Danish Facilities Management Association (DFM) and the former Dutch association, NeFMA. The DFM now claim 115 members; NeFMA and two other Dutch associations formally merged in 1995 to form FMN (Facility Management Netherlands), now the second largest group in Europe with 1500 members.

Facility management is seen by the NFM as 'the integrated planning, realization and maintenance of building, equipment and services'. The DFM does not define facilities management but suggests that it is 'more comprehensive than building operations'.

The Hungarian Facility Management Association (HUFMA), despite claiming facility management activity to be not as widespread as in other EuroFM countries, date their foundation to 1990. Members are reported as mainly coming from the National Association of Building Maintenance. The Technical University of Budapest includes FM in a post-graduate building management course.

The Finnish Facility Management (FIFMA) started in 1993 as an IFMA Chapter and stays true to the classic IFMA definition, claiming its special viewpoint as the working environment. They now have over 80 members. Facility Management Austria started in 1995 and affiliated as an IFMA Chapter in 1998; they have 'over 50 members'. They claim to represent suppliers, professionals and educational institutions and assert that in Austria, FM is 'not cost cutting but adding value'.

In contrast, the German Facility Management Association (GEFMA: Deutscher Verband für Facility Management eV), in a 50 billion euro internal market (outsourced or in-house), aims to 'allow organizations to profit from cost reductions of up to 30%'. With

280 current members, their main objective is stated as growing the market for FM. In the German-speaking world the contrasting traditions seem stark.

Norway claims a network, linked to IFMA and chaired from their Building Research Institute, that seeks to respond to firms switching from 'Operations and Maintenance' to total FM. Three research projects are being promoted in life cycle assessment, space management, and adaptability and condition assessment. In Spain the Sociedad Española de Facility Management, founded in October 1998, has so far attracted 70 members. Interestingly their executive is constitutionally balanced between education providers, facility managers and service providers: a formal attempt at harmony between FM's three strands that does not appear to have been replicated elsewhere. Corporate sponsorship has yet to materialize.

Swiss (Facility), French (Facility) and Irish (Facilities) Associations have either not yet developed sufficiently to announce a structure and direction on their respective sites or have had other priorities.

There seems to be no detectable pattern in the development of the various national bodies: each association seems to emphasize a part of the total mosaic that is FM. Each seems to have organized around the particular nuances of its founders, and with the exception of NeFM, none has become the broad church that BIFM represents and only two have stuck with the classic IFMA line. David Rees (pers. com., 2001) does suggest, however, that EuroFM is currently adopting the BIFM rather than the IFMA framework for its future competencies model.

3.5.5 Asia Pacific

Facility Management Association of Australia (FMAA)
According to the FMAA website the association was founded as the International Facility Management Association of Australia in 1988 and changed its name in the early 1990s. The name change suggests that an IFMA chapter had been established but this was not the case. Brian Purdey (pers. com., 2001) suggested that the name change may have been partly due to local resentment of American domination and money being extracted from Australia to the USA, and partly to the separate markets. Australia, with a government shifting to the right and a surplus of under maintained property, especially in the public sector, was more akin to the UK than the US. He further added that some strong personalities active in Australia at the time may have contributed to clashes. Waddell (pers. com., 2001), however, provided a somewhat different view:

> There was never a split. What happened was that several guys got together to form an Australian association. It was an incorporated company that was established with no direct links to IFMA. The name was International Facility Management Association of Australia and was a properly constituted company under Australian law. We have on file a letter from Ed Rondeau, IFMA Chair at the time, giving us the right and authority to use the IFMA name for this purpose. Various IFMA Chairs have seen a copy of this letter. So, there was no split as there was never a joint association.

Apparently, despite the letter, some 'pressure' for a change was exerted, and the FMAA board decided a title which reflected their national identity would better serve their association's and their members' needs.

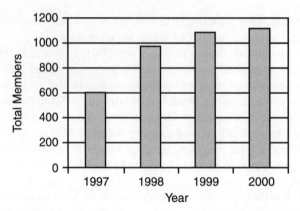

Figure 3.3 Membership growth by year: FMAA.

Membership statistics available date back only as far as 1997 (Figure 3.3) but record growth rates comparable to those of other associations. Indeed the 60% increase in membership from 1997 to 1998 is the largest annual increase (in percentage terms) that any of the associations analysed has recorded. A subsequent slowdown, again mirroring trends in IFMA and BIFM, is apparent.

Japan Facility Management Association (JFMA)

In 1999, trade press reports indicated completion of the third certification process for the qualification of CFMJ (Certified Facility Manager Japan) through a certification examination which 770 candidates had successfully completed. The hiring of a CFMJ as a bidding condition for tenders for a new government building provides further evidence of the rising recognition of FM in Japan. Unfortunately approaches to the Japan Facility Management Association in 2001 did not elicit a response[12] nor was it possible to find an English translation of the JFMA website. That site does, however, contain a section[13] with the English heading 'What's FM' in which a diagram appears to locate FM within the context of the well known Deming 'plan-do-check-act' cycle. Given frequent reports of the reception Dr Deming received in Japan, whose most prestigious quality award carries his name, the page offers a fine example of FM's diverging definitions being influenced by particular national managerial and business concerns.

Singapore

Andrew Green (pers com., 2001) reported that as of June 2001:

> The Singapore Chapter of IFMA is chartered and its registration application is currently being considered by the Singapore authorities. If the usual time applies for these things, we should be fully registered by end-August! We currently have some 58 members – remembering that we are allocated any member in our catchment. Therefore we are the nearest Chapter before Hong Kong for members living in Malaysia, Indonesia, Thailand, etc.

Previously the discipline in Singapore was largely thought of in terms of building maintenance so this effort to lift the profile may, perhaps, mark a new departure.

3.5.6 South Africa

The South African Facility Management Association (SAFMA)'s website records that their founding in June 1997 occurred because '. . . certain large South African institutions wanted to improve and enhance their facilities management operations'. The South African Property Owners Association and Property Council were heavily involved, and by that date, three years after the introduction of majority rule and the end of economic sanctions, many large facilities suppliers were actively promoting FM and outsourcing in the country, up to and including officials at ministerial level. In the opinion of two people who were active in South Africa at the time (Fari Akhlaghi and Gerry Baron-Fox), the introduction of the concept was 'supplier led' to an unusual extent, and many investors were willing to back sale and FM enhanced lease back of properties. This may explain both the SAFMA definition of facilities as 'the physical entities where products or services are created, stored and distributed' (i.e., anything that could be sold) and FM as 'the management of specific physical entities to enable the business to carry out its core functions' (a definition which veers towards outsourcing). Three years later, however, the SAFMA and IFMA concluded an interim agreement on collaboration.

3.5.7 CIB W70: FM and Maintenance

The International Council for Research and Innovation in Building and Construction (CIB) chose its conference in Brisbane (November 2000) to announce the renaming of its W70 working commission as *Facilities Management and Maintenance*. At the same conference, Jones *et al.* (2000) provided a useful summary of W70's history including an analysis of the themes of papers presented at previous biennial conferences. It seems the commission was established in 1977 to examine 'emerging issues surrounding the utilization of buildings' with three themes: maintenance planning, modernization and tenant participation in building management. Its early focus seems to have been social housing. Subsequent development may be traced through the growth and diversification of topics presented as conference papers. Facilities management appears on the conference agenda in 1990, overtakes 'maintenance planning', in terms of numbers of papers, in 1996, and becomes the largest single topic in 1998 (the year when 'facilities management' was formally incorporated into its remit). By 2000, FM dominates the conference, with, it must be said, an emphasis on FM as a frequently outsourced maintenance process making claims to a more strategic role in integrated asset management.

Grimshaw (pers. com., 2001) confirmed much of the above:

> The appearance of FM papers at W70's meetings in the early 1990s reflected the fact that a number of people working in maintenance were becoming aware of FM and W70 was a natural outlet for such papers. Lee Quah, the then W70 co-ordinator, took the view that FM was a technical function allied to maintenance and not a management discipline linked to organizational development (the line being promoted by Keith Alexander). Therefore she saw FM as a natural development of W70's work (not everyone in W70 agreed with her especially the maintenance technologists).

In 1996 (or thereabouts) Keith Alexander (and EuroFM) proposed a new CIB working commission to deal exclusively with FM. This was initially supported by the CIB Board but support was withdrawn when Lee Quah kicked up a fuss (saying W70 was already doing it). A number of other commissions also said they had a legitimate claim on FM. At a meeting in Rotterdam of interested parties it was agreed that FM should reside in W70 but that liaison would be maintained with other interested commissions. So the idea of a dedicated commission for FM was dropped. However, a task group on FM was formed chaired jointly by myself and George Cairns – this is still in existence but has not been very active. The 1998 and 2000 W70 symposia have seen the consolidation of FM in W70 to the point where this is becoming its main focus and the maintenance technology aspects are withering. Its title is now 'FM and maintenance'.

Again the competition between operational, strategic and architectural views of FM is evident. By the time of the 2000 symposium, while FM may have become a main focus of W70, it had, in the process and with a few notable exceptions, remained determinedly operational.

3.6 Conclusion

In just 30 years a diverse range of building and workplace related activities have collected together under the banner of facility (or facilities) management: a banner many claim as a new professional discipline. Like 'jazz' or 'modern art', or for that matter any business fashion, the term embraces many quite different interpretations, yet they are all linked by some common characteristics or concerns. Meanwhile the 1990s have also seen the rise, and increasing global spread, of the facilities management industry.

Clearly there is no single definition that satisfies all concerned, and no one focus (strategic, design, operational) that is universally accepted as being the core of FM. What is evident, however, is that many people who are involved in areas of activity that are commonly considered to be part of the spectrum of FM are becoming increasing assertive about the validity of their title of 'facility manager' and their right to exist in a separate and identifiable profession. Questions remain, especially the confusion between the professionals and those selling facilities management (Green and Price, 2000) and the insecurity of FM's ethical dimension (Grimshaw, 2001). There is also a case to be made (Price, 2002) that FM must develop the evidence and confidence to express its contribution in 'business' relevant language if it is to truly reach its potential.

Endnotes

1 Santayana, G. (1905) *The Life of Reason*, vol. 1, ch. 12. A remark usually repeated as 'Those who ignore history . . .'.
2 Honda and Fujitsu recently released the results of computer simulations suggesting that a car manufacturing plant built not on the classic production line model, but as a system of autonomous agents emulating a Complex Adaptive system, would be 89% more effective. http://www.wmg.warwick.ac.uk/mcn/issue5.pdf

3 Don Young, currently vice president communications for IFMA confirms this and adds that the early NFMA participants had been attendees at workshops organized by Herman Miller's 'Facility Management Institute' whose researchers he credits with developing the original concept of FM as 'people, process and place'.

4 *Facilities* was launched in the same year but its contents are not indexed.

5 One interpretation of its future history is a selective competition between IT, Real Estate and FM for the corporate high ground.

6 However, seven years were to elapse until Cornell University could announce that its Facility Planning and Management Program was the first in the world officially recognized by IFMA (24 September, 1996) http://cunews.cornell.edu/releases/Sept96/facility.program.ssl.htr.

7 Duffy being one of the founders.

8 Former chairman of British Steel and Chancellor of Stirling University. From his remarks it is clear he was referring to what the Americans called Facility Management. This is the earliest reference located that indicates the source of subsequent confusion.

9 The selection and the comments in the headings are the author's.

10 There seems to be no evidence that can cast further light on this incident, however, the contrast between the 'strategic' designers and the 'hard-grafting' practitioners appears stark.

11 The first sign of the conservative privatization of service agenda that was to drive much of the industry in the UK during the 1990s.

12 E-mail inquiries were addressed to the association's contact address and their vice president. A fax enquiry was transmitted. No response was forthcoming.

13 www.jfma.or.jp/fm/fm_gaiyou.html. The page is copyrighted so the diagram cannot be reproduced here.

References and bibliography

Abrahamson, E. (1996) Management fashion. *Academy of Management Review*, **21** (1), 254–85.

Alexander, K. (1994) A strategy for facilities management. *Facilities*, **12** (11), 6–10.

Anonymous (1972) Facilities management is here to stay. *Administrative Management*, **33** (7), 16.

Anonymous (1977) Facilities management comes to the mailroom. *Modern Office Procedures*, **22** (3), 85.

Anonymous (1978) Task/ambient lighting: often the answer but not a panacea. *Administrative Management*, **39** (6), 26.

Anonymous (1989) IFMA's model for FM degree provides guidance to institution. *Facilities Design & Management*, **8** (9), 37.

Becker, F. (1990) *The Total Workplace: Facilities Management and Elastic Organisation* (New York: Praeger Press).

Cairns, G. and Beech, N. (1999) Flexible working: organisational liberation or individual straitjacket? *Facilities*, **17** (1/2), 18–23.

CFM (2002) Centre for Facilities Management. www.cfm.salford.ac.uk

Clarke, T. (1986) Time to change to Facilities Management? *Management Accounting*, **64** (3), 16–17.

Duffy, F. (1993) In: Hannay, P. (ed.) *The Changing Workplace* (London: Phaidon Press).

Graham, M.C. (1988) FMs rising star in the land of the rising sun. *Facilities Design and Management*, **7** (2), 62–63.

Green, A. and Price, I. (2000) Whither FM? A Delphi study of the profession and the industry. *Facilities*, **18** (7/8), 281–92.

Grimshaw, B., (1999) Facilities management: the wider implications of managing change. *Facilities*, **17** (1/2), 24–30.

Grimshaw, B. (2001) Ethical issues and agendas. *Facilities*, **19** (1/2), 43–51.

Hannay, P. (1993) Frank Duffy (ed.) *The Changing Workplace*, London: Phaidon Press.

Haynes, B., Fides Matzdorf, F., Nunnington, N., Ogunmakin, C., Pinder, J. and Price, I. (2001) *Does Property Benefit Occupiers? An Evaluation of the Literature*. occupier.org Report 1. www.occupier.org/reports.htm

Heavisides, R. (2001) *Output based Facilities Management Specifications in the National Health: Service: Literature Review and Directional Outcomes*. Occupier.org Working Paper 1. www.occupier.org/working_paper1.pdf

Hennessy, E. (1983) The way we work now. *Director*, **37** (4) November, 82–5.

IFMA (2002) *Facts About The International Facility Management Association*. www.fmkp.cz/fm/fakta_ifma_en.htm

Jones, K.G., Mureithi, S. and Then, D.S.S. (2000) Facilities management & built asset maintenance: a research agenda for the 21st century. In: Then, D.S.S. (ed.) *Moving towards Integrated Resources Management: Proceedings of the Brisbane 2000 CIB W70 International Symposium*, 609–15.

Kaufman, M. (1985) Who should be running the office? You or the experts? *The Office*, **102** (1), 172–3.

King, S. (1988) Daring to get that FM degree. *Facilities Design and Management*, **7** (8), 44–7.

Magnus, M. (1981) Work environment: its design and implications. *Personnel Journal*, **60** (1), 27.

Makoto, T. (1990) Facility management enhances usefulness of buildings. *Business Japan*, **35** (10), 97–102.

Mandell, M. (1990) Cost cuts are in the mail. *Nations Business*, **78** (1), 30–1.

Mills, D. and Davidson, N. (1982) Say hello to the facility manager. *Executive*, **24** (11), 52.

Nutt, B. (1999) Linking FM practice and research. *Facilities*, **17** (1/2), 11–17.

Nutt, B. (2000) Four competing futures for facility management. *Facilities*, **18** (3/4), 124–32.

Nutt, B. and McLennan, P. (eds) (2000) *Facility Management: Risks and Opportunities* (Oxford: Blackwell Science).

Price, I. (2002) Can FM evolve? If not what future? *Journal of Facilities Management*, **1** (1), in press.

Price, I. and Shaw, R. (1996) Parrots, patterns and performance (the learning organisation meme: emergence of a new management replicator). In: Campbell, T.L. (ed.) *Proceedings of the 3rd Conference of the European Consortium for the Learning Organisation*, Copenhagen.

Price, I. and Shaw, R. (1998) *Shifting the Patterns: Breaching the Memetic Codes of Corporate Performance* (Chalford, UK: Management Books 2000 Ltd).

Proquest (2002) www.il.proquest.com/

Rees, D.G. and Clark, E.V. (2000) Professional facilities management in public sector organisations. *Facilities*, **18** (10/11/12), 411–20.

Scarborough, H. and Swan, J. (1999) Knowledge Management and the Management Fashion Perspective. In: *Proceedings of British Academy of Management Refereed Conference, 'Managing Diversity'*, Manchester. http://bprc.warwick.ac.uk/wp2.html

Schein, R.H. (1990) Organizational culture. *American Psychologist*, **45** (2), 109–19.

Scott, C.R. (1971) Why Facilities Management. *Bankers Monthly*, **88** (10), 38.

Spooner, P. (1983) Why Smiths Industries decentralised its DP. *Chief Executive*, April, 47–8.

Thomson, T. (1988) *Facilities management consultancy: the development of a plan for its proposed implementation in an architectural company*. MSc Dissertation, University of Reading.

Tickle, K. (1986) Facilities for the future. *Interior Design*, July, 34–6.

Trayer, G.T. (1984) The information revolution for the facility manager. *The Office*, **99** (1), 159.

Yee, R. (1986) The 1986 Corporate Real Estate Executive Survey. *Corporate Design & Realty*, **5** (2), 56–7.

Wilson, S. (1987) *Making Offices Work Management Today*, London, October, pp. 91–3.

4

The development of facility management

Ilfryn Price*

Editorial comment

The facility management (FM) industry can basically be divided into three categories: facility managers, specialist consultants and service providers. Facility managers are responsible for particular facilities either for one organization or on behalf of a number of organizations and function largely at a *strategic* level. Specialist consultants provide targeted expertise in areas as diverse as architectural, structural, fit-out, services and landscape design, cost management, project management, environmental assessment, due diligence, energy planning and dispute resolution, and function largely at a *tactical* level. Service providers include cleaning contractors, insurers, furniture suppliers, security, construction, catering, fleet management and a range of other support services, and function largely at an *operational* level.

This distinction is important. There needs to be a balance between the various roles that collectively combine to form the FM industry. Professional bodies recognize this diversity through levels of certification and career development; however, there is an imperative that the higher strategic aspects are allowed to flourish as it is at this level that the maximum value-adding occurs. Too much of a focus on operational aspects will have the effect of slowing the growth of the profession, or worse still, minimizing its relevance to society.

In its widest sense, the FM industry is the largest contributor to gross national product on a worldwide scale. The importance of the industry to global issues such as the need for sustainable development, use of renewable energy sources, the minimization of environmental impact and habitat destruction, reduction of greenhouse gas emissions, and so on, is undeniable. Buildings alone consume more resources than any other infrastructure type, including transport. The quality of facilities is directly proportional to perceptions of living standards and economic progress, and indeed underpins the very fabric of our society.

* Sheffield Hallam University, UK

The events of 11 September 2001 have changed the world forever. The way future facilities will be designed and managed must take account of external threats that can seriously damage societal processes. Business continuity and disaster recovery have become key considerations alongside financial return and operational efficiency.

Workplace strategies are at the heart of the FM discipline and represent the greatest opportunity for value enhancement. The FM industry needs to remember who are its customers, and what products and services they are looking for. The bottom line is providing value for money, but the definition of value is broad and the possibilities for optimization are virtually limitless.

4.1 Introduction

After reaching the UK in 1983 and Japan in 1985, facility management (FM) went global during the 1990s, becoming more diverse as it did so. In the previous chapter the FM literature in 1990 was left with a single reference to cost cuts being 'in the mail' (Mandell, 1990), yet with hindsight it is clear that both the storm clouds and the opportunities were gathering. Core competencies (Prahalad and Hamel, 1990) were just entering strategic thinking with the corollary that if you were not world class, you should 'buy it, don't make it'. Information technology, especially the network, was becoming ubiquitous (even if IBM had two years to enjoy its old hegemony) bringing with it business process re-engineering (BPR) and the call to not just make work more efficient but to eliminate the unnecessary. White-collar productivity overall had stopped rising in 1975 and 'white collar value was there to be extracted'.[1] Markets, relishing new and faster information, wanted to spread risk by investing in separate business entities rather than paying rent to the managers of diversified conglomerates to spread it for them. That, and the lower transaction costs enabled by IT, progressively reduced the rationale for doing everything in house.[2] Globalization was becoming a reality but hardly anyone had heard of the World Wide Web.

4.2 The UK FM market

With cost reduction dominating many boardroom agendas, real estate, facilities and any 'non-core' activities came under the spotlight. Engineering or business service providers who regarded such services as their own core competency were quick to respond. IBM outsourced its former in-house property and facilities operations as Procord in 1991 (acquired by Johnson Controls four years later). Facilities management as an industry had arrived with a vengeance.[3] In 1992 Drake and Scull, an FM firm with roots in general engineering services, signed two deals to provide operations, maintenance and miscellaneous support services to BP Exploration's Wytch Farm oilfield and to British Airways' engineering operations at Heathrow. Their marketing director at the time, Gerry Barron-Fox, had spent part of the 1980s working with Herman Miller and brought back to the UK a mix of their history in facilities management and the tradition of facilities outsourcing.[4] Other engineering services firms plus cleaning and catering companies were not slow to join in. Meanwhile the construction industry was badly affected by the recession of the early 1990s and several of the major players created facilities management subsidiaries.

Politics favoured the industry. The Conservative government in the UK had spent the late 1980s divesting 'fully privatizable' entities (e.g., steel, gas and telecoms) and then turned its attention to outsourcing. Some outright sales went through, including the privatization of AEA technology and sale of their facilities arm to Procord, but most attention switched to public services. Health trusts were created to bring 'market forces' into the National Health Service and required hospitals to submit their support services to compulsory competitive tendering (CCT), a discipline also mandated for local authorities.[5] Significant opportunities for single and multiservice FM providers developed. National government departments and especially defence establishments were likewise under pressure to 'market test' and most non-military support work was outsourced by the middle of the decade. The Property Services Agency (PSA), which had provided most public sector property support and a significant service to defence establishments, was privatized through a mixture of sale and buyout in 1993 (Cameron, pers. com., 2001).

As these measures matured over the second half of the decade into the private finance initiative (PFI), the FM industry became even more attractive to construction firms. In order to bid on PFI projects they either had to develop their own FM arms or develop sustainable working relationships with integrated service providers. By the decade's end those who had succeeded in gaining re-listing on the London Stock Exchange as business service providers (a classification which attracts a premium in expectation of higher returns) were generally reported in the trade press as 'more successful'. As the decade closed corporate PFI or infrastructure management was being written about as the new arena, though commercial deals have proved slow to close. Only three of 1993's top players survive in the top 20 players in the industry;[6] the others have all entered FM in the last seven years.

4.3 Academic and professional developments

FM was not lost as an opportunity for 'cash-strapped universities' (Leaman, 1992). In the USA this seems to have manifested itself in undergraduate courses. Six were accredited by the IFMA in 1996 (with Cornell boasting of theirs being the first) and the emphasis has tended to be on the design and planning of the workplace. In the UK (where undergraduate courses are state funded) the emphasis has been on part-time, Masters level programmes for those in employment. In the UK, Strathclyde's Centre for Facilities Management (CFM) and University College London were joined by Herriot-Watt, Sheffield Hallam,[7] the University of the West of England, Salford and Leeds Metropolitan. In 2002 an MSc in Corporate Real Estate and FM is reportedly under development at Reading and in Infrastructure Management at Salford, where the CFM moved in 2001. A new Centre for Advanced Building Research (CABER) has been opened by Glasgow Caledonian and is also entering the MSc market. There is no common pattern. FM has been launched in design and planning, property, construction, and business schools, a range as broad as the range of professions and activities attracted into FM. Accreditation started in 1997 but, given that the BIFM was then only three years old, it emerged in parallel with, rather than before, the associated academic centres, and there was not such an established professional line to dictate what could and could not succeed as academic FM. Australian universities seem to have been slower to develop FM, with only two Masters level programmes locatable online in 2001, one an MSc and one an optional specialization within an MBA. Competition does, however, appear to be increasing, and in 2002 several postgraduate courses, including at least one going to doctoral level, have appeared.

4.4 Developing practice

In the UK, the BIFM has conducted three surveys of members' responsibilities (in 1997, 1999 and 2001). The first two showed, in decreasing order, members' backgrounds as estates/property, office administration, service engineering, general management and surveying. Only 2% had a hotel services background.[8] Members were mainly employed in financial services, education/training consultancy, communications and property management. Outsourcing, surprisingly, declined slightly between the first two surveys, perhaps because political pressure for it had diminished. In 1997, members saw lifting FM to a strategic level as an issue for five years hence. In 1999 it was seen as an issue for now.

Rees (1997, 1998) conducted two surveys of FM in the UK National Health Service. Between 1995 and 1997 (when the surveys were done) integrated structures with all estates and hotel services in a single management structure, usually named a facilities directorate, expanded from 74 to 82% of the survey while the representation at senior management levels increased from 10% of the sample to 24%. The service has since gone further and currently has an initiative designed to ensure a cadre of suitably qualified personnel to occupy 'Facilities Director' roles in the future. Developments have, however, tended to be led by the former 'estates' side of the support services spectrum and it is their association that rebadged itself as 'estates and facilities'. At least until recently separate associations still existed for caterers and 'domestic services'. In universities a similar trend is apparent: five years ago only two had appointed a Facilities Director, now there are six but the estates role remains strong. In Local Government the picture is even more confused and only 30% of authorities have a single manager below Chief Executive level with unified responsibility for the range of what would be considered 'facilities services' whether carried out in house or procured. That individual is most likely to have a property or finance background. In commercial organizations an accurate picture is harder to obtain but unpublished research by students at Sheffield Hallam shows most organizations preferring to source services on single, relatively short-term contracts. The FM industry certainly see conservative, and even protectionist, policies as relatively rife and 'intelligent clients' as rare (Green and Price, 2000).

The RICS, as the dominant professional body for traditional property specialists was relatively slow to respond to the rise of FM,[9] but now has an FM faculty and has recently been a lead sponsor of occupier.org. There remains, however, a body of opinion that suggests that FM should concern itself with operations and maintenance and leave property and design to the respective specialists, and certainly operations and maintenance still dominate the FM market place.

It is harder to report on any other country's experience. Don Young, Vice President of Communications for the IFMA, kindly provided a view (2001) on the state of play in North America and painted a picture similar to that found in the UK:

> Today, IFMA's classical definition of facility management is again having to assert itself from the word-coining exercises of related professions, such as real estate, that would like to claim FM responsibilities as their own (spending levels of new real estate projects have dramatically decreased and some companies are treating the

acquisition part of it as a commodity to be outsourced). Some say their real estate jobs/duties increasingly are being questioned and targeted for elimination. Groups representing that profession are trying to stretch real estate to encompass the classical people, place and process focus of FM. In effect, with some success, real estate is repackaging FM into 'corporate infrastructure resource management' (CIRM) and similar terms. But, when you look at what composes CIRM, it's all FM. Additionally, the interior design profession, led by the professional membership organizations, also is stretching itself to encompass people and process issues. This is evident in reviewing interior design-directed research on 'productivity.'

4.5 Defining FM

Eleven members of a 1999 Delphi group (Green and Price, 2000) offered 11 definitions of FM (Table 4.1). As part of research for this chapter contributors to the volume were polled and asked to rate their agreement on a scale of 1 (strongly disagree) to 5 (strongly agree). Table 4.1 shows the average and individual response. Every definition bar one has at least one disagreement and all have at least one supporter, normally at the strong agreement level. The sample cannot be claimed as statistically valid but it reproduces the disagreements that were found, which could not be resolved, in the earlier research (Green and Price, 2000). One participant, an academic, found no definition absolutely to his liking and offered the following comment:

> Bernard Williams Associates defines facilities[10] as 'the premises and services required to accommodate and facilitate business activity'. In other words, facilities are the infrastructure that support business. This is a wide definition and is intended to cover not just land and buildings but other infrastructure such as telecommunications, equipment, furniture, security, childcare, catering, stationery, transport and satellite work environments (home, car, client office, airport lounge, hotel lobby, etc.). Facilities should not be seen as an overhead, but rather as an integral part of an 'ecosystem' necessary to enable people to perform at their best. The most valuable asset of the organization is its people. The two main support activities for any organization are therefore human resource management and facility management.

The respondents in the same sample were also given an opportunity to rate their agreement with views of FM's impact on the business planning process given by the original Delphi panellists. Again Table 4.1 summarizes. Perhaps not surprisingly, given the survey, only one respondent was prepared to agree with the view that FM actually does not impact on the business planning process, though cost optimization was the impact that received most agreement. In practice the results seem to support the observation that, despite claims to a strategic role, FM remains largely operationally based.

4.6 Developing knowledge

The pattern of knowledge development in FM, or perhaps the lack of it, as revealed through academic and professional literature seems to support that last observation. This evaluation was made in the occupier.org review (Haynes et al., 2001; Heavisides, 2001). The occupier.org database references some 300 books, papers, articles and web sites,

Table 4.1 Responses to a survey question regarding as set of definitions of Facility Management (FM)[11]

Definition	Average	Individual Scores												
1 Total management of property, plant and human resources to improve service quality, reduce operating costs, and increase business value to provide competitive advantage.	4.31	3	5	2	5	5	4	5	2	5	4	5	5	5
2 Facility Management is a business, which derives profit in exchange for managing risk transfer between the user and the supplier of a service infrastructure.	2.00	2	2	4	3	2	1	1	2	2	1	2	2	3
3 The development, co-ordination and control of non-core business services managed to deliver competitive advantage to the supported business.	2.85	3	2	3	2	1	3	5	2	2	3	3	3	5
4 The creation of an environment within which core business can be undertaken.	3.46	2	5	2	5	1	4	4	4	4	4	2	2	5
5 A process of continuous improvement in the workplace and its technology to achieve beneficial improvement in human, social and business outcomes.	3.69	3	2	4	2	1	3	5	5	4	4	5	4	4
6 The management of spatial environments to actively support the business.	3.92	4	3	3	4	3	3	5	4	5	4	4	4	3
7 Integrating operational needs with property and plant assets.	3.85	4	2	5	2	4	2	5	4	3	5	5	4	3
8 Management of all facilities in the built environment (except frame and construction).	2.92	2	1	4	1	4	3	5	4	2	4	1	2	2
9 Any service, or range of services, we can provide to a client more economically than they can provide themselves, whilst delivering an acceptable margin for our business.	1.85	2	1	3	1	1	2	1	2	4	2	2	1	1
10 Supplying support for integral non core activities.	2.46	3	4	2	4	1	2	3	4	2	2	3	2	2
11 Property related support service.	2.85	3	2	2	2	4	2	4	4	2	2	3	2	2
The Impact of Facility Management														
1 Very little/None/Not much – except for a few major corporates.	2.25	3	1	3	1	2	2	1	3	4	1	3	2	2
2 I believe that FM is a business in its own right. Hence in an FM business it plays a full role in the planning process.	3.25	3	5	4	5	3	5	5	3	1	1	1	5	3
3 It is often seen as the first port of call for disinvestment and reduction in costs.	3.27	4	2	2	2	4	4	3	4	3	3	3	3	4
4 Optimization of the cost and quality of services that support the core function of an organization.	4.18	3	3	4	3	5	4	4	4	4	5	5	4	5
5 An essential element. Gives practicality to planning.	3.83	4	3	3	3	4	2	5	4	5	5	3	4	4

evaluated in an attempt to establish what was known about the impact of property and workplace initiatives on the business performance of the occupying organization. Heavisides (2001) evaluated a largely separate 100 works on performance specification. Most of the references in both studies were written during the 1990s. The search was not confined to property literature – indeed it was hoped that more evidence would be found in business publications. There was very little. By way of example, one search of the Proquest database, which has a US bias, found 22 recorded citations of facility management in article titles since 1 January 1998, largely in building related professional journals. In the same period 304 hits appear for knowledge management, with a healthy proportion of them coming from periodicals aimed at financial managers and other board level decision makers.

Judged by the criterion of the number of new titles published, interest in the 'new' workplace is reaching fad status; it is showing exponential growth. However, most of this interest appears to be pushed by practitioners, advisers or professionals rather than pulled by line managers or even business and organizational theorists. Another sign of the explosion is the plethora of new terminology, as property or real estate specialists, facilities managers and workplace designers all lay claim to a (or often the) strategic role, claims for which evidence is frequently lacking.

4.6.1 The impacts of property

In the UK in particular, despite new forms of service offering being advertised to the market and despite shortening lease terms (it is unclear whether this is for reasons of supply and demand or as a true sign of a shift in the business ecology), traditional approaches to property procurement still dominate. A case has been proposed that they are more deeply embedded and discouraging of innovation than in the USA or Scandinavia, although comparative research is hard to find. Information on true occupancy costs, and especially whole life costs, is frequently not available and standard codes for comparative costing of true occupancy costs have only recently appeared on either side of the Atlantic.[12]

More disconcertingly, the influence of property on the feedback from customers' perceptions of business income is poorly understood (McLennan, 2000), apart from sectors such as retail and leisure facilities where it has always been more immediately obvious. A theoretical underpinning for occupation decisions does not exist. The new market emphasis on intellectual capital and the growing gap between market values and asset values is leading to questions of whether, on the one hand, markets properly value property assets held by non-property companies, and, on the other, whether such companies need to hold property. There is recent research (Deng and Gyourko, 1999) suggesting that, in higher technology industries, firms in the USA with lower property holdings derived superior stock market value over the period 1983–94. The same evidence was not, by 2001, publicly available in the UK.

Property acquisition and operational costs are the second largest expense, after salaries, for most office-based organizations, yet they are not necessarily gathered to any standard. Moreover, IT costs are frequently considered and managed separately, a factor that is bound to impede decisions such as workplace investments that may involve a trade-off between physical and virtual space. Despite these deficiencies, space charging is the most usually recommended method of allocating costs to individual business units or product lines. Whether it is effective in persuading departments to make more efficient or effective use of their space is unclear. Evidence of falling space densities with occupation time suggests not.

4.6.2 Measuring success

Service level agreements and output specifications for hard and soft FM services are an operational practice much recommended in theory (again with a bias towards advisers or service providers in the recommending group), but their effectiveness in practice has again received little attention. Many organizations, whether or not they outsource or manage FM in-house, have a tendency to prefer a significant measure of control on inputs, staff numbers and budgets. Some have found that in practice the effort devoted to the construction of service level agreements fails to justify the return and have abandoned them in favour of benchmarking (properly applied) and demonstrations of year on year improvement. Others have found that service levels expressed in terms of failure to comply with minimum standards fail to encourage a customer focus.

The focus of much operational FM on costs and technical measures is in any case misplaced. A variety of schemes for assessing staff perceptions of workplace exist but often have only indirect links to measures of productivity. Again there is a knowledge gap. The functional, as opposed to the physical or financial, obsolescence of buildings has been little studied and even less attention has been paid to the changing nature of work in the post-Fordist economy (Laing, 1991 – another pioneering work subsequently largely overlooked).

Better FM measurement practice would seem to lie in the development of more holistic, balanced scorecard-style measurement systems where business relevant measures of customer impact are included. In some cases, say retail sites or hotels, customer footfall provides an obvious indication of business impact. In others, say higher education, hospitals (in the UK) and perhaps call/service centres, the feedback is beginning to be appreciated. Student recruitment, patient satisfaction and service or retention levels, respectively, provide prospective measures in each sector. For mainstream offices it tends to remain invisible despite some evidence in practice of well-designed workplaces facilitating faster knowledge creation and dissemination.

4.6.3 The new workplace

A theoretical basis of new workplace design, notably Duffy's model of cells, dens, hives and clubs, each matched to different work demands, is well established. Recent examples supporting completely mobile workers suggest that the model needs a mobility dimension and the 'café' may supply a better metaphor than the 'club'.

The evaluation of new workplace environments – especially the claims for the benefits of what still tend to be termed open-plan, non-hierarchical environments – is split. Some high-profile failures, such as Chiat Day's reversion to traditional executive offices, have delighted the 'flexophobes' (Berger, 1999). Meanwhile the 'flexophiles' can point to St Lukes (a Chiat/Day spin off) as an example of a company whose non-territorial and creative workplace environment is integral to their thinking and is perceived as a contributor to dynamic business growth which has recently earned them the label of 'the most frightening company on earth' (Coutu, 2000). To some extent the promotion of new ways of working and new workplace styles may have ignored differences of individual psyche. Two people doing essentially similar jobs may have genuinely different needs from their working space. The argument is, however, obscured by questions of status and organizational culture. Management attitudes appear to have a large part to play in the success or failure of new forms of working.

Apart from surveys of occupants of individual offices, the evidence base on new workplaces is mainly journalistic and biased towards interviews with successes and failures. Companies that have successfully implemented teleworking and new spatial arrangements do report, in addition to reduced costs:

- reduced absenteeism
- easier recruitment
- reduced turnover
- improved morale and customer service
- faster development of new products and ideas
- higher knowledge worker productivity in terms of case loads/sales visits/innovation
- improved profitability.

Others report the exact opposite. The reasons why have not been explicitly researched but it is generally assumed that the successes come from cases where any initiative was either built into a business from day one or undertaken with a view to increased output rather than simply reduced cost. The point is made that the rhetoric of the former frequently obscures an intention that is much more focused on cost (Loftness, in Vischer, 1999).

There are signs of change. The business value of workplace initiatives is beginning to be considered as part of the wider question of managing and measuring knowledge work. The link to organizational culture, widely made in the knowledge management arena, is beginning to be appreciated in the workplace design arena.[13] The term 'process architecture' (Horgen et al., 1999)[14] has recently been suggested to indicate the interaction of the designer with the culture and unwritten design rules of the organization.

4.6.4 The workplace and organizational culture

If occupiers are to be able to make correct decisions on property they, or their advisers, need to understand how it contributes, not only to costs but also more importantly to the knowledge and service processes of the organization. The cultural reactions to space, while undoubtedly being influenced by national cultures, also have a significant generic element. There are, however, particular factors of the operation of the property market in the UK that may make the lack of understanding more of a problem. No specific research has been found but experience of the Scandinavian market, where owner occupation is more common, suggests a more direct involvement of the occupying organization in how their premises are designed and built. In the USA shorter lease terms and greater movement of businesses may make experimenting with new forms of both financing and design easier. Conversely the argument that key staff will leave if they are not given private offices is made more strongly in American literature.

4.6.5 The outcomes

It may be argued that the report has not produced answers; however, that was not the objective of the occupier.org project, which was to summarize what is and is not known about the impact of property and workplace on occupiers' businesses. It is clear that there are significant gaps in both the professional and business literature. It is also clear that the issue, if it is to be understood, needs to be considered from a business perspective. The question is less to do with how property benefits occupiers, and more about how occupiers secure maximum benefit from property. In other words, the workplace has to become a management issue rather than just an issue on which professionals offer advice, but it is up to the professionals to earn their place at the top table by showing how the transition can be achieved and what the benefits are.

The occupier.org report (Haynes *et al.*, 2001) concluded with a brief speculation. Changes in workplace may enable changes of culture but only, perhaps, if they are accompanied by changes in managerial thinking and belief systems. If the modern school of management thinking is correct in the assertion that new managerial paradigms are needed in the new economy, or to the extent that that it is true, then they may also be needed to make a success of new workplaces. Conversely, the creation of physical (and perhaps virtual) space may be the most under-utilized managerial tool of the knowledge era (Ward and Holtham, 2000), a claim that is only beginning to be investigated.

4.7 Development paths for FM

The rise and fall of management fads (Pascale, 1991; Schapiro, 1995) has become a subject of scholarly analysis under the perhaps less pejorative title of management fashions. Quality circles (Abrahamson, 1996), knowledge management and learning organizations (Scarborough and Swan, 1999), and benchmarking (Price, 2000) have all been analysed. Citations of a particular term in journals or wider trade publications provide an indication of its spread or decay. With the advent of Internet databases the method becomes easier to apply, if only as some proxy of the spread of a particular professional discourse. While the limitations of the approach as a research method are manifest and acknowledged the above works indicate classic S-curve dynamics (peaking and decaying) for quality circles, benchmarking and learning organizations. Knowledge management and complexity are still displaying exponential increases. FM can, by these standards, lay claim to considerable longevity. If the origin point of current FM is taken as the 1978 Ann Arbor meeting or the 1980 decision to found the IFMA (see Chapter 3) then it has a history of roughly the same duration as benchmarking, whose origin is generally attributed to Rank Xerox in 1979 (e.g., Price, 2000), but FM is both a larger industry and more formalized in terms of professional and educational activity. It has been interpreted as both BPR (Alexander, 1993a) and a quality circle (Alexander, 1993b) but has seen both rise and fall without securing a permanent niche in the language and practice of management.

4.7.1 Fads, genes and memes

Abrahamson (1996) and Scarborough and Swan (1999) have both drawn attention to the property of management fashions to act as self-perpetuating discourses. More recently Abrahamson and Fairchild (1999) have argued for a selection process operating on fashions in the conversations of 'knowledge entrepreneurs'. My theoretical preference goes further and argues that we should interpret such discourses, and the organizations which emerge to support them as an example of, quite literally, a selection process between not genes, but memes (Price and Shaw, 1996, 1998), a term coined (Dawkins, 1976) to convey the idea of auto-replication of 'ideas' by 'broadly speaking a process of imitation'. Biological evolution acts on individual organisms (phenotypes) but ultimately selects for 'successful' genes, and genetic success equals getting copied. Memes are conceived as replicating themselves not through biological inheritance but through the effects of language and cultural artefacts on individual minds and collective organizational paradigms (Hull, 1988; Dennett, 1995; Price, 1995; Blackmore, 1999). In doing so they may enable organization, but the measure of success of a meme is the number of minds it affects. Considered from the meme's perspective (a rhetorical device and not a suggestion that memes are in anyway conscious) if multiplicity of meaning speeds replication then 'so what?' (Price and Shaw, 1996, 1998).[15]

Multiplicity of meaning is common in successful 'fads'. Benchmarking began its life as the specific practice of lateral comparison of performance with a view to improvement, but is now used in some areas as a term for standards setting, quality assurance and the downwards measurement of performance. Small wonder that there is managerial obfuscation when the team from corporate headquarters announces 'we are here to benchmark you'. It sounds nicer than 'audit' or 'evaluate' (Jackson and Lund, 2000; Price, 2000). Learning organization has had a similar range of meanings attached to it (Price and Shaw, 1996) and knowledge management, its partial successor in popularity terms (Scarborough and Swan, 1999), carries a range of meanings from IT to organizational development. FM is no exception to what seems to be a widespread phenomenon if not a general rule. While confusing to users, diversity is a good trick in meme replication space.

If this evolutionary perspective is correct, FM is firstly engaged in a Darwinian struggle to secure its niche in a wider business 'ecology' and secondly it is a system in which competing interpretations, and the institutions that hold them, are themselves engaged in such a struggle.[16] Some will lock into secure positions in the FM ecosystem, others will not. As Thomson (1988) notes (quoting Armstrong, 1982), the point was made early in FM's development by one of the IFMA's founders:

> How do new disciplines behave at first? They fight for attention, identity and respect. The established ways and their associated support systems form barriers. Some jockeying takes place. Proof that the new is good is tough. The new needs three to five years to establish. The new is conceptual, descriptive; the old is analytical, full of case studies. All new disciplines begin with the basics, usually describing and counting things. In the 1800s, forestry began by counting trees, entomology began by identifying bugs.
>
> In some sense, this is where we are today [1982] with facility management. We are arguing for its use, crying loudly to be heard above the dins created by both

change and status quo. We are gathering basic data, even counting facility managers.

We know there is a need to manage the physical environment in concert with people and job processes and as the management science becomes more mature, the debate and headlines will diminish. That will not mean the need has diminished, it will only mean that facility management will be then well established and well accepted – part of the new status quo.

The history revealed here, of general expansion accompanied by emergent institutions each propagating different variants on a theme, seems explicable as such a selection process and there seems to be no more attractive explanation.[17] In parallel with FM's largely successful effort to establish its niche in managerial/business/professional discourse, there are continued competitions, as reviewed above, between major professional FM bodies, between the professional patterns of other professions (architecture, surveying, project management and engineering), between operational practitioners and designers/consultants/property managers with boardroom ambitions, and, not least, between the business providers and in-house facility managers.

Such an evolutionary perspective does not incidentally deny FM its status as a profession. Professions, from such a stance, are no more than systems of belief and practice that have managed to capture a sufficiently permanent niche in a social ecosystem, equivalent to adherence to particular paradigms in science (Hull, 1988) or bodies of religious belief and practice (Price and Lord, 2000). Surveying, for example, currently holds the niche in the UK for exchanges involving assessments of value. In continental Europe the term surveyor has no meaning and other professions occupy the equivalent niche (Adair *et al.*, 1996). Nor does the perspective deny the need for professions to have a code of ethical practice (*cf.* Leaman, 1992; Grimshaw, 2001). To the extent that they need a code to gain social legitimacy and replicate, a code is another useful adjunct to replication.[18]

An evolutionary perspective does place FM within the wider body of understanding of Complex Adaptive Systems (Waldrop, 1992, is the classic introduction) which is offering an explanation of order, generated and preserved, in both biotic and non-biotic systems in terms of agents interacting according to particular rules or schemata (e.g., Gell-Mann, 1996). A strong evolutionary metaphor already exists within FM as a result of Becker's descriptions of offices as ecosystems, and consideration of workplaces in terms of CAS theory may offer the elusive foundation for FM as a body of knowledge of workplaces as a management tool. It has been argued elsewhere (Price and Akhlaghi, 1999) that management who conceive of organizations as more akin to living systems than to machines deliver better results in the field of FM practice. For them management is more concerned with designing environments in which a certain outcome is likely than with the traditional processes of planning and control. Complexity theory forces management toward design (e.g., Stacey, 1995; Price and Shaw. 1998; Pascale *et al.*, 2000) and the built environment may indeed be the most under-appreciated tool for the process (Ward and Holtham, 2000) available to the manager.

However, the perspective also carries a warning: organizations that are enabled by a particular pattern of beliefs and practices, a certain meme system, are also limited by them. If external environments change faster than the replicating code, genetic or memetic, then the carriers and propagators of that code become extinct. In the biotic domain, over geological time, it has happened to 99.9% of species that have existed. The time scales of organizational extinction are generally much shorter. FM may already be locked in to a pattern system that effectively stops it evolving. If so, the future is bleak (Price, 2002) and a different body of discourse will emerge to capture the strategic ground that is waiting to be taken. There are warning signs in the rise of 'Corporate Real Estate Management', 'Infrastructure Management' and 'Real Estate Asset Management'. Many of the significant players in FM's development, including, recently, Duffy (2000) have been writing articles bemoaning its failure – a sign, in the rise and fall of other fashions, of their impending demise (Abrahamson and Fairchild, 1999).

4.8 Back to the future

Nutt and his colleagues have recently identified four 'trails' or directions down which FM might develop (Nutt, 2000). The Human Resource trail sees FM contributing to the management of increasingly flexible and diverse working practices. The Financial Resource trail sees the increasing separation of property ownership and business management as serviced infrastructure provision becomes a business in its own right. The Physical Resource trail is essentially concerned with the physical fabric of buildings. The Knowledge trail sees a convergence of design and management strategy so that, in essence, the built environment is seen as a strategic management tool.

In a recent survey (Price, unpublished results of an informal Internet survey) respondents were asked whether FM can embrace all four, or is it more likely that separate disciplines will emerge to cover each. The replies reveal, yet again, the different mental models of FM that are now out there. For example:

> FM may seek to embrace all four, but in my opinion the four and others will inevitably need to become individual disciplines requiring detailed competencies and capabilities.

> I think FM should embrace all four trails.

> Yes and no depending on the sector and organization. HR and Financial involve potentially different skill sets, but the Physical and Knowledge are more commonly/ more likely to be covered by FM.

> FM can and should embrace all four. However FMers will need the appropriate skills and face the challenge of moving FM closer to or preferably into the Boardroom.

> A convergence of design and management strategy. An object-oriented approach.

> FM has to continue to embrace all four streams – FM has to deliver suitable premises that meet business goals, reflect a company's brand values and provide a stimulating and productive environment in which people can work – to do this FMers have to understand the needs of the business, the real-estate market, personnel needs, physical building resources – in order to plan the company's property strategy.

> Facility management will always remain an umbrella discipline due to the nature of its title and the inherent meanings therein. With all umbrella disciplines, and these must include asset management, property management, corporate real estate and quality assurance, it is hard to escape the generalization of the discipline and to force a

direction down one of its many tributaries. The emphasis on one particular discipline will change with time as the prevailing economic, business and social trends change.

My view is that if FM is to be a well-established profession (as opposed to an emerging one), it must be elevated above matters of physical stewardship of facilities to holistic corporate strategic management. The FM team must be represented at the Board of Directors level and involved in business-wide decision-making processes. This may, in the future, mean that FM encapsulates human resource management as well as responsibility for built assets, systems, technology and planning processes. While there will of course be specialization of companies in various areas, such as financial management and advice, the 'knowledge trail' described by Nutt appears to best reflect the goals of the FM profession. It is possible for FM, therefore, to embrace all four areas, but I believe the latter is the most important.

The human resources trail is the one that I prefer as it focuses on people and business. The rest is only dressing, and finance is only part of being in business. I hope there will not be a split as there will be no parity and that is already evident in this profession. It will also mean that people are the losers.

FM is ideally placed to be the integration point between HR, Finance, the built environment and Knowledge Management. It is up to the profession to grow out of its infancy and embrace all four rather than languish in its role as the 'too hard basket'.

It is very much dependent on the type of facility being managed [emphasis added] – is it a hospital, a warehouse and distribution centre, an oil rig, or a corporate office environment. In any facility there would be combination of two or more of the trails mentioned above and unique management skills will be required depending on the environment being managed. Unfortunately due to the very proactive involvement and influence of systems furniture manufacturers, all of FM thinking is so 'office centric' that people tend to lose the bigger picture.

The emphasized sentence which starts the last comment, while perhaps trivial, makes a statement that FM ignores at its peril. FM has enjoyed an evolutionary bloom radiating from its original niche into a range of others by virtue of the fact that most organizational activities still need a facility. Perhaps the future needs to be less concerned with what FM is and more with how different organizations are impacted. Here perhaps we may see where the four trails, and others, viewed as sub species of FM may find their optimal niches.

4.9 Conclusion

There will be businesses for which knowledge generation is an absolute business priority. Here classic FM may finally find its strategic outlet. If it can rediscover some of its lost or ignored roots and finally demonstrate that effective office design can indeed speed up organizational learning and innovation in the workplace FM may find its home on the knowledge trail. To do so it may have to shed much of its operational association or reassert the IFMA's 'classical facility management'.[19]

There will also still be facilities in which knowledge generation is of lesser importance. Many, especially where customer impact of the facility is high, will find facilities services important. Hotel and leisure facilities know the importance of 'back stage' services and some retail sites and hospitals are finding business benefits (or their public sector equivalent[20]) from upskilling auxiliary staff as part of the customer experience (Price and Akhlaghi, 1999). Such facilities, where the customer impact is immediate, seem the natural home of the HR or service trail.

In other businesses the physical estate itself may influence customers, or staff, through the image it conveys. Certain university campuses, for example, are positive assets in attracting students (Price et al., 2001). In such cases the physical resource trail beckons, but it is only part of the story – the design, grounds and location of a building all contribute, and there is perhaps an estates trail.

Then there are facilities where the operational risk of failure is high – acute hospitals, offshore oil rigs and nuclear reprocessing plants spring to mind. Here, potentially, maintenance achieves its highest profile.

One can perhaps then see four distinct areas where different bodies of specialized FM knowledge will be most needed. Some at least still carry the potential for transfer of risk, and appropriate reward, to a specialist provider. Whether FM can brand itself up market in four different areas at once, while still embracing its large body of operational practice, remains to be seen. The challenge is certainly significant and the history suggests the subject has colonized mass niches rather than specialist ones. The danger of a failure to learn from history remains real.[21]

Endnotes

1 Comment to the author by the then Finance Director of BP Exploration early in 1991.
2 To the extent that the boundaries of a firm stabilize where internal administration is cheaper than the market place in determining transfer costs between different divisions (Coase, 1937).
3 Eley (1993) illustrates the then makeup of the new industry just as outsourcing was developing. The six largest players were Serco (founded in 1987 with 8000 employees), Drake and Skull (1987 – 3000), BET FM (1989 – 300), Haden (1990 – 300), P&O FM (1989 – 140) and FPN (1985 – 50).
4 Personal recollections and files of the author. Gerry and I were members of the same network at the time. I was then involved in change management and Business Process Review in BP Exploration where we believed we were leading the UK field in business outsourcing.
5 The degree to which CCT succeeded or even yielded a net benefit has not been researched. Avoidance tactics were much discussed in same trusts and authorities and some unpublished results suggest that CCT decisions followed the political map of the country more than any market factor. The point is that it was a significant opportunity to FM providers.
6 http://www.i-fm.net/
7 Who entered the market offering an MBA and MSc and also boasted of being first to receive BIFM accreditation early in 1997. Subsequently (2002) the MSc has been dropped and a DBA (in Facilities and Property Management) validated.
8 The statistic is interesting. Large catering/cleaning companies have sought, during the decade, to become broader FM providers and the Hotel Caterers and Institutional Managers Association represents a number of those who manage 'soft FM' services but their professional traditions have by and large failed to infect FM.

9 In 1996 Tom Kennie and I facilitated an action-learning group of partners from small and medium surveying practices in a project to develop a set of practice management guidelines (Kennie and Price, 1997). FM was seen by most of the group as a threat during a SWOT exercise. Subsequently some larger practices have sought to develop FM divisions. Anecdotal evidence suggests that most have found it difficult to switch from a traditional property perspective,

10 This definition appears in the introduction to book called *Facilities Economics in the European Union* by Bernard Williams and Associates (published by BEB in 2000), it also defines facilities management as 'the process by which the premises and services required to support core business activities are identified, specified, procured and delivered'.

11 Responses of authors in this volume and others in their networks to the question: 'Below are 11 definitions of FM provided by a British Delphi panel of leading FM and property people (academics/professionals, procurers and suppliers). On a scale from 1 (strongly disagree) to 5 (strongly agree) can you express your opinion of each definition.'

12 The OPD Occupancy code is more precise on property and FM costs. The US Government Office of Real Property's cost per person model incorporates a better treatment of IT costs with a view to evaluating flexible working developments in an integrated manner.

13 Franklin Becker made the link in 1990 but the property professions as a whole still struggle.

14 The term 'process' is used in the sense of that organizational development consultants use it, i.e., the human and group process, rather than in the sense of business process. Elizabeth Shove (1991) made a similar call largely ignored in subsequent literature.

15 Adaptation to the Deming cycle in Japan (see Chapter 3) could be seen as a classic example of a new meme hitching itself to an established success.

16 It also predicts that FM will show properties such as fitness landscapes, identifiable patterns and a log normal distribution of size. Research to test that proposition is currently in progress.

17 Several variants on the theme of evolutionary selection exist in sociology, organizational theory and economics (see Aldrich, 1999, for a useful review).

18 This may be interpreted as cynicism, but that is not the intention – to paraphrase Santayana, scepticism, like chastity, should not be abandoned lightly.

19 See quote from Young in Section 4.4 regarding 'classical facility management'.

20 A major 'spring clean' initiative for British hospitals, begun in 2001, arose from the belated political realization that food and cleanliness impact taxpayers perceptions of hospitals. In these 'customer critical' facilities FM ignores the skills of the 'soft services' managers at its peril.

21 Many people have supplied data and recollections relied on in this and the preceding chapter, especially fellow authors and others who contributed to the 2001 survey. Dennis Longworth, Don Young and Kim Hadden (from the IFMA) and Lionel Prodgers, Valerie Everett and Karen Ward (from the BIFM) supplied data on the founding of their respective institutions. Brian Purdey and especially Duncan Waddell illuminated the early history of the FMAA and Bob Grimshaw that of CIB W70. Colleagues and students have provided more details, especially Barry Haynes, Andrew Green, Shaun Lunn, Nick Nunnington and David Rees. Tony Thomson was an invaluable guide to the

early years of FM in the UK. Craig Langston and Gerard de Valence introduced Australian contacts. The interpretations I have made are, however, mine.

References and bibliography

Abrahamson, E. (1996) Management fashion. *Academy of Management Review*, **21** (1), 254–85.

Abrahamson, E. and Fairchild, G. (1999) Management fashions: lifecycles, triggers and collective learning processes. *Administrative Science Quarterly*, **44** (December), 708–40.

Adair, A., Downie, M.L., McGreal, S. and Vos, G. (1996) *European Valuation Practice: Theory and technique* (London: E. & F.N. Spon).

Aldrich, H. (1999) *Organizations Evolving* (New York: Sage).

Alexander, K. (1993a) Facilities management as business re-engineering. *Facilities*, **11** (1), 6–10.

Alexander, K. (1993b) Facilities management as a quality cycle. *Facilities*, **11** (2), 25–7.

Armstrong, D.L. (1982) DEAR DAVE series – is facility management a fad? Why all the hoopla now? *Interiors*, June, 44.

Becker, F. (1990) *The Total Workplace: Facilities Management and Elastic Organisation* (New York: Praeger Press).

Berger, W. (1999) Lost in Space. *Wired*, **7** (2). www.wired.com/wired/archive/7.02/chiat_pr.html

Blackmore, S. (1999) *The Meme Machine* (Oxford: Oxford University Press).

Coase, R.H. (1937) The Nature of the firm. *Economica*, **4** (16), 386–405.

Coutu, D. (2000) Creating the most frightening company on Earth: an interview with Andy Law of St. Luke's. *Harvard Business Review*, **78** (September/October), 142.

Dawkins, R. (1976) *The Selfish Gene* (Oxford: Oxford University Press).

Deng, Y. and Gyourko, J. (1999) Real estate ownership by non-real estate firms: an estimate of the impact on firm returns. http://reserver.wharton.upenn.edu/workingpapers/321.html

Dennett, D.C. (1995) *Darwin's Dangerous Idea: Evolution and The Meanings of Life* (New York: Oxford University Press).

Duffy, F. (1993) In: Hannay, P. (ed.) *The Changing Workplace* (London: Phaidon Press).

Duffy, F. (2000) Design and facilities management in a time of change. *Facilities*, **18** (10/11/12), 371–5.

Eley, J. (1993) One-stop shopping for one-stop shopping. *Facilities*, **11** (1), 20–2.

Gell-Mann, M. (1996) *Complexity, Global Politics, and National Security – The Simple and the Complex*. Address to the US National Defence University. www.dodccrp.org/comch01.html

Green, A. and Price, I. (2000) Whither FM? A Delphi study of the profession and the industry. *Facilities*, **18** (7/8), 281–92.

Grimshaw, B. (2001) Ethical issues and agendas. *Facilities*, **19** (1/2), 43–51.

Haynes, B., Fides Matzdorf, F., Nunnington, N., Ogunmakin, C., Pinder, J. and Price, I. (2001) *Does Property Benefit Occupiers? An Evaluation of the Literature*. occupier.org Report 1. www.occupier.org/reports.htm

Heavisides, R. (2001) *Output based Facilities Management Specifications in the National Health: Service: Literature Review and Directional Outcomes*. Occupier.org Working Paper 1. http://www.occupier.org/working_paper1.pdf

Horgen, T.H., Joroff, M.L., Porter, W.L. and Schön, D. (1999) *Excellence by Design: Transforming Workplace and Work Practice*. (New York: John Wiley).

Hull, D. (1988) *Science as a Process* (Chicago, IL: University of Chicago Press).

Jackson, N. and Lund, H. (eds) (2000) *Benchmarking for Higher Education* (Milton Keynes: Open University Press).

Kennie, T. and Price, I. (1997) *Practice Management Guidelines* (London: RICS).

Laing, A. (1991) The post-Fordist workplace: issues of time and place. *Facilities*, **9** (8), 13–18.

Leaman, A. (1992) Is facilities management a profession. *Facilities*, **10** (10), 18–20.

Mandell, M. (1990) Cost cuts are in the mail. *Nations Business*, **78** (1), 30–1.

McLennan, P. (2000) Intellectual capital: future competitive advantage for facilities management. *Facilities*, **18** (3/4), 168–71.

Nutt, B. (2000) Four competing futures for facility management. *Facilities*, **18** (3/4), 124–32.

Pascale, R.T. (1991) *Managing on the Edge* (New York: Simon & Schuster).

Pascale, R.T., Milleman, M. and Gioja, L. (2000) *Surfing on the Edge of Chaos: The New Art and Science of Management* (New York: Crown Publishers).

Prahalad, C.K. and Hamel, G. (1990) The core competence of the corporation. *Harvard Business Review*, **68** (3), 79–91.

Price, I. (1995) Organisational memetics?: organisational learning as a selection process. *Management Learning*, **26** (3), 299–318.

Price, I. (2000) Benchmarking higher education and UK public sector facilities management. In: Jackson, N. and Lund, H. (eds) *Benchmarking for Higher Education* (Milton Keynes: Open University Press).

Price, I. (2002) Can FM evolve? *Journal of Facilities Management*, **1** (1), 56–69.

Price, I. and Kennie, T. (1997) Punctuated strategic equilibrium and some leadership challenges for university 2000. *Proceedings 2nd International Conference on the Dynamics of Strategy*, SEMS, Guildford, 335–49.

Price, I and Lord, A.S. (2000) Isomorphism in biotic and abiotic complex adaptive systems. *International Conference of Complexity and Complex Systems in Industry*, Warwick.

Price, I. and Shaw, R. (1996) Parrots, patterns and performance (the learning organisation meme: emergence of a new management replicator). In: Campbell, T.L. (ed.) *Proceedings of the Third Conference of the European Consortium for the Learning Organisation*, Copenhagen.

Price, I. and Shaw, R. (1998) *Shifting the Patterns: Breaching the Memetic Codes of Corporate Performance* (Chalford, UK: Management Books 2000 Ltd).

Price, I. and Akhlaghi, F. (1999) New patterns in facilities management: Industry best practice and new organisational theory. *Facilities*, **17** (5/6), 159–66.

Price, I., Matzdorf, F. and Smith, L. (2001) *Where to Study: Understanding the Importance of the Physical Environment to Students in Choosing their University*, Sheffield, occupier.org Working Paper 3. http://www.occupier.org/working_papers/working_paper3.pdf

Rees, D.G. (1997) The current state of facilities management in the UK National Health Service: an overview of management structures. *Facilities*, **15** (3/4), 62–5.

Rees, D.G. (1998) Management structures of facilities management in the National Health Service in England: a review of trends 1995–1997. *Facilities*, **16** (9/10), 254–61.

Scarborough, H. and Swan, J. (1999) Knowledge management and the management fashion perspective. *British Academy of Management Refereed Conference:, Managing Diversity*, Manchester. http://bprc.warwick.ac.uk/wp2.html

Schapiro, E. (1995) *Fad Surfing in the Boardroom* (New York: John Wiley).

Shove, E. (1991) Letraset zombies? *Facilities*, **9** (8), 9–12.

Stacey, R.D. (1995) The science of complexity: an alternative perspective for strategic change processes. *Strategic Management Journal*, **16**, 477–95.

Thomson, T. (1988) *Facilities management consultancy: the development of a plan for its proposed implementation in an architectural company*. MSc Dissertation, University of Reading.

Vischer, J.C. (1999) Will this open space work? *Harvard Business Review*, **77** (May/June), 28–36.

Waldrop, M.M. (1992) *Complexity: The Emerging Science at the Edge of Order and Chaos* (New York: Simon & Schuster).

Ward, V. and Holtham, C. (2000) *Physical Space: The Most Neglected Resource in Contemporary Knowledge Management?* www.sparknow.net/cgi-bin/eatsoup.cgi?id=602

PART 2

5

Strategic management

Danny Shiem-Shin Then*

Editorial comment

Strategic management is a core competency for the facility manager. It refers to an involvement in decision-making processes that affect the organization as a whole. Ideally the facility manager would have a seat on the executive board, but where this is not the case it is critical to exercise strategic thinking in relation to all aspects of the facility portfolio. Failure to do so undermines the ability to deliver significant value enhancement.

Strategic management is based on goals that are identified in a number of strategic facility plans (SFPs). These cover key aspects of the facility portfolio and reflect organizational directions and performance benchmarks. It is obviously critical that SFPs are aligned to organization goals, but what is more difficult yet highly useful is to predict future facility needs and develop strategies that will enable a timely response. Strategic thinking is therefore all about anticipating and managing change.

Spatial need is a good example of the importance of strategic management. Most organizations will require suitable workspace for its people, but the amount of space will grow and shrink over time as a result of economic conditions, business performance, new initiatives and changes in technology. Additional space may involve new construction, acquisition, leasing, renovation or relocation, but which choices are best and when to make the decisions determine levels of productivity and govern a whole raft of other actions that must be successfully co-ordinated.

Change is both inevitable and necessary. With change comes the opportunity for progress and innovation, without it comes vulnerability and stagnation. Change costs money in terms of both transitional expenditure and loss of productivity, and this must be weighed against potential benefits in order to determine whether a particular change proposal is worthwhile. Strategic management brings with it issues of value identification,

* The Hong Kong Polytechnic University, Hong Kong

cost–benefit analysis, risk assessment and stakeholder negotiation. It is a high level activity and one of considerable importance for any organization.

In relation to workplace practices, strategic management offers the opportunity to re-engineer processes to make them more innovative and efficient. This can lead to quantum leaps in value and, through the introduction of new technology, open up new business options. This is why strategic facility decisions must have the support of the organization's top management and be co-ordinated with other divisional units, particularly human resources and information technology. There is much to commend such integration. Enterprise resource planning (ERP) is a contemporary example of how information can be effectively shared within the organization through adoption of supporting technology and a process re-engineering approach.

Treating a facility manager as a non-strategist is akin to treating an interior designer as a painter and decorator. It is important to elevate the role above its operational base, and in fact this is part of the difference between a trade and a profession. The ability to link and integrate a diverse range of critical issues under the control of a single representative is of great advantage when looking for ways to improve quality, reduce cost and minimize risk.

5.1 Introduction

In today's business environment, business support infrastructures are no longer static. The pace of change promoted by rapid technological developments and the dynamics of globalized supply chains and markets has altered the traditional assumptions and economics of workplace provision and management. As part of strategic management, workplaces must now be planned, designed, created and managed to accommodate the continuous change faced by today's enterprises. There is growing acceptance by corporate management that the blending of the physical environment, information technology and employee issues is central to the development of effective management policies in an organizational setting. Strategic management is aimed at achieving organizational effectiveness through dynamic alignment of the required infrastructure support of buildings and services to meet changing business needs. Strategic inputs are vital ingredients for success in effective workplace management.

In the corporate world there is growing acceptance of the suggestion that, in order to achieve the much-needed alignment between business strategic direction, organizational structure, work processes and the enabling physical environment, an organization's strategic intent must clearly reflect the facilities dimension in its strategic business plans. What is also clear from published literature (Then, 1999a), however, is that there is a lack of an integrating framework for considering the likely impact and implications of business management trends and corporate management decisions on the provision, and subsequent ongoing management, of the workplace environment.

In this respect, current literature points to three prerequisites for a strategic approach to the effective management of workplace provision and management (Then, 1999a):

- the need to link real estate/facilities decisions to corporate strategy
- the need to proactively manage functional workspace as a business resource
- the need for a framework to integrate business resource management with the provision and management of the corporate operational assets, and associated facilities support services, in their business settings.

The above themes, in turn, lead to at least three requirements in corporate real estate and facility management in an organizational setting:

- the requirement for an appropriate linking mechanism for considering the facilities implications of business decisions by promoting meaningful dialogue between business corporate planners and real estate/facilities personnel
- the requirement for management processes to monitor the strategic relevance of facilities requirements and monitor their performance-in-use over time
- the requirement of appropriate skills and competencies within the real estate/facilities function to monitor and continuously review procurement policies and strategies to take advantage of advances in technological development and market offerings from the supply side.

It is in the context of the above requirements that this chapter will explore the strategic role of corporate real estate and facility management in providing appropriate facilities solutions to meet emerging business challenges.

5.2 The context of real estate and facility management (RE/FM)

For the purpose of this discussion RE/FM is defined as the interface between strategic business planning and operational asset management. The term has the advantage of encapsulating three key aspects (real estate, facilities and management) which, when taken together, influence the perceptions of the main stakeholders that bring the physical assets (i.e., buildings) into being and manage them over time.

Both published literature and practice reviews (Then, 1999a) support the need to consider facilities provision plans in conjunction with the other business resource plans relating to finance, manpower and technology. A pro-active management model of the corporate real estate resource necessitates a constant two-way dialogue: from strategic management, i.e., the strategic intentions and direction of where the company is going, and from operational management, i.e., the best way of achieving the desired outcome in functional facilities and services and their ongoing management. The conceptual framework that justifies the need for a constant dialogue between strategic management and operational management is illustrated in Figures 5.1 and 5.2 (Then, 1999b).

Figure 5.1 is a conceptual representation of current practice of RE/FM in many organizations. A key requirement of a pro-active approach is not only to raise the awareness of the two sides to the strategic importance of closely aligning the real estate resource with the corporate strategic intent, but also to establish channels of formal communication that keep both parties fully informed of the external market and its likely implications for the operational real estate asset base. Figure 5.2 illustrates the use of the *strategic facilities brief* (SFB) and the *service levels brief* (SLB) as the instruments for promoting and maintaining this crucial interface between strategic management decisions and operational management decisions.

A critical barrier to the success of RE/FM is access to the right information from business units, as well as from the facilities operations. Establishing the existence of useful information within the business units, and gaining access to this information on a

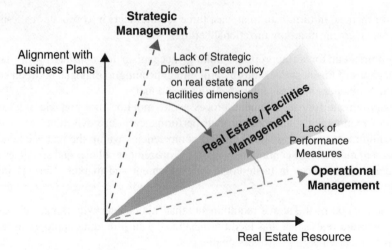

Figure 5.1 Conceptual representation of current practice.

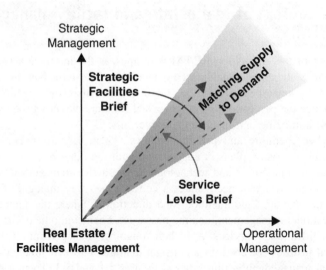

Figure 5.2 Justification for RE/FM.

regular basis, is often problematic but is critical if effective actions are to be taken. In this respect, one of the key roles of RE/FM is to act as the informed interface (Carder, 1995) where the overlapping concerns from strategic management and operational management (Cole, 1994) can be reconciled to provide an informed solution to apparently conflicting goals when seen in isolation from either perspective. The effectiveness of RE/FM relies on the regular flow of information, i.e., localized information built up from the bottom of the operational structure, and business information brought down from the core business end of the organization. A schematic illustration of the proposed model is illustrated in Figure 5.3.

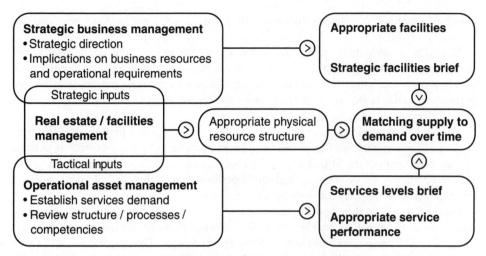

Figure 5.3 The role of the strategic facilities brief and the service levels brief.

RE/FM provides the strategic link between strategic business planning and operational asset management. In the context of any organization or business unit, the focus of RE/FM is to reconcile the demand for, and supply of, the real estate asset base and associated facilities support services essential for the delivery of its core products or services. Simply expressed, the principal role of RE/FM is to support the core business of the organization it is serving. In economic terms RE/FM provides a suitable platform for:

- defining and quantifying the *demand* emanating from strategic business direction expressed as operational space needs of facilities and support services to support core business activities
- defining the *supply* in terms of the necessary physical asset base and appropriate service levels from the service delivery perspectives and their management over time
- *matching supply to demand over time* as a continuous process of maintaining relevance in terms of an appropriate physical resource structure to support the corporate strategic intent.

RE/FM fulfils a much-needed intermediate role between strategic management and operational management in any organization. Representation in RE/FM should comprise strategic inputs from executive management staff responsible for strategic business planning, and from senior staff from the facility management or property services management department.

The derivation of a corporate strategic choice, without integration of the real estate and operational dimensions, clearly contributes to sub-optimal solutions in many organizations, reducing the role of the real estate/facilities function to one of reacting to the demands of individual business units.

From the above discussions it is clear that an integrating framework for RE/FM must be built by creating a continuous dialogue between the strategic management of core business development and the operational management of business resources. As illustrated earlier, in Figure 5.3, two key instruments and processes are essential to bring about a formal and continuous dialogue – they are the *strategic facilities brief* and the *service levels brief*.

The SFB is an output statement or document that defines the operational needs emanating from the organization's business plans. The principal purpose of the SFB is to define a corporate guide that outlines the key facilities attributes and physical service performance criteria that are required to fulfil the organization's deliverables in facilities and services, as dictated by the business plans.

The SLB represents the definition of acceptable environmental performance levels in respect of the physical asset base and facilities support services requirements as defined by the deliverables of the SFB. The principal purpose of the SLB is to define and quantify the appropriate support services and their performance within the physical working environment that supports the core business activities.

Figure 5.4 proposes a feasible expectation of a desirable outcome that integrates decisions at the level of making strategic choices, with the economics of facilities provision and management as an important business resource. The outcome is a set of well thought out and thoroughly evaluated supporting facilities strategies that are continuously monitored and adjusted to meet business requirements through flexibility in supply and fitness for purpose.

The strategic inputs are the expressed intentions of the corporate strategy for core business development in the short to medium term. The role of the SFB is to attempt to ensure that any investments in physical resource terms result in the delivery of appropriate facilities to support the fulfilment of the business plans. The source of the strategic inputs is *business information*. The role of the SLB is to ensure that the appropriate facilities support services are delivered for the level of operational assets as defined by the SFB. Similarly, the tactical inputs from operational management should contribute to the strategic evaluation of potential capital investments that result in altering the capacity of the physical assets. The source of the tactical inputs is *facilities information*.

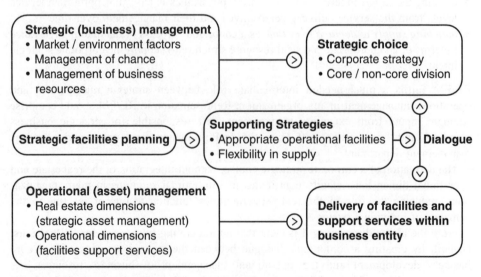

Figure 5.4 RE/FM as a response to generating supporting strategies.

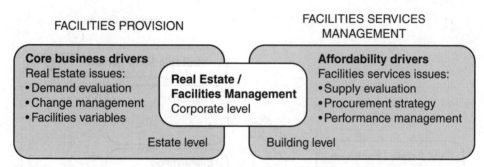

Figure 5.5 Business drivers and affordability drivers.

The interactions between the SFB and SLB are necessary to ensure proper matching of demand for facilities to support strategic business development and supply of appropriate facilities-related support services to enable the implementation of the desired business plans. Pro-active management of the corporate real estate resource demands clear strategic direction from senior management and clearly defined and measurable deliverables from operational management.

The critical interfaces between *facilities provision* and *facilities services management* within an organizational context are illustrated in Figure 5.5. The objective is to identify the crucial interfaces and to identify the drivers motivating actions at three identifiable levels in terms of real estate and facility management, i.e., corporate level, estate level and building level.

The corporate level is concerned with the adequacy of the real estate assets, as a business resource, in fulfilling strategic objectives. The estate level then interprets this strategic intent in terms of implications for the current operational real estate portfolio, i.e., facilities provision. The building level is primarily concerned with meeting users' requirements on an ongoing basis, while at the same time minimizing disruptions while action is taken to adjust to the next 'steady state' as a consequence of the strategic response initiated at the corporate level.

It is important to point out that underpinning any decisions in facilities provision is the constant interplay between the pulls from three key resource drivers: those of people, technology and the workplace environment (property).

5.3 Key organizational variables

The preceding section describes the dynamic business environment within which the role of the operational real estate assets in organizations must evolve in order to accommodate changes in the market place. The desired outcome from RE/FM, in terms of real estate resources, is an appropriate portfolio structure that is aligned with the organization's business operational requirements. Buildings and land, as physical assets, are relatively static products. Effective matching of demand for, and supply of, functional accommodation and associated support services to meet operational requirements in a dynamic business environment demands the management of the real estate assets as a dynamic integrated process.

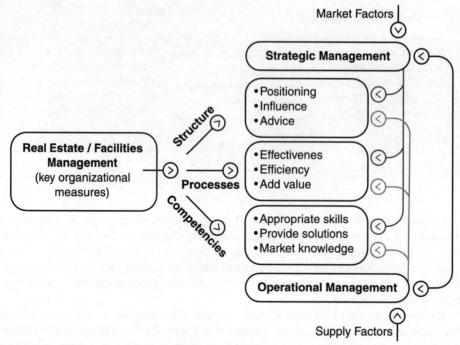

Figure 5.6 RE/FM – key organizational variables.

Research has indicated that three measures are critical in evaluating the performance of RE/FM within an organization (Then, 1998). The choice of the measures is driven by the need to evaluate how organizations respond in managing their operational real estate assets as a business resource. Figure 5.6 illustrates a model that proposes that the performance of RE/FM can be evaluated in term of three key organizational variables:

- *Structure* – the organizational set-up for operational real estate provision and facilities services management
- *Processes* – the systems and procedures for the management of the delivery of operational assets and their associated facilities support services
- *Competencies* – the necessary skills required for an efficient and effective delivery system – both in-house and bought-in expertise.

The performance of RE/FM is seen as the outcome in which the three organizational variables of structure, processes and competencies work in concert.

5.4 An integrated resource management framework

The dynamic capabilities of the model framework to *review* strategic relevance against the competitive realities of the business environment within which it operates, as well as *measuring* operational management effectiveness and efficiency against best conceivable practices externally, are illustrated in Figure 5.7. The business imperative, from a resource

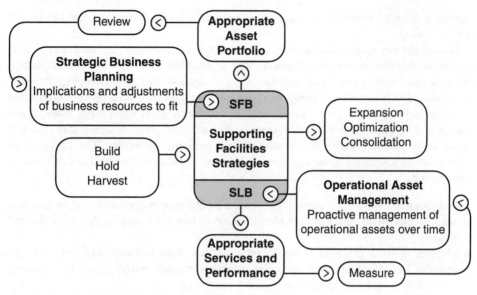

Figure 5.7 An integrated resource management framework (Then, 1999a).

management perspective, is to align the appropriate supporting facilities strategies with the current strategic business plans.

The realization of an integrated approach necessitates a formal planning framework that must cater for the cultural, procedural and existing knowledge base of the organization concerned. Ultimately, the practice of RE/FM is concerned with the delivery of the enabling workplace environment – the optimum functional space that supports business processes and human resources.

The *raison d'être* of RE/FM is to meet the business challenges that confront the organization it is supporting, as an enabler in the first instance. In the long term, a more sustainable role must be to build upon an aspiration to continuously add value by providing appropriate and innovative 'facilities solutions' to business challenges through the skilful manipulation of all business resources – the optimum balance between people, physical assets and technology.

5.5 The crucial link between work and the workplace

In providing facilities solutions to meet emerging business challenges it is becoming increasingly important to show the correlation between work and the workplace, between workplace provision and the effectiveness of people, and the correlation between the way the workplace is configured and serviced and effectiveness of businesses in meeting the needs of their customers. The rapid pace of technological development, particularly in information technology, has had, and continues to have, a considerable impact on the design and subsequent use of buildings. Far from being regarded as a necessary evil, for an increasing number of organizations there is a growing acceptance that buildings (as operational assets) must now be managed as a valuable business resource, just like people and technology.

One of the key responsibilities of facility managers in managing demand over time is to ensure that the original assumptions that formed the basis of the accommodation strategy are still valid, and remain so, set against the ever-changing business environment. All too often facility management is reduced to maintenance, engineering or office services, but space management is a strategic role. Used well, space is not purely a cost, but a generator of revenue in its own right. Well-designed and managed space motivates staff as well as attracting and retaining customers. The provision of workspace is a strategic issue because of the need to manage change that can occur at three levels:

- changes to the business – operational changes, expansion or contraction of the business and its constituent parts, and the identification of business needs as they vary through time
- changes to work practices – the adoption of new work methods and processes, the introduction of technology to automate or obviate manual activities, and the changing expectations of the people in the business
- changes to workspace – the process of reconfiguring existing workspace, relocation to new premises, and changes to the way workspace is allocated.

The process of maintaining the strategic relevance of the accommodation strategy is a continuous one. Any review of an accommodation strategy today must consider the implications of the above factors on the existing operational portfolio and the likely future need for other buildings.

5.6 Conclusion

The strategic management of RE/RM is not only about fitting employees to desks and rooms, but about supporting their value-adding tasks and engendering a corporate culture that supports creativity and innovation.

Decisions relating to real estate investments and facilities operations are *strategic* because they:

- are not only about transactions and operational management
- are about strategic asset management (alignment, location and life-cycle management) and space management
- are going to be, in the future, more about workplace strategies, not on a national, but on a global basis, driven by branding of corporate image and culture.

The changes that are confronting businesses are increasingly demanding workplace solutions that are 'employee-person centric' rather than 'building-place centric'. The primary concern is the delivery of a productive workplace environment as a total serviced package – an integrated system as a 'bundle of services'.

There is a need to rethink the workplace, not as a place but as a supporting system of services that follows and supports individual workers and their teams, no matter where

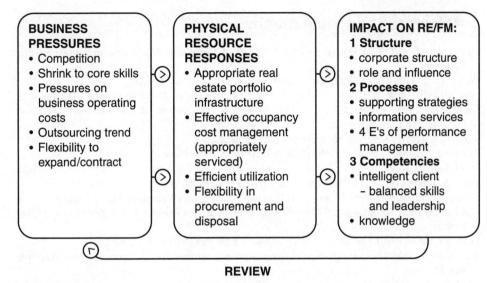

BUSINESS PRESSURES	PHYSICAL RESOURCE RESPONSES	IMPACT ON RE/FM:
• Competition • Shrink to core skills • Pressures on business operating costs • Outsourcing trend • Flexibility to expand/contract	• Appropriate real estate portfolio infrastructure • Effective occupancy cost management (appropriately serviced) • Efficient utilization • Flexibility in procurement and disposal	1 Structure • corporate structure • role and influence 2 Processes • supporting strategies • information services • 4 E's of performance management 3 Competencies • intelligent client – balanced skills and leadership • knowledge

REVIEW

Figure 5.8 Context of RE/FM – summary.

they work. The shift in focus for RE/FM is from managing the physical assets and equipment to helping people to be productive. The RE/FM response will be governed by a thorough understanding of business operational demands, and providing appropriate facilities and tailored services within affordable costs.

The conceptual models and frameworks presented in this chapter represent the consolidation of influences from theory and practice, the objective being to understand and explain the nature of the problem and logically justify the crucial need for an informed interface between strategic business management and operational asset management.

The role of RE/FM is predicated on change. If facility managers are to make an impact in adding value to the business delivery process while minimizing cost, then they must aspire to a position of influence via their ability to offer facilities solutions that address business challenges. Such a position of influence can only be attained by actions that are in tune with the strategic intentions of the organizations they are serving. Facility managers must develop methods that can be employed to gain an understanding of the needs of the business (current and future), and the capabilities and constraints of the current portfolio of buildings, and to reconcile the supply and demand equation. In fulfilling their management roles it is fundamental that they understand the role of facilities in supporting business delivery processes.

The potential scope of activities that facility management embraces may be diverse, but they are essential services for the delivery of an enterprise's core business. It is the clear definition, appropriate procurement and cost-effective management of operational buildings (as assets) that demonstrate the value-adding role of RE/FM in an organizational context.

Figure 5.8 represents an attempt to summarize the 'causes–responses–impact' relationship that accounts for the justification and conceptual development of real estate and facility management.

References and bibliography

Arthur Anderson (1995) *Wasted Assets? A Survey of Corporate Real Estate in Europe* (Arthur Andersen).

Avis, M. and Gibson, V. (1995) *Real Estate Resource Management* (GTI).

Avis, M., Gibson, V. and Watts, J. (1989) *Managing Operational Property Assets* (GTI).

Carder, P. (1995) Knowledge-based FM: managing performance at the workplace interface. *Facilities*, **13** (12), November, 8.

Cole, G.A. (1994) *Strategic Management* (DP Publications).

Debenham Tewson Research (1992) *The Role of Property – Managing Cost and Releasing Value* (Debenham Tewson).

Ernst & Young (1993) *The Property Cycle – The Management Issue* (Ernst & Young).

Gallup Organization (1996) *Shaping the Workplace for Profit*. (Workplace Management).

Graham Bannock & Partners Ltd (1994) *Property in the Board Room – A New Perspective* (Hillier Parker).

Henley Centre (1996) *The Milliken Report: Space Futures* (Milliken Carpet).

McGregor, W. and Then, D. (1999) *Facilities Management and The Business of Space* (London: Arnold).

O'Mara, M.A. (1999) *Strategies and Place* (The Free Press).

Oxford Brooks University and University of Reading (1993) *Property Management Performance Monitoring* (GTI).

Then, S.S. (1998) Corporate leadership in real estate and facilities management. In: Quah, L.K. (ed.) *Proceedings of CIB W70 International Symposium on Facilities Management and Maintenance – The Way Ahead into the Millennium*. Singapore, A9–16.

Then, D. (1999a) An integrated resource management view of facilities management. *Facilities*, **17** (12/13), 462–9.

Then, D. (1999b) Bridging the gap between strategic business planning and facilities provision and management: implications on competency and training at management levels. In: *Property and Construction Education and Research, Proceedings of the 3rd and 4th International Electronic AUBEA Conferences 1997–98*, 73–81.

6

Space management

Alison Muir*

The main purpose of most facilities is to provide suitable workspace, yet this is not an easy task, as it involves the determination of requirements, design, construction, fit-out, adjustment as people come and go and as business demand changes, and eventually new space acquisition or disposal. Space is often measured as net rentable area, but attached to this are other essential spatial needs like circulation/travel, engineering, amenity, car parking and utility areas that support the primary function. In some cases this non-rentable component can be as high as 50% of the total building area.

Given the importance of suitable workspace, effective space management is a vital competency and a major source of value optimization. Most benchmarks for facility performance are expressed as a ratio to floor space, so by reducing floor space everything suddenly becomes more efficient. Strategies for reducing space requirements include the creation of non-territorial workspaces, shared resources, telecommuting, hot-desking and open planning. There are trade-off costs associated with any of these approaches, but if improvements in worker productivity can be gained, then it is sure to be a worthwhile initiative.

Space management relies on a clear understanding of organizational need and direction. A needs analysis is used to determine the number and size of functional spaces when a facility is designed. Subsequently, space usage must be monitored and ownership reallocated to ensure that this expensive resource is not wasted. The facility manager should be constantly concerned with whether available space is sufficient and where additional space, if necessary, can be readily acquired.

While floor area is a significant consideration, so too is the quality of the space. This is a key factor during the design of new facilities and must remain so during the period of occupancy. There is a direct relationship between quality and productivity whereby

* Muir and Muir, Sydney, Australia

increases in quality will increase productivity and vice versa. Access to natural light, visual privacy, noise reduction, circulation flow, storage availability and workstation design are all examples of influencing factors. Job satisfaction is closely linked to productivity, and is therefore affected by facility quality and the successful management of space.

Space configuration changes are the result of human resource movements, technology upgrades, specific project tasks, promotions, new business ventures and refurbishment work. The impact of these changes, known as *churn*, can be a critical cost centre. Strategies that minimize churn impact are therefore valuable and will directly reduce operational budgets and unproductive activities. One such strategy is to purchase workstation furniture that is modular, flexible and adaptable to different uses. Another strategy is to endeavour to move people, not technology. Churn costs are not routinely recorded and form a hidden expense for most organizations. It is areas such as this that enable improvements in performance to be realized, leading to greater value for money and better use of limited resources.

6.1 Introduction

Space management and space planning are separate but related processes. Space planning is the process of optimizing the layout of a building to suit a business's needs; ideally this is done within the context of a business plan and a facility management plan for the business. Conceptually the process is simple: data about the business, the legislative/legal environment, the building, the needs of the business and those of the employees is gathered. This information is then analysed and translated into a space plan, through the 'design' process. The success or failure of the space planning solution can be assessed using various indicators at any time through the plan's life.

Modern economic and social considerations almost inevitably lead to demand for more flexibility in space planning. As a result, non-physical solutions, commonly called space management, are increasingly being used to provide flexibility; examples include hotelling and job sharing. The efficiency of these management strategies can be also be assessed using indicators similar to those that are used to evaluate space planning solutions.

This chapter will use the commercial office as an example of space planning and space management, but factories, schools and tertiary institutions, hospitals and even homes may be the subject of the same discipline. Definitions of terms are found in the Glossary at the end of the chapter.

6.2 Information gathering

As this is the beginning of the process, and the rock on which everything else is built, the thoroughness of information gathering is critical to the success of the process. Information gathering is easily divided into the three main headings of corporate context, legislative/legal requirements and the building.

6.2.1 Corporate context

The corporate context is a collection of business plans and operational instructions that are particular to the business, and must be interpreted as part of the space plan. Figure 6.1 shows the corporate context in perspective. The business will be expected to provide a number of items that describe the goals, organization and structure of the business and its operations.

Figure 6.1 The corporate context.

The business plan (BP)

This should show projections and established deliverables across time for the core business, such as sales and delivery targets and new product development/factory expansion. The BP provides the context for the whole business at a strategic level for the life of the plan and the business.

The facility management plan (FMP)

This is a strategic document that will address space performance, space management, the base building specification, financial performance, procurement, workplace relations, communications, risk assessment, project management, how to manage change, and, if a public sector organization, public sector guidelines and reporting requirements. The FMP provides a consolidated view of property that supports the BP.

Organizational understanding

This document is the introduction to the BP and an overview of global context. It shows internal and external influences, interpretation of business goals and understanding of the physical infrastructure of the business.

Organizational structure

This shows, in diagrammatic format, the whole business and its component parts, the links between them, and particularly the unit/part needing the space plan.

Strategic space plan

This is likely to be part of the facility management plan but defines space standards and the expected overall image and quality of space planning as part of the whole business.

Marketing policy and operational systems

These are business decisions defining the how, what, when, where and why of sales and marketing. Such decisions will have an impact on how the office is managed and the expectations of employees.

Core business policy and operational systems

These are descriptive: they tell the employees how, what, when and where to perform their role in the business. This information will have an impact on adjacencies to other business units, business unit forecasts and expectations, and management styles.

Human resources policy and operational systems

These will have an impact on the expectations of employees and any unions, social justice within the business, and, if in the public sector, public sector guidelines and reporting requirements, such as equal employment opportunity (EEO) and occupational health and safety (OH&S). Private sector reporting requirements will need to be advised as well.

Environmentally sustainable design (ESD) policy and corporate attitude

ESD strategies, often referred to as 'green design', involve issues related to extraction and supply of raw materials, manufacturing, distribution, use and the end life (disposal) of products. Sustainability means that economic, social and environmental development should meet the needs of present generations without compromising the ability of future generations to do the same. A number of manufacturers, especially workstation and chair suppliers, are producing products that comply with ESD principles. In Australia, annual reporting by businesses must now include a 'green' component. The corporate world is recognizing that green is not just good for the planet, but good for business as well.

Brief for space plan

This includes needs analysis, staff numbers, blocking and stacking and the like and will need to include:

- total number of full-time employees, employee total by division and working relationships and adjacencies
- gross area per employee in m^2
- number and size of cellular offices and other rooms such as utility rooms, libraries, quiet rooms, reception and computer rooms or server cabinets
- number and size of boardroom, conference, meeting, interview and training rooms, and adjacent support facilities
- records and storage requirements

Relationship Matrix		1	2	3	4	5	6	7	8	9	10	11	12	13	14	15	16	17	18	19	20	21
Executive	1	▓																				
Director, Marketing Operations	2	S	▓	S	S	S	S			X												
Australian Marketing and Distribution	3		S	▓	C	C	M											M	S			
International and Sydney Marketing	4		M	M	▓	M	S			C								S				
Regional Tourism and Development	5		S	S	S	▓	S															
Marketing Research	6		S	S	S	S	▓															
Director, Policy and Development	7							▓						S								C
Planning and Co-ordination	8				S	S	S		▓	M				S								C
Development	9				S	S	S		M	▓				S								C
Finance	10							S			▓											
Employee Services	11									S		▓										S
Information Systems	12												▓									
Information Systems, Library	13													▓								
Director, Marketing Communications	14														▓	M	M	M	M	M		
Ministerial and Board Liaison	15	S							S						M	▓	M	M	M	M		
Special Events	16														M	M	▓	M	M	M		
Destinational Promotion	17														M	M	M	▓	M	M		
Industry Promotion	18														M	M	M	M	▓	M		
Video Images / Desktop Publishing	19														M	M	M	M	M	▓	S	
Common Facilities	20																				▓	
Australia Day Council	21														C							▓

Legend: M   Must be close to
         S   Should be close to
         C   Could be close to
         X   Must not be close to

Figure 6.2 Sample affinities matrix.

- security requirements (internal and external)
- power, data, lighting, communications and media requirements
- air-conditioning, ventilation and acoustic requirements
- fire services
- plumbing and hydraulic requirements
- building automation requirements.

An affinities matrix (Figure 6.2) identifying 'essential', 'important', 'occasional' and 'never' adjacencies will assist with block planning and stacking, e.g., the affinities matrix will show that it is essential that the Board Room be close to the CEO's office, but never adjacent to the mail room.

Anthropometric/ergonomic analysis

This is a physical analysis of the employees who will use the result of the space planning. An ergonomic assessment will take a critical review of the job to be performed, the organization and the associated workplace design in terms of the workforce, their efficiency and productivity, and their health. Comparisons should also be made between the job demands and the employees' capabilities and limitations. At this time the emphasis is not on the built environment but only on the physical human needs. The elements of the ergonomic assessment that are required for office design include general physical activity, lifting and handling, work postures and movements, accident risk, job content, job restrictiveness, worker communication and personal contacts, decision-making, repetitiveness of the work, the need for attentiveness, lighting and vision, thermal environment, and noise. Ergonomic assessment of at least the parameters shown in Figures 6.3 and 6.4 are required:

1 Seat height with feet flat on the floor
2 Height to top of thigh when seated (for desk height)
3 Eye height above seat level (for use of a computer)
4 Height from seat to underside of elbow (for arm rest height)
5 Near reach when seated and far reach when seated, affecting spread of equipment and functions on a desk and adjacent surfaces.

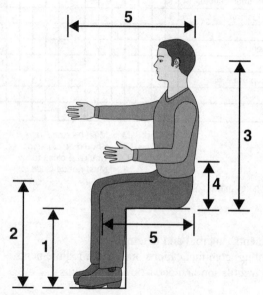

Figure 6.3 Ergonomic parameters – male.

Two other items are vital: an overall programme with times for staging and for delivery of the completed office project, and a budget for delivery of the completed office project. To develop a complete cost plan a number of factors are involved and are generally divided into building works, loose items, engineering services, fees, contingency and taxation. Figure 6.5 provides a basic format for the preliminary budget estimate for an office plan.

As with all documents, the information on which the budget is based is important. What is also important is what is not included, typically items such as land costs, legal fees, holding costs and interest charges, approvals and permits, council contributions and payments, escalation, prolongation and time extension costs, real estate and marketing

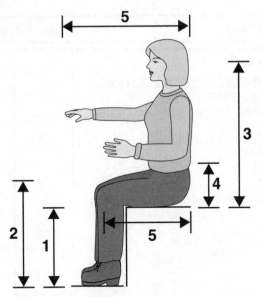

Figure 6.4 Ergonomic parameters – female.

fees, commission, site allowances, computer hardware or software, PABX, audio visual equipment, out of normal hours work, relocation costs, making good to existing premises (if relocating), cutlery and crockery, artworks, televisions, staff security cards, or other items – all that are applicable should be noted in the documentation. This is not an exhaustive list and other items may need to be added in particular situations.

The combination of the documents described above will provide the professional designer with a clear picture of the business, and a well-defined starting point.

6.2.2 Legislative/legal context

Legislative and legal constraints on space planning will vary from country to country. In the following discussion the situation in Australia is used as an example; however, similar codes and statutes exist in most places, governing many aspects of the physical planning of space, and designers should investigate the particular requirements of the authorities in their region.

Every employer, manager and supervisor in Australia must ensure the health and safety of employees, through compliance with the *Occupational Health and Safety Act*. In NSW this is an explicit and unambiguous legally binding duty of care for employers to protect employees against injury or illness related to their work. The duty of care provisions in all OH&S legislation in Australia includes foreseeable risks. Employers and managers are expected to take all reasonable and practical steps at the workplace to eliminate or minimize these risks, and thereby avoid employee injury and illness. The duty of care provision extends to the workplace and includes:

● design of work areas and spaces that addresses concerns such as lighting, noise, thermal comfort, equipment, furniture and tools

TRADE DESCRIPTION	Cost/m^2 $	Subtotal	Total
BUILDING WORKS			
ADJUSTMENT TO BASE BUILD/DEMOLITION			
INTERNAL WALLS & PARTITIONS			
INTERNAL DOORS			
WALL FINISHES			
FLOOR FINISHES			
CEILING FINISHES			
BUILT-IN JOINERY			
WORKSTATIONS			
Sub-Total of Building Works			
LOOSE ITEMS			
BLINDS			
SIGNAGE			
WHITE GOODS			
ARTWORKS			
STORAGE			
LOOSE FITMENTS & CHAIRS			
Sub-Total of Loose Items			
ENGINEERING SERVICES			
HYDRAULICS			
MECHANICAL SERVICES			
BMS/DDC			
FIRE PROTECTION			
ELECTRICAL INSTALLATION			
COMMS/DATA			
SECURITY			
A.V. EQUIPMENT			
BWIC			
Sub-Total of Eng. Services			
FINAL CLEAN			
CONTINGENCY (5%)			
PRELIMS & C.M. FEE (9–13%) say avg 11%			
PROFESSIONAL FEES (11%)			
ESCALATION			
Sub-Total (Excl Tax)			
TAX (10%)			
Sub-Total (Incl Tax)			
INFORMATION USED			
EXCLUSIONS			

Figure 6.5 Typical format for a preliminary budget estimate.

- work and task design including what the employee is required to do in terms of work content, work demands, restrictions and time requirements, working with other employees, and the responsibilities of their position
- work organization including how tasks are sequenced, patterns of work, shift work, rosters, work and rest, consultation and feedback, supervisory and management practices, who the employee reports to and/or works for, how the tasks are organized and why, the nature of the organization, and its problems
- employee health and other problems either unrelated to the work, or as a result of their work, the matching of skills to the demands of the tasks, the need to adjust to the work conditions, and the use of personal protective equipment.

The space planning and management of an office must typically address issues such as walkways, floor coverings, stairs and steps, and the space around work and storage areas. Lighting should provide appropriate artificial light for tasks, and control the impact of natural light (e.g., the elimination of glare from windows). Protection should be provided against annoying or harmful noise, and the need for privacy and quiet recognized and accommodated. Air quality concerns, including humidity control, exhausting of harmful fumes, protection against dusts, gases and vapours are all controlled by statutory regulation. Thermal comfort, including consideration of people performing active and sedentary tasks, can affect productivity. Storage spaces including location, access, adequacy, flexibility and personal versus general all affect space management and health and safety. The maintenance of the building, plant, equipment and furniture affect an employee's occupation and ability to perform. The employees' needs for visual privacy, personal space and flexibility of spaces are in continual conflict with space management if not clearly defined as an OH&S strategy rather than simply being tied to the financial bottom line.

In 1992 the Australian Building Codes Board introduced the *Building Code of Australia* (BCA) BCA90, which produced a uniform set of regulations for the design and construction of buildings across Australia. The original BCA was fully prescriptive and buildings were 'deemed to satisfy' the code if they matched the specific prescribed requirements contained in the code. In 1997, BCA96 introduced performance alternatives for the Code, thus giving designers greatly flexibility in their solutions.

Compliance with the BCA enables the achievement and maintenance of acceptable standards for structural sufficiency, safety (including fire safety) and health and amenity (including such things as adequate ventilation), for both the base building and the office fit out. Specific concerns addressed by the BCA include structural stability and loads, type of construction, fire compartmentalization and separation, protection of openings, access and egress, access for people with disabilities, provision of fire fighting equipment, smoke hazard management, lifts (elevators), emergency lighting, and provision of exit signs and warning systems.

Requirements for special provisions for people with disabilities are covered by a different code, the *Disability and Discrimination Act, 1993* (DDA). This Act addresses such concerns as access to buildings for people with disabilities, which, under the DDA, must be provided through the principal public entrance, not via a back door or loading dock.

The Australian Standard AS1428.1 – 1998, *Design for Access and Mobility*, defines the circulation spaces at door ways for swinging doors, stairways and handrails for buildings with no lifts, heights for switches, general purpose outlets (GPOs) and door handles, tactile indicators at stairways, escalators and ramps, and the construction of sanitary compartments.

There are separate standards that cover many other things such as lighting, air conditioning, workstations, car parking, computer accommodation and general access. Compliance with all the relevant codes and standards is obviously a fundamental part of successful space planning. Compliance may be assessed by local government authorities or in some cases (as is the case in Australia) by private certifiers.

The influence of insurance companies and an increasingly litigious society have contributed to the growing importance of insurance considerations in the office environment. Professional indemnity, public liability and workers' compensation insurances may have an impact on space planning and management, e.g. if an employee chooses to become a telecommuter, and is injured while working at home, how will the employer's insurance situation be affected? Planners and managers need to be aware of these concerns and seek information that will allow them to make adequate allowance for them in their work.

Legislation relating to company directors in the private sector needs to be understood, analysed and considered. Such legislation defines the responsibilities of senior positions within the private and public sectors and therefore it affects the space planning. If, for example, a company director were to agree to the permanent locking of a door that is a required exit under an applicable code or regulation, then the directors are individually responsible for any misadventure that may result. The worst example is a locked fire door that causes the death of someone in a fire stair. A clear understanding of the legislative/legal responsibilities of senior management is therefore important when space planning an office for an organization.

6.2.3 The building consideration

Once the building requirements have been defined as part of the corporate context and the strategic space plan, information about the physical 'building' is collected and compared with the requirements. A number of consultants will be useful during this gathering of information phase including the architect and the service engineers (for electrical, mechanical, fire services, communications, hydraulics, structural and acoustic data). A building surveyor/ordinance surveyor, quantity surveyor and a public or private certifier will contribute other data to the total package of information about the building.

The involvement, at an early stage of the 'design' process, of a building surveyor to check dimensions and area calculations, and a private certifier to check code compliance is early insurance against time and money being wasted later in the process. With the growth of compliance consultants, the architect and interior designer are less likely to have an interest or competency in this field, and are more likely to specialize in 'design' alone. Both the building surveyor and the private certifier will use 'as built' drawings as the basis for checking compliance and provide a report to the project team of the findings. Typical problems with 'as built' and archived CAD drawings are the level of inaccuracy and lack of checking prior to issue. Figures 6.6 and 6.7 illustrate some of the salient points when considering a building for lease or purchase.

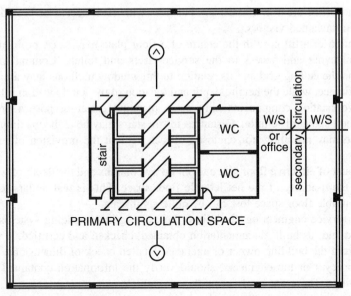

Figure 6.6 Central core building, no columns.

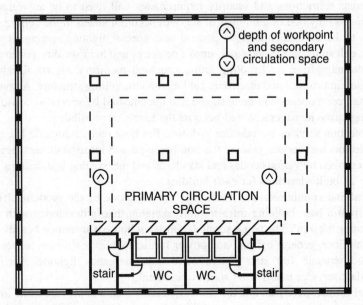

Figure 6.7 Side core building with columns.

When considering a building's suitability for a business, the architect will consider the building floor shape. The dimensions and overall floor shape (preferably rectilinear), and the location of the core and its geometry, impact on the effective internal usable space. The effective floor depth, i.e. the distance from the windows to the lift core, should be a multiple of the workstation dimension plus required circulation spaces. Arrival points from lifts to office areas can assist the business with clear access or provide a needed barrier from unwanted visitors.

The location of a lift core in the centre of a floor plate may cause problems with the workstation layout and access to the service risers and toilets. Column location and spacing, and the ceiling grid and its relation to the window mullions may assist planning or be a hindrance, while the height, depth and location of the windowsill in relation to the standard workstation components may make planning easier or pose a problem that requires a non-standard solution. Distances to fire stairs may be at the maximum for open plan and so may not allow for enclosed offices without the provision of extra escape corridors.

The amount of effective floor space taken up by columns and the depth of window sills will affect the usability of the net lettable floor space. This is real dollars lost, and the amount of usable floor space lost can be quite significant.

With all service engineering disciplines the existing base building systems should be investigated and 'as built' documentation obtained, checked and certified, if not already available from the building owner or agent. A detailed check of dimensions, either by a building surveyor or the engineers, should verify the information contained on the 'as built' drawings. These are rarely accurate or complete.

For air conditioning and other mechanical services, questions about supplementary air conditioning, exhaust air, ventilation, separate air-cooled systems, air conditioning zoning, and any noise restrictions and security requirements will need to be answered. The air conditioning is only one of a number of services accommodated in the ceiling space and will need to be co-ordinated with general and special lighting, emergency lighting, sprinklers, exit signs, EWIS speakers, smoke detectors and loud speaker systems, security systems, drainage from air handling units, structural members, electrical cabling, data cabling, communications and electronic cabling, and the ceiling structure. A complex web of issues and components will be designed and documented by specialist consultants, but the ongoing space management will become the user's responsibility.

Fire protection services include fire hydrants, fire hose reels, automatic fire sprinklers, smoke detectors and service pumps; the number, type and location of such equipment is usually prescribed by various codes and standards and the existing installations should be shown on 'as built' drawings for each building.

Electrical and communications cabling is the backbone of the modern office and is designed from a base building infrastructure skeleton that is developed with the office space planning 'design'. The house services consist of house distribution boards, generally one on each floor, general lighting and power for lift lobbies, corridors, toilets and plant rooms, and separate fire stair lighting and exit/emergency lighting. Vertical risers accommodate services mains from roof or basement.

The brief for the space plan and the strategic facility plan will provide some information about the requirements for the initial feasibility and 'design', but will need to be clarified and expanded as the space planning progresses. Design work for electrical and communications will include the tenant distribution board, metering panel, light circuits and light fittings, cable management, light switching, electrical capacity (dependent on loads), and uninterruptible power supply (UPS) and stand-by diesel generator, if required.

The reflected ceiling plan is the major tool for co-ordination of the services. Following the electrical consultant, who will provide a lighting layout to suit the space planning, the mechanical engineer will check that the lighting load will not exceed the standard allocation, and the electrical engineer will provide circuitry and switching. Switching that includes master, zonal and local switches provides the most energy efficient operation. Exit and emergency lighting is added to meet code requirements.

Most office buildings will provide low-brightness, one or two tube fluorescent light fittings, with air slots, mounted in an acoustic tile ceiling as standard. Overall lighting level is likely to be 250 lux (lumens per square metre).

Location of power and data circuit cables is a bigger concern than that of lighting circuits as access ways need to be found both horizontally and vertically to allow for reticulation of cables to the workpoints. Generally the tenancy distribution board allocated to the floor will suffice, although some larger offices with computer rooms will require additional boards to deal with the number of circuits and to manage circuit cabling volume.

Cable management is generally achieved using a mixture of hard (fixed) and soft (movable) wiring, with an interface between the two systems either in the ceiling or at the perimeter using ducted skirting. Soft wiring works best in the workstation environment where cableways and basket systems allow reticulation of cables and associated outlet boards. Offices and other fixed partition spaces are generally still served by hard wiring, since churn of offices is less frequent than that of workstation clusters.

One form of horizontal reticulation popular with heavy cable users is the raised floor where an 80–600 mm high underfloor space is provided and cable management is facilitated through removable floor tiles. A number of design issues have to be addressed when this design option is being considered, such as sill height, ceiling height, door head heights from public spaces, lift doors, ramps and access points, accessibility, cost, and cable access from vertical risers.

Communications, like electrical cabling, has base and tenant components and generally includes data, master antenna television (MATV), security systems, fire intercommunication and a building management control system (BMS). Communications is divided into *passive* and *active* networks, and includes the cabling installation, workstation outlets, communication racks, test point frame, patch leads and fly leads for the passive, and hubs, routers, switchers, UPS, PABX, modems, servers, PCs, and printers in the active. Unlike the power network, data networks operate on a Cat 5E data cable in star configurations and require each RJ 45 outlet at each workstation to be individually wired from a central location on the floor, within a maximum distance for quality of signal. This generates quite high concentrations of cables at the board, at the start of a run of workstations or in ducted skirting. Within the ceiling space cables are tied, grouped and secured in multiple runs rather than installed on cable trays up through an existing ceiling grid.

Generally, as a minimum, three communications outlets are provided at each workstation, for telephone, data, and a spare for a network printer. If telephone quality cabling (Cat 3) is still used for traditional voice (telephone) cabling another cable is run to the workstation and only two data outlets provided. It is imperative that outlets are permanently labelled and that complete 'as built' records are maintained.

MATV facilities generally only need to run co-axial cable from a take-off point located in the communications cupboard at the core of the building to the outlet provision in the space(s) required. Communications rooms and antennae dishes on terraces have become quite common additions to modern office buildings where tenants require alternative routing for their communications services to minimize risk of outages, especially when 24/7 operation is required for the business.

Security can take a number of forms and will depend on the risk assessment of the business. Common forms are physical barriers, electronic access control, intruder detection, video surveillance and duress alarms, but may include CCTV surveillance, vibration sensors, bio recognition systems and pressure mats. Most commonly used access control systems involve 'proximity' cards, which utilize a credit-card-sized ID card with an embedded coil that is detected or read by a transmission/receiver device at key entry points to the building and the tenant's secure access points. Security doors require co-ordination of door hardware, door thickness, door jamb, electrical connections and the correct swing for egress in the event of an emergency.

Knowledge of the company's business needs in all the above areas is vital if an early analysis of the suitability of potential premises is to be accurate and so enable the elimination of inappropriate buildings from those under consideration.

6.3 Analysis and synthesis of 'design'

Analysis and synthesis is a 'design' process that creates proximity diagrams using the information from a number of sources to give answers, often preliminary, to all or part of the problem. For example, with information on the type of activity, overall numbers of people and a budget, it is possible to check the feasibility of potential buildings and select a suitable one. The analysis and synthesis or 'design' phase will interpret the collection of information described above and produce the space plan. The process of analysing the data fairly rapidly identifies what data is critical and more importantly what data is missing from the process.

It is not part of this chapter to describe the philosophy of the design process; suffice to say it is iterative, like most design, and is best undertaken by experienced professionals. The current space planning approach, however, is to place cellular offices and enclosed rooms inboard from the windows/natural light and place the employees in workstations close to the windows. This concept allows employees, who do not have the flexibility in their job description that would allow them to leave the office, access to the outside world and natural light. In the past, cellular offices were located near the windows and were seen as a reward that came with promotion, but now with progression based on performance this tradition is disappearing, with the adoption of a team based approach built around shared common facilities such as storage and utility rooms.

The most common form of planning for an office is a mixture of open plan workstations and enclosed offices and meeting spaces. The open plan standard workstation concept arose out of financial imperatives to reduce space requirements, and therefore costs, in the office. One principle is that costs can be reduced when change is required as the staff are rearranged and not the workstations or offices.

Generally, a standard workstation is designed for the largest group of users, based on ergonomic and anthropometric data and becomes the office 'standard.' The standard workstation generally is in the range of 1800–1800 up to 2400–2700 mm^2 or multiples of 300 mm in between.

The storage of files and stationery is likely to be the cause of the greatest waste of floor space, and should be addressed independently so that the most suitable and efficient method can be identified. Lateral filing systems are a more cost and space effective method of storing records than traditional drawer style filing cabinets. Lateral filing can be achieved in standard-sized, space-efficient storage units that can also be adapted for the storage of manuals, books, products and stationery.

Many office design concepts are changing, and some of the current practices include hotelling, hot-desking, satellite offices and systems furniture. The different work practices should be clearly matched with the form of office accommodation that best supports the work practice, and should offer more than one format to give variety and choice in the space plan.

The design process used to be carried out by traditional methods, using a drawing board with a pencil and scale ruler. Now there are sophisticated computer programs that can optimize layouts, work more accurately, with greater speed to change and analyse more varieties of layout in a given time. These techniques track the changes, quantify materials, fixtures and costs, and document the required physical works using the one medium. The main benefits of computer aided design, or CAD, are speed and accuracy of design options. The quality of the space plan is dependent on the analysis and synthesis of the information gathered, but it is also dependent on the quality of the information and the interpretation provided by the business.

With efficiency checks, discussed below, integrated into the design process, the outcome should be more effective space planning. These are the same tools and techniques that have been developed for military and aerospace technology, used to produce greater efficiency in the office.

6.4 Space indicators

Space planning and space management solutions must be monitored at various times and their performance included as part of routine financial reporting procedures. The focus will be on return on investment for both the initial development and during the ongoing management of the space.

There are three main indicators of space performance that are useful for financial reporting purposes:

- efficiency – this indicates how well the space is apportioned as a component of the total space
- flexibility – this is indicative of the changeability of the physical space
- space utilization – this measures the use of space over time, and the potential for use in future time.

These indicators should be used to check progress at various stages throughout the planning and occupancy phases, most importantly at the following points:

- preparation of the building surveyor's report
- preparation of the initial space planning diagram
- before commitment to a lease or purchase
- after final space planning is completed
- as part of post occupancy evaluation after construction is complete
- as part of ongoing space and churn management
- as part of maintenance management.

These indicators can be evaluated quantitatively using the following formulae.

6.4.1 Efficiency

$$\frac{\text{Workpoints } (m^2)}{\text{Number of employees}} = \text{net } m^2/\text{employee}$$

This calculation measures the total area occupied by work points (enclosed offices and workstations) and divides it by the number of employees. The result is a net area per employee in square metres, which excludes any other facilities. This should not be below 4 m^2 to be efficient and includes tertiary circulation space.

$$\frac{\text{Workpoints } (m^2)}{\text{Circulation space } (m^2)} \times 100 = \% \text{ circulation space}$$

This calculation compares workpoints, as above, in relation to primary and secondary circulation space, and should not be less than 30% to be efficient. Building codes will generally have an impact on this calculation by defining access and egress widths.

$$\frac{\text{Net lettable area } (m^2)}{\text{Number of employees}} = \text{gross } m^2/\text{employee}$$

Net lettable area, as discussed previously, is the area of the lease or purchase and when divided by the number of employees gives the gross area per employee. In Australia this area should not be less than 10 m^2 per employee, as the air-conditioning system for the floor will have been designed to suit, including an average for employee and computer equipment heat loads. If the gross area is less, supplementary air conditioning may be required to meet the code requirements.

$$\frac{\text{Net lettable area } (m^2)}{\text{Gross building area } (m^2)} \times 100 = \%$$

The efficiency of the net lettable over gross building area provides a useful assessment of the return on investment whether calculated when considering the purchase of a building or used as part of the calculation of building outgoings for a lease.

$$\frac{\text{Usable floor space } (m^2)}{\text{Net lettable area } (m^2)} \times 100 = \%$$

The higher the percentage result here, the more real space is used efficiently.

$$\frac{\text{Usable floor space } (m^2)}{\text{Number of employees}} = m^2 \text{ per employee}$$

This calculation provides a measure of 'real' square metres per employee rather than net lettable area per employee.

6.4.2 Flexibility

$$\frac{\text{Cellular office area (m}^2)}{\text{Total workpoint area (m}^2)} \times 100 = \%$$

Total workpoint area includes cellular offices and workstations, and should include corridors and passageways, but not meeting and other support rooms. This calculation will show high flexibility when the percentage is low and will also indicate workforce flexibility.

$$\frac{\text{Cellular office area (m}^2)}{\text{Net lettable area (m}^2)} \times 100 = \%$$

This calculation will show the flexibility of the space plan as a whole and again the higher the percentage, the lower the flexibility. The flexibility of space can often be more efficiently achieved by moving, or churning, the employees and not the workstations or cellular offices.

6.4.3 Space utilization

$$\frac{\text{Meeting space use (hours)}}{\text{Total time available (m}^2)} \times 100 = \%$$

This formula is a useful tool for calculating the number of meeting rooms needed in the space plan. Total time available should be the time the office is open for business, e.g., 7.30 a.m. to 6 p.m., 5 days per week or 7.00 a.m. to 9.00 p.m., 7 days per week.

$$\frac{\text{Time at workpoint (hours)}}{\text{Total time available (hours)}} \times 100 = \%$$

This calculation is useful for hot desking and hotelling space management, and should be calculated on the total time the office is open, up to 24 hour/7 day operation.

$$\frac{\text{Net lettable area (m}^2)}{\text{Number of employees}} = \text{m}^2 \text{ per employee}$$

In Australia space utilization is generally targeted at 10 and 16 m^2 per person in the public sector, with slightly lower figures in the private sector.

$$\frac{\text{Records space (m}^2)}{\text{Net lettable area (m}^2)} \times 100 = \%$$

The calculation of space used for storage multiplied by rent/m^2, versus the cost of time to retrieve records from elsewhere and the cost of rent/m^2, will determine the efficient and cost-effective utilization of storage space.

The practice of space management basically comes down to the management of churn, and the success of flexible thinking and space planning at the earliest stage of an office's conception.

6.5 Space management

The excesses of the 1980s and the devolution of power to staff in the 1990s has seen the transplant of factory and assembly line terminology to the 'design' of offices. Re-engineering and restructuring are commonplace, and are often used when a business needs to respond to a changing marketplace. *Workplace re-engineering* is the fundamental rethink and redesign of business processes to achieve dramatic improvements in cost, quality, service and speed. On the other hand, *workplace reform* includes the full range of industrial relations and human resources issues affecting the occupation of facilities by employees.

Successful management and implementation as the result of workplace reform and re-engineering can guarantee the survival of a business. A 'redesign' of the office space plan may be required as a result of workplace re-engineering and negotiated workplace reform, or simply a churn of the office. The identification of total churn costs, including employee rates for down time, is imperative if churn, as part of space management, is to be successfully financially managed, delivered and reported.

Space management is the delivery of space services and the management of the completed space plan. It is a tactical activity responding to the strategy set out in the facility management plan. Space management includes alternative officing, free addressing, group addressing, 'just in time', virtual office, satellite offices, shared spaces, hotelling and hot-desking, and 'on and off premises' management. It is not for the tactical operation of space management to increase or decrease efficient and effective use of office space – the role is one of maintenance of space services to the employees.

The long-term strategy of space management is rightly the providence of the facility management plan and may include business re-engineering and outsourcing.

6.6 Conclusion

In summary, space planning and management both provide a supportive work environment that delivers the business' products, within the business' budget and time. Space planning is a discipline used to produce a working environment suitable for a business to produce a return on its investment; with the right tools and information anything is possible. Space management, on the other hand, provides a range of working environments and space services for the business' employees.

Glossary of terms

Alternative officing – a collective term used to describe alternative accommodation strategies that influence the design of the workplace and how people work. The quiet zone and quiet rooms are examples of alternative officing where workstation occupants can withdraw to a thinking space.

Anthropometrics – the study of the human form and the measurement of the dimensions of that form compiled into tables for comparison and analysis.

Blocking – the process of determining and illustrating the location of each business unit on the floor of a building. This will depend on affinities with other business units and specific physical aspects of the space such as views and daylight.

Brief or *Space Needs Analysis* – the process undertaken to determine the amount of space needed by an organization, both now and in the future, based on projected employee numbers and space use guidelines for the range of different functional requirements. The space and needs analysis forms part of the facility plan strategy and tactics for space management.

Cat 5E – a 4 pair data quality cable with high performance data transmission capabilities, the most commonly used cable in office tenancies. A *Cat 6* cable has been developed but is not yet commonly used in Australia.

Cellular office – a room with a door and four walls occupied by one individual who has a need for isolation.

Changeability – the ability for space to be adapted to respond to change.

Churn – the internal accommodation re-arrangements undertaken in response to changing organizational and functional requirements which includes the following:

● *primary churn* is the result of a major strategic change in facility utilization by an organization and includes a relocation to new premises
● *secondary churn* is the result of the need to respond to deviations in the organization's business environment or the result of internal operational constraints and includes the need to respond rapidly to new business opportunities or change to work flows
● *tertiary churn* is an internal phenomenon and relates to the competition for resources between different groups within the organization and includes minor re-arrangements undertaken to resolve inequities in facilities utilization.

Circulation space – the space within a building assigned to the movement of people, goods and/or vehicles, and from which access is gained to other functional rooms and spaces. Circulation space can be categorized into three levels:

● *primary circulation space* consists of those parts of the floor or building such as aisles, walkways, entrances, exits, foyers and space required for access to stairs, lifts, toilets or common rooms – statutes define the widths of primary circulation space
● *secondary circulation space* is the area of a building required for access to some sub-division of space (open or enclosed) that does not serve all employees on a floor
● *tertiary circulation space* is the movement space within an individual employee's workpoint.

Ergonomics – the design of equipment, processes and environments so that tasks and activities required of humans are within their limitations, but also make the best use of their capabilities. Simply, ergonomics is designing for people.

Environmentally sustainable design (ESD) – when developing the environmental sustainability of an office a number of products, services and the base building itself, must be considered as part of a whole, rather than as separate and unconnected component parts. Most manufacturers will provide information on their approach, attitude and processes related to being a 'Good Green Citizen'. The business using the products must have a similar attitude to see value in green design. A good example is the use of timber in office design, which was seen in the 1980s and 1990s as sophisticated and appropriate to the image of the business. The image may still be appropriate in the twenty-first century but now we have endangered species, and some timbers that are so expensive in solid form that they are only seen as veneers. Most sustainability-conscious timber suppliers are supplying plantation timber from a list of species grown for the purpose; a species list is available from most forestry organizations.

EWIS – emergency warning and intercommunication system.

Facility plan – the outcome of the process of planning an organization's present and future operational requirements and translating them to the working environment in the most effective way.

Free address – a workpoint shared on a first-come, first-served basis, usually situated in a large open space with many workstations.

Group address – a designated group or teamwork space allocated for specified periods of time.

Home base – a dedicated individual workpoint.

Hot-desking – the practice of sharing dedicated workpoints between two or more employees.

Hotelling – the practice of allocating available workpoints, on demand, to employees and others when required.

Just in Time – a management practice of not assigning offices to specific individuals on a permanent basis.

Master antennae television system (MATV) – a passive cabling network for distribution of 'free to air' commercial broadcasting TV plus FM radio.

Net lettable area (NLA) – the Property Council of Australia defines the equation for NLA, which generally includes any column footprint, windowsill depth to the glass line and other potentially contentious component areas within the leased area of an office floor, but excluding common areas, toilets, lift halls and the like.

Office – a room or place for the transaction of business, discharge of professional duties, or conduct of clerical or administrative activities.

Off-premises workplace – a workplace remote from an organization's head office, e.g., satellite office, virtual office or telecommuting site.

On-premises workplaces – part of an organization's head office.

Open plan – the design of interior building spaces with the minimum of dividing partitions between areas designed for different uses.

PABX – an automatic telephone switchboard.

RJ 45 – an outlet with four pairs or eight conductors that is the most common outlet used for structured cabling system networks.

Satellite offices – provide administrative support for employees living close to the satellite site.

Shared space – a single assigned workpoint and work tools shared by two or more employees either simultaneously or on different shifts/schedules.

Space audit – the surveying and recording of the amount and use of the various functional space types currently within an organization.

Space planning – the action of translating the space needs of an organization onto the floor plate(s) of a building and at the same time taking into account the defined adjacencies between business units.

Space utilization (rate) – the estimation of the number of people that use the space in relation to its potential capacity and for what proportion of the available time. Space utilization is usually expressed as a ratio.

Stacking – the process of determining and illustrating the location of business units on the floor of a building adjacent or relative to other business units based on predefined proximities or affinities.

Support areas – part of the office not assigned or dedicated to a specific task or function, including meeting rooms, waiting areas, storage, computer rooms, utility areas and libraries.

Systems furniture – interconnecting demountable screens, desks, shelves and storage components forming a workstation. Systems furniture usually includes provision for integrated electrical, data, communications and lighting. When assessing the most suitable systems furniture for a space plan, the checklist in the Appendix will be useful. Each of the items listed should be given a weighting for the specific project and the systems compared using that rating. A numerical result will be useful for approvals and assessment.

Uninterruptible power supply (UPS) – a power supply often from a back up battery system or back up diesel generator as security against power failure.

Usable floor space – the net lettable floor space minus the footprint area for columns, windowsills and other structures where furniture and staff cannot be placed.

Virtual office – a workplace concept in which employees are free to work anywhere (home, car, hotel and similar) through the use of portable technology.

Workpoint – a dedicated area for an individual to work, although it may be shared with others.

Workstation – a workpoint at which a single user or multiple users operate screen-based equipment (SBE) mostly from a seated position, but sometimes standing. A workstation includes a desk or counter and associated storage furniture, and the facilities which accommodate the visual display unit (VDU), input devices (keyboard), documents and other equipment associated with the SBE. The work surface height of a workstation should be adjustable or a user provided with an adjustable chair and footstool.

References and bibliography

Bondin, P.C. (1999) Space planning considerations. In: *Proceedings FMAA Conference*, Brisbane.

Connolly, S. (1999) *Building Code of Australia and New Approval System under the Environmental Planning and Assessment Act 1998*. Unpublished course notes, Facility Planning, University of New South Wales.

Erlich, R. (2000) Sustaining Industry. *Qantas Club Magazine*, Spring, 6–10.

Heard, K. and Meadows, W. (2000) *Service Engineering for Facility Planning*. Unpublished course notes, Facility Planning, University of New South Wales.

Máté, K. (1995) *Sustainable Design*. Catalogue for 'M.A.D.E Accountable' exhibition (Sydney: Society for Responsible Design).

McPhee, B. (2001) *Ergonomics in Occupational Health and Safety*. Unpublished course notes. Facility Planning, University of New South Wales.

Standards Australia and Facilities Management Association of Australia (2000) *Glossary of Building Terms – Facility Management Terms* (Standards Australia).

Tutt, P. and Adler, D. (1979) *New Metric Handbook* (London: The Architectural Press Ltd).

Appendix. Systems furniture assessment

1. Function

- Is the workstation panel based, desk based, a tile system or freestanding?
- Is under shelf lighting provided?
- Are the storage units lockable?
- Is a master key system provided?
- What desk sizes are available?
- Is the desk surface one level or more?
- Is the desk level adjustable between 680 and 720 mm from floor?
- What accessories are available? Are they lockable?
- What is accessible within near reach? And far reach?
- Is a power pole provided and what is the cross section?
- On the screen, where is the access to services? Floor? Midrail? Or other?

2. Component parts

- What is the total number of component parts?
- Can the supplier provide a drawing showing the components of a screen and separation of services in the cable duct?
- How is a screen relocated?
- What tools are needed to relocate component parts of the whole system?

3. Safety and OH&S

- Where are the sharp edges?
- What leg interference is under the desk?
- Are power outlets safe from spillage, pens, and dust build up?
- Show compliance with statutory requirements
- Can cables be unhooked by a foot or create a trip hazard?
- Is anti-tipping provided on the mobile pedestal?
- Is the under hamper light switched or hard wired?

4. Changeability

- Can the system be rearranged without a technician?
- What tools are required to rearrange the system?
- Is a monitor arm provided as a standard item or integral to the system?

5. Screens

- How do the shelves and hampers attach to the screen?
- What weight does the screen carry?
- What sizes and increase increments are available for standard screens?
- What is the acoustic rating of the screen?
- Is the screen a pinable surface?
- What standard metal and fabric finishes are available for screens?

6. Cabling

- Where is excess cable stored?
- What is the maximum riser reticulation capacity?
- What GPO, data and communication plates are available?
- What is the minimum radius for multicore optical fibre in the ducting?
- Is the screen hard or soft wired?
- Where is the location of the cable riser?
- What provision is made for power and data separation?

7. Guarantee

- What is the cost of the system for a single unit and a cluster of six?
- What is the sales life of the system?
- How long will the components of the system be available for additions?
- What are the manufacturer's warranties on paint, fabric, laminate, timber, glass, castors, metalwork and hydraulics?
- What time is required to respond to defects on site?
- Is the guarantee for unconditional replacement of the components at no cost to the purchaser?
- What is the delivery time of the system to site?
- How long does the system take to install on site?

8. Appearance

- What is the aesthetic affect overall in a large open office area?
- How durable are the finishes?
- Is the system easy to keep clean?
- Does the system image suit our staff and our image to our clients?
- Where can we see an installation of the same size/type and similar finishes to the proposal for our office?

7

Information management

Stuart Smith*

Editorial comment

One of the most significant changes affecting the way facilities are managed is *e-FM*. The term is used to cover a wide range of technology-based innovations that are automating processes and integrating management systems. The concept includes intelligent buildings and contents, monitoring sensors linked to computerized management systems, graphic-based software tools, Internet-based information databases, and direct business-to-business and customer-to-business procurement services. Facilities can be monitored and corrected remotely, multiple facilities can be linked via centralized management control, and service requirements can be requisitioned and tracked with minimal human intervention.

New built facilities are becoming more complex, mainly through the sophisticated services that are being designed into them. Many of these are specialist services aimed at improving performance and efficiency with low operational demand. In order words, capital is being invested in systems initially in order to provide value savings later in the facility life cycle. However, with complexity comes the potential for unforeseen problems, failures and lost productivity, so the financial viability of embracing new technology may not be easily determined.

Computer-assisted facility management (CAFM) tools are popular as a means of improving facility control and customer service, monitoring planned expenditure against actual performance, and recording vital information in a single database. Maintenance work schedules, prioritizing activities and budget reconciliation are well handled by even the most rudimentary CAFM tools. The advantage lies in making them Web-enabled and linking them to other, traditionally separate, management functions. Where such efficiencies can be found, the in-house FM team can devote more attention to areas of strategic business importance that have higher value-adding potential.

* KIBT Consultancy, Sydney, Australia

The future of FM is likely to be viewed in the context of sophisticated command cent. with highly trained managers responsible for a wide range of assets within an organization, or even across organizations. These control rooms will monitor the performance of all environmental systems, adjust capacities, shut down services to unoccupied areas, and look after security, technology, personnel needs and other requirements while simultaneously recording data and evaluating performance against established benchmarks. However, beyond this largely operational façade will be found tactical and strategic processes aimed at keeping the facilities closely aligned with organizational needs and priorities. The facility manager of the future will be an information-charged, connected, business-orientated strategist with a team of educated people properly versed in all the core competencies of the profession.

Therefore e-FM offers enormous opportunity for value improvement through the elimination of labour-intensive functions and repetitive processes, and the concentration on matters of more strategic significance to the goals of the organization in a changing and highly competitive market.

7.1 Introduction

This chapter is about knowledge – specifically how, why, where and when technology augments the function/actions of facility management (FM) to create knowledge. It is a journey through the development and application of technology. It is not an expose on technology and FM systems.

The types of technology used in FM span database development and web technology to the use of virtual private networks (VPN) for location-independent FM and three-dimensional computer-aided design (CAD) modelling that enables a 'walk through' of space.

Technology in itself is not a panacea. Many organizations have viewed technology in this way to their detriment. Technology is a tool – either stand alone or systems based – that gathers and creates data or information but not knowledge. It is the enabling capabilities of technology that is the key to knowledge creation. It is knowledge that influences the ability to establish sustainable facility performance, value-added efficiency in organizational workflow or better strategic facility planning, and not technology.

This chapter explores a number of aspects of FM that are influenced by or predicated on technology. They are strategically focused even though most FM activities are directed at operational issues. (It can be assumed that operational processes are developed as a result of strategic organizational planning although in practice some FM professionals may think this is stretching the truth.)

The themes are: what is FM technology and where does FM technology add value? How does technology impact on people and organizations? What are the key technology solutions used in FM? And, how do FM systems deliver sustainable competitive advantage for the organization and/or enhance long-term performance.

Finally, within the framework of the ideas presented, a few questions that arise are addressed. These include, where is FM technology heading? Are facility managers to be victims of information overload? How will ongoing change, in all aspects of life and business, from greenhouse emissions reduction compliance to radical transformations in organizational shape and structure, influence what is important to know and what is not?

...ate information confusion or information clarity? Does technology ...ledge complexity or is simpler work the outcome and indeed a worthy ... In other words, are all the bells and whistles that come with the ...lly needed?

2 What is FM technology?

A number of words recur throughout this chapter. They are important words. They are words that underscore what technology is in a general business sense as well as in an FM context. The important point to note is that if the driver of the technology is not business performance, then there can be no reason for facility performance. These words are:

- *Facility Management* – the Facility Management Association of Australia (FMAA) defines FM as 'a business practice that optimizes people, process, assets (place) and the work environment to support the delivery of the organization's business objectives'. There are other definitions, all expressing more or less the same ideas.
- *Technology* – encompasses building systems, architectural structures, office automation, information technology, 'plug 'n' play' furniture systems, management practices and processes. This is not a traditional definition – it is a facility centric definition that integrates many FM applications.
- *Efficiency* – may be defined as a standard of performance internal to the organization that is measured by the ratio of resources utilized to the output produced. What this means is that managers are fully rational and have a clear expectation about future costs.
- *Value* – the difference between what a customer is willing to pay for a good or service and the cost incurred in performing the activities that secure the sale. Note that this is different to profit as each customer 'values' the good or service differently.
- *Effectiveness* – may be defined as the relationship between an organization's outputs and its objectives. The more the objectives contribute to the organization the more effective it is. While it is not as easily defined or measured as efficiency, effectiveness is more important because it is applied to the value created by organizations rather than the costs imposed on organizations. Effectiveness is an important element of sustainability.
- *Substantive Value* – this takes the definition of 'value' a step further. Substantive value (or value adding efficiency) is a link between efficiency and effectiveness. However, it still sits on the efficiency side of the fence and does not contribute to sustainability.
- *Sustainability*, or more precisely sustainable development, has been defined by the World Commission on Environment and Development (1987) as 'development that meets the needs of the present without compromising the ability of future generations to meet their own needs'. In a facility context sustainable development is defined by the FMAA 'as the theoretical effective balance between economic progress and environmental conservation embodying intergenerational goals of maximizing wealth and utility and minimizing resource use and impact over the life of the development'. Sustainability is much more. Sustainability must include social equity, corporate citizenship and relationships with all of an organization's stakeholders (management, employees, the community, unions, government, investors, suppliers and customers).

Without this, maximizing wealth and utility, minimizing resource and development impact are only words. Technology will add significantly to ensure the sustainable approach to the future is a success.

FM is in a strategic position to shape future directions for organizations in relation to culture and accommodation, to be 'sustainable', and to deliver effective outcomes. Sustainability has primarily focused on reducing energy consumption in buildings. This is not 'energy efficiency' as the functionality of the building may not have been considered.

If a building can be defined as a 'product' then the next major step toward sustainable growth is to improve the value of the product (building) and services (offered by the building) per unit of natural resource.

It is the synthesis of these words, or concepts, which provides the basis for the use of technology in FM. Technology will assist in creating and measuring future minimum benchmarks for effectiveness, efficiency and sustainability in the pursuit of better facility performance. Note that the classical definition of 'efficiency' does not fit with the 'whole-of-life' focus of FM. All of these terms do, however, underscore the value that can be obtained from the use of technology.

A fundamental aspect, not considered here, is the impact the implementation of new technology has on the people in the organization. A purely rational view is that technology relates to improving efficiency and creating substantive value; what is not clear is how the systems, processes, work practices and intra-organizational relationships between different functional groups are affected.

An equally important consideration in the application and implementation of specific hard and soft technologies is the associated implementation of management strategies, organizational philosophy and work processes. In short, what is the level of change that is required, as a parallel process, to the implementation of the technology. This is discussed in more detail below.

7.3 Technology development

Technology systems used in FM have developed along a similar timeline to the development of technology in finance and accounting.

Figure 7.1 is a diagrammatic interpretation of the fundamental components of an FM technology system. The arrows indicate the communications pathways between the various components of the system. Where each component is physically located is unimportant to the user of the system.

There are four aspects that are fundamental to all information technology applications. They are relational databases, networking infrastructure, computer-aided processes and data communications technologies. In present FM systems data communications technology is equivalent to Internet technology. Additionally, FM technology tracks advances and applications of technology to management in the manufacturing sector and to changes that have occurred in the management of the workplace. Figure 7.2 illustrates the relationship between FM technology and management innovation.

Over time the focus of FM has changed. These changes, not unexpectedly, align with major management innovations. Phase 1 is characterized by an acceptance and application

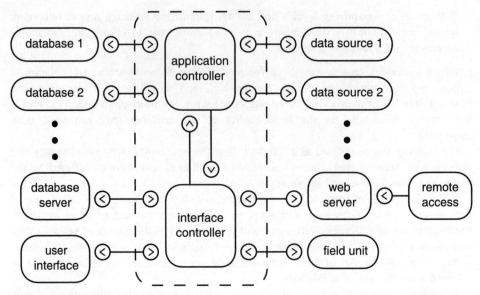

Figure 7.1 Basic components of an FM system.

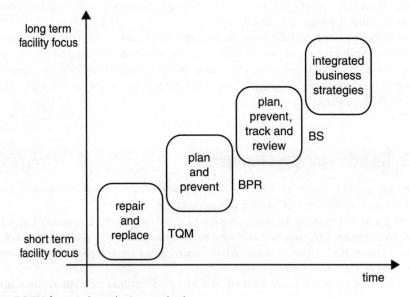

Figure 7.2 FM focus and new business technology.

of total quality management (TQM). The focus of this phase is internal, operational and short term. The next phase is business process re-engineering (BPR). The focus of this phase is still internal and operational; however, there is a shift toward learning and whole of life analysis. The final phase recognizes that there exists a strong relationship between facility operation and the balanced scorecard (BS).

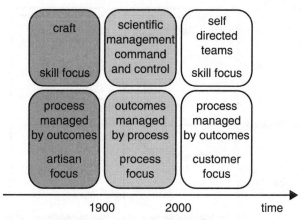

Figure 7.3 Changes in the management of the workplace.

Importantly, the use of technology has both shaped and been shaped by these transitions. Figure 7.3 illustrates how the view of the management of the workplace has changed in the same time frame.

7.4 Efficiency and effectiveness – end result or starting point?

The driver behind the development of TQM, BPR and many of the management processes and systems is a focus on efficiency. Efficiency is most usually defined as a standard of performance that is internal to the organization, measured by the ratio of resources utilized to the output produced. It disregards those factors that are difficult to measure such as social wellbeing and the aspirations of external stakeholders. In the case of facilities management, efficiency has been equated to operating a facility at the lowest possible cost. This usually means there is a focus on physical assets only, or more simply plant and equipment.

As with other developments in technology this has been the focus of the introduction and development of technology in FM. Interestingly, since the introduction of the PC, employee productivity has not increased even though computers allow mundane tasks to be completed more quickly. Technology has not made us more efficient, but it can make us more effective. Database technology has allowed FM to capture information over extended time periods and so provide 'whole of life' analysis not possible prior to accessible computing. This focus raises an interesting question. If FM focuses on the operation of the building, and hence implies long-term operation, and the facilities managers are charged with optimizing people, process and function over the life of the building yet efficiency is defined in the short term, then a paradox exists. Resolution of the paradox will require a wholesale change in values and understanding of how technology (as an FM tool) is assessed. Figure 7.4 relates sustainability, efficiency and effectiveness with the assessment of technology. Note the **r** and **K** on the matrix in Figure 7.4 – the notation relates to mechanistic and organic organizational structures respectively.

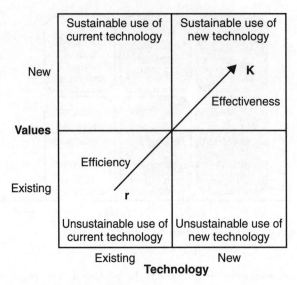

Figure 7.4 New values for technology and effectiveness.

7.4.1 The efficiency conundrum

Supporting the long-term/short-term paradox is the use of the term 'efficient use of floor space'. This is evaluated by dividing the total floor space by the number of people in the facility. A simple example is an office building that has 2000 people working in it and a total usable area of 30 000 m^2. The efficiency factor is 15 m^2 per person. However, this does not imply that each of those 2000 people has 15 m^2 to call their own – the 15 m^2 is distributed throughout the facility against different space types that might include individual open plan workspace, team space, collaboration space, circulation space and enclosed office space.

Note that the distribution calculation is based a wide ranging analysis that incorporates organizational culture, strategy, the type of work activities people undertake, and levels of interaction and collaboration in the organization. In this sense it is more useful to talk about 'effective use of floor space'. Unfortunately, effectiveness is much more difficult to measure as it implies the measurement of intangibles.

Efficiency is easy to measure and hence FM technology systems have focused on it. While this focus has established the performance baseline of FM technology it is not where the future of FM technology lies. The greatest impact of FM technology is not on efficiency but on developing quantitatively strategic solutions based on the interpretation of data and the translation to knowledge (i.e., effectiveness). This can be illustrated by a simple example:

A premium class office building can attract a rent of $320/m^2. If the building occupies 30 000 m^2 a 2% improvement in space utilization (i.e. 600 m^2 less space is used) can return $192 000 if the extra space can be sub-leased. This does not mean that the employees are squeezed into less space. It means that by quantitative analysis of the data available better space utilization has been achieved. Greater savings are afforded if the extra space is used as 'swing space' – this is extra space that allows for short-term fluctuations in space requirements. Using swing space is far more economical than attempting to find and fit-out 600 m^2 off site, and this does not consider the business, organization and social dislocation that may occur. (Note that 600 m^2 is around half the area of a floor in a typical office tower.)

Further to this there is a strong link between optimized space utilization and employee productivity – if the average salary is $30 000 and there are 2000 people in the organization then an increase in productivity of just 1% will return $600 000.

Technology can help make enhanced space utilization and productivity improvements a reality by filtering collected information and applying the appropriate analysis tools to target opportunities.

There are other examples where a focus on effectiveness rather than efficiency is delivering improved facility performance: they include improved energy usage, water usage and churn management. An improvement in each area supports a better use of resources and hence supports the philosophy of sustainability.

The moral of the story is that FM technology has dismantled the financial disincentives (through measurement) to developing sustainable buildings and at the same time provided FM professionals with more opportunities to shape the facility, with a focus on effectiveness rather than efficiency, and hence add value.

7.4.2 Value loops

There are other factors that will determine how FM will utilize technology in the pursuit of effectiveness. For example, traditional business processes are founded on value chains, i.e., systems of interdependent activities that are connected either by function or technology. Activities are connected when the performance of one activity interrelates with another. Connections exist in such a way that they minimize 'leakage' and so optimize the process and minimize waste (cost).

Organizations attempt to match each stage of a production or service process (linear chain) to optimize the value (efficiency) that occurs in the process. Most FM functions exist in this environment. A simple example is the accommodation procurement process. At each stage of the process 'leakage', or small changes to the building's specifications and performance, occurs. It is important, therefore, to ensure that leakage is minimized to ensure that maximum value is derived from the process. More insidious is 'innovative filtering'. In this case key performance conditions are changed due to technical limitations of the products that form the fabric of the building (examples are the façade, mechanical services, telecommunications and space planning). FM technology plays a substantial part in how procurement activities are carried out – by developing new information flows (networking, web or integrated processes) FM can exploit linkages between activities, further reducing leakage and innovative filtering.

In a traditional building acquisition process each action has a linear relationship with others in the chain. The value of the building is compromised as leakage and innovative filtering occurs over time. Figure 7.5 illustrates this concept.

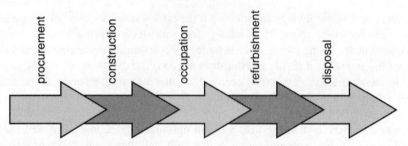

Figure 7.5 A traditional linear building life cycle.

In a sustainable environment 'value chains' are reinterpreted as 'value loops'. Value loops consider the relationship (effectiveness) of all components in a circular process that links the final outcome (usually an end customer) tightly with the input to the process. FM has a natural affinity with value loops due to the multidisciplinary nature of the profession. Figure 7.6 illustrates the same building life cycle process. There is one major difference, and it is the notion that the building, or building components, can be recycled, refurbished or reprocessed that is critical to the procurement process. This assists in changing the FM focus from efficiency to effectiveness.

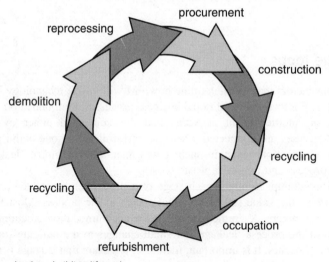

Figure 7.6 The value loop building life cycle.

7.5 Technology, people and strategy

In the introduction it was stated that this chapter is about knowledge. It is for this reason that before discussing the very technology that underpins the information used by FM professions it is important to understand the drivers of technology and the effect on the people who are using it. It needs to be made clear from the outset that business drives technology and not the other way around. Any technology used that does not add value to a process or outcome or does not create knowledge delivers no value to the organization.

Research indicates that 80% of technology system implementations fail to deliver the expected outcomes. This is not because of a fundamental flaw in the technology or that the systems purchased are not appropriate but because in implementing the system scant regard is given to the people who are expected to use the systems. There is little evidence to suggest that strategic alignment is considered and/or that strategies for maintaining the data are developed. This applies to all types of technology, not just information dependent technology, or FM technology.

Facility management is susceptible to this outcome. In many cases the implementation and use of technology completely changes the processes and systems that were in operation prior to the use of technology.

A full understanding of these relationships requires an understanding of change management. While that is beyond the scope here it is worth noting that technology is a part of the facility infrastructure along with the social, organizational and physical elements. It is an enabling tool. This view can be summarized, in the context of FM, as follows:

- FM technology enables strategic decision making to be supported by statistically significant information derived from real data
- FM technology supports the day-to-day operation of a facility by providing relevant and real time data to FM professionals
- the implementation of technology systems impacts on all aspects of the organization
- the capabilities afforded by using FM technology are only achieved if integrated with strategic and change management issues
- FM technology has extended the value loop to include organizational customers through web interfacing – this has meant that the FM community now includes stakeholders with different requirements, and in that sense technology is both a stimulus for the development of new organizational cultures and a facilitator of those cultures.

All of this is academic, however, if the way that information is managed is not considered.

7.5.1 The line of invisibility

The line of invisibility is an imaginary line between the customer and the organization, with the customer in this sense being either internal (i.e., within the organization) or external. The line has been widened by the use of technology, not narrowed as is anecdotally understood. Technology has enabled all people who use it to distance themselves from real human contact – consider the way that E-mail has removed the need for face-to-face contact, or voice mail has allowed people to avoid voice communication, or even the telephone answering systems that invite us to 'Press 1 for accounts, 2 for complaints', and so on. These 'progress' technologies, in a business sense, have distanced the organization from the customer and not the other way around. The challenge is to incorporate technology in a way that ensures that the organization regains the confidence of the customer and stays 'in touch'.

FM in a traditional sense is primarily an operation that occurs on the side of the line of invisibility that deals with managing the organization's property and as such it is removed from the customer. FM professionals have tended to deal mostly with plant and equipment. A dichotomy exists here – while we have become 'less personal', technology now allows faster data transfer and has made us more connected. Technology is changing the dynamic of the relationship and is pushing FM and the customer together. Success therefore is predicated on understanding how to use technology to cross the line of invisibility and create value.

7.5.2 The management of information

Organizations collect data and collate information in a number of different ways. Mostly the collection of data is the result of planned actions. In other cases it is the result of a passing comment. Much of the information collected by organizations is acted upon, or changed, and returned to others in the organization without knowledge being created. In FM it is no different.

Facility managers have a responsibility to make sure that value is added to the information that they collect, and this can be done through their management of that information. In most cases the knowledge obtained about an organization's systems,

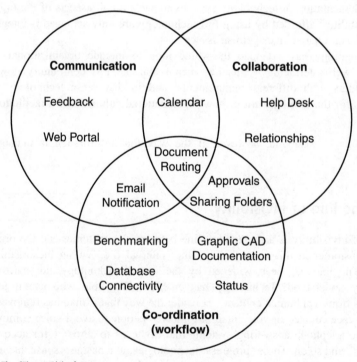

Figure 7.7 The 'Three C's' of FM.

processes and work history is embedded in the memory of individuals. Web-based documentation is changing this, providing a focus for the alignment of organizational policies and procedures with technical developments. Figure 7.7 outlines the 'Three C's' of FM where FM systems have been enhanced through the integration of technology.

7.5.3 Mechanisms for information translation and knowledge creation

The conversion of information to knowledge is vital to the value adding strategy of the use of technology in FM. There are a number of ways information can be transformed. These are:

- reports
- statistics, analysis and benchmarking
- maintenance and service filters or other established communications protocols
- frequent reviews of work practices and management policies to align with information flows and vice versa.

There are other outcomes arising from the implementation of technology systems into FM. Rather than deliberate on why this has occurred, it is more important to know what has occurred. This can be summarized as:

- there has been employee empowerment and attitudinal change among people working in FM – technology has enabled people to see clearly what, and how, the technology assists them in their work (tasks) – technology has enabled people to understand why the work they do is important and where it adds value (and fits) within the organizational context
- FM does not only manage the physical infrastructure, it influences the experience of people in the facility and contributes to the success of the organization
- customers (and other internal/external stakeholders) will become part of the FM equation by interacting with the technology at their disposal – this is especially relevant in the case where occupants maintain the data that reflects their own space
- FM technology will function interactively with people – people will define their environment, the technology will respond to optimize the conditions for individuals and groups – this is already happening within the Internet through the use of software agents
- integration with stakeholders creates a feedback loop that promotes maintenance of the data and hence greater accuracy in predicting future facility needs
- data warehousing is being used to counteract the condition when the amount of data accessible becomes unwieldy, and to prevent data from becoming obsolete.

These concepts are discussed further below.

7.6 Building systems focus

The ideas expressed so far have explored the influences and outcomes of the use of technology in FM but not the applications that support this view. In this section various

types of systems will be discussed. Most of these systems fall within the scope of computer-integrated FM (CIFM). These systems differ from traditional FM technology applications – used as computer-aided planning tools – in that CIFM places greater emphasis on the management of information rather than the planning process.

7.6.1 Building management systems (BMS)

Building management systems are sophisticated FM tools. They work at the leading edge of FM technology, incorporating high-speed digital networks and their use is being driven by a greater need (and use) for monitoring of building systems as the controlled zone size of the building decreases. There are three catalysts for this – each is discussed below.

The development of communications networking technology that embraces a common protocol

The Building Access Control Network (BACnet), an Open Systems Interconnect (OSI) protocol developed to create a communications environment where different system components could interface together is such a system. Other systems include LonWorks, an open source protocol, built around a LonWorks network agent. The important point about communications systems is that, in the case of building management systems, they have embraced the technology and language of information technology, whether Ethernet or packet switching. It is now hard to distinguish where IT ends and BMS communications start.

Distributed computing

The development of distributed computing has enabled greater flexibility (but at the same time greater complexity) of systems control algorithms. This is a good thing. The complexity of the algorithm, established at the systems level, enables more simplicity at the human level. We are now able to adjust the microclimate in our space on the fifteenth floor without unduly affecting the rest of the building and have a lift on the floor in 15 seconds. The BMS will understand and learn what our environmental wants and needs are. Given that most complaints to the facility manager are about heating, air conditioning and vertical transportation, then accessing an individual's environmental history – through technology (if the opportunity to ask directly does not present itself) – aids the facility manager in improving facility performance.

Sustainability

The focus on sustainability and energy effectiveness has seen greater emphasis placed on technology integration in order to enhance the capabilities of building management systems in delivering environmental benefits. Sustainability and energy effectiveness have provided the launching pad for facility managers to influence the organization in more strategic ways than merely implementing fluorescent light replacement programmes.

7.6.2 Computerized maintenance management systems (CMMS)

As the name suggests, these systems support the maintenance operation of a facility. Historically, this has been viewed from a production-based environment with each element of the maintenance task acting in isolation from other maintenance focused tasks. The introduction of computers – and more specifically, information technology – has enabled maintenance, and hence the management of maintenance, to be more strategically focused. A number of developments underscore this advance. The most innovative has been the introduction of business-to-business (B2B) information systems, using e-commerce that enables vendors and suppliers to interact via the Internet and to integrate supply systems. This applies to the procurement of maintenance consumables and to the delivery of preventative maintenance procedures. The outcome of this is greater operational alignment between the suppliers of products and services, the FM professionals and the customer.

7.6.3 Computer-aided FM (CAFM)

These systems are based on the interaction between CAD drawings of rooms or other definable spaces and the attributes of those rooms or spaces stored in a relational database that is dependent on a stable data hierarchy. In most CAFM systems this is based on a campus, building, floor and room hierarchy. CAFM systems are the most important of the family of systems in the FM technology household. CAFM is used to predict (forecast) and optimize facility related activities such as churn rate, growth requirements and space allocation, and capital works program management. Each of these aspects of CAFM (as stand-alone systems) is enabling facility managers to do what they do best, i.e., manage facilities now and into the future. Unfortunately all is not what it seems as more information has not meant more interactions between managers or systems.

The relationship between systems and work, as information that resides in the minds of people, is now being translated and adapted for use with technology. What was once a simple discussion between colleagues is now a complex relationship in Boolean logic. It is either a one or a zero, and undefined between these two options. In the future, fuzzy logic will make this interchange simpler and at the same time more challenging. Driving much of this is the convergence of the Internet, database technology, CAD and software based intelligent agents. These have made it possible to maintain all CAFM information

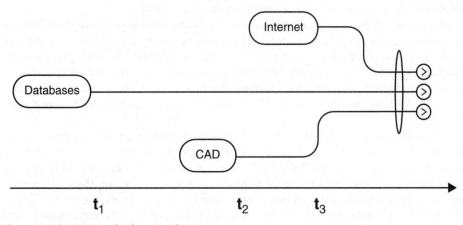

Figure 7.8 The CAFM technology time-line.

Table 7.1 A not so comprehensive list of building assets and their attributes

Asset	Attribute	Asset	Attribute
Light	fluorescent halogen location of switches/dimmable natural light augmented natural light blinds/blackout curtains	*Data*	number of networking points and data rate type of networking location networking administrator responsible for the room
Seating	number of seats benches individual seats moveable	*Power*	number of power points location of power points
Ambience	finish images quality of equipment	*Utility services*	toilets nearby water access catering

in real-time, offsite and online. Figure 7.8 illustrates the timeline convergence of CAFM technologies.

By far the most common use of CAFM technology is the management of space. Site, building, floor, room, item – it is possible to define all of the attributes of the space to the most finite element. Table 7.1 illustrates the breadth of attributes that may be associated with a space. The list is far from complete. These are some of the definable attributes for a teaching space.

Space management is not the only application of CAFM and it barely touches the surface of what is possible. The most useful application of CAFM is the management of construction and the support of project management (based on the number of CAFM system suppliers) that support this application. If the focus of using technology is to build better and manage better, then using CAFM systems can assist in achieving this as an end result. Figure 7.9 illustrates the range of FM applications (databases) that can be incorporated into a CAFM system. The rating beside the application indicates the level of importance of the application to the management of the facility.

The list of attributes highlighted in Table 7.1 is the only way of understanding the functional aspects of space. It is more appropriate, however, to visualize space through CAD. The most powerful aspect of CAFM systems is the ability to 'drill' down through precinct (campus), site, building and floor plans to describe specific details about a person's workspace, and to be able to manage the data at each level in real time. The plans in Figure 7.10 illustrate this concept. Each plan, 1 through 4, drills down to reveal more detail.

Impressive as this may seem, it is absolutely dependent on the accuracy of the information in the database. Eternal vigilance is required; the client will spot the inconsistency between audits. Technically this is unfortunate. It takes time to thoroughly audit the organization, noting new additions, modifications to space, the movement of people as project teams and department name changes. Unfortunately, the creditability of the system – a system that may or may not have been imposed on the organization – and the credibility of the FM office is in question.

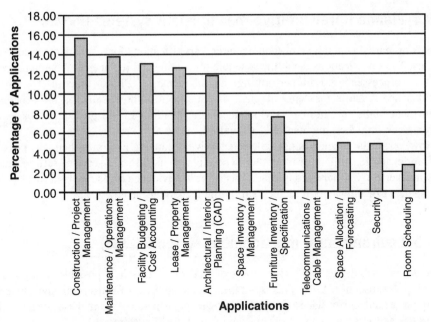

Figure 7.9 Distributions of CAFM applications.

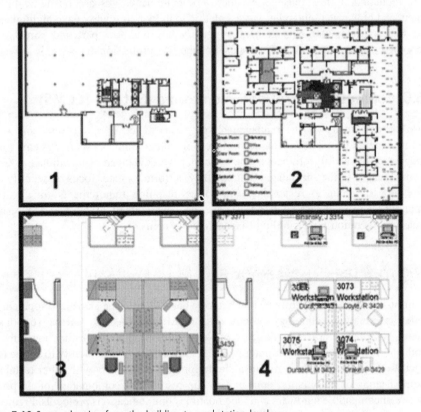

Figure 7.10 Space planning from the building to workstation level.

7.6.4 Planned preventative management systems (PPMS)

This is primarily a catalogue of manufacturers' specifications for all products used in the facility. The specifications detail the schedule for routine and replacement maintenance as well as processes used in performing the maintenance. PPMS function by issuing maintenance requests on items of plant or equipment automatically, as the scheduled maintenance falls due. In many FM maintenance situations PPMS provides an easy way of remembering when maintenance is required.

However, if the process is not co-ordinated with planning activities then it is possible that, for example, an item of plant due to be replaced as part of a services refurbishment may inadvertently be serviced instead. This is further evidence that supports a focus on the management of information rather than on the technology itself.

7.6.5 Environmental management systems (EMS)

It could be argued that EMS is a subset of CAFM, just as it is possible to argue that CMMS is a subset of CAFM. This is not the case as the focus of EMS is on environmental compliance and performance. It is the knowledge generated from these systems that integrates and influences the performance of the CAFM system.

The reduction of greenhouse gases, control of legionnaires' disease, onsite treatment of effluent and grey water, and other issues relating to indoor air quality have highlighted the need for environmental management systems. They are now part and parcel of the technology systems arsenal at the disposal of the FM professional.

7.6.6 Computer control and monitoring systems (CCMS)

These systems focus on the management of security, access and egress, and related building operations. They provide a level of integration between systems in the management of security protocols. The systems have developed as an adjunct to CAFM systems because of the specialized function they perform. Greater focus on security using technology, in order to respond to workplace planning requirements for flexibility, adaptability and openness, will necessitate the development of CCMS and result in increased integration with building management systems.

7.7 The IT infrastructure for FM technology

The infrastructure that supports FM technology has followed the same development path as other information technology systems. Early CAFM or CMMS systems were discrete devices, based on small mainframe technology supporting a single function only. The introduction of the personal computer enabled more aspects of the facility to be measured in a cost-effective way albeit still as a single function device. As with other technology systems islands of automation appeared as the result of a combination of proprietary software, incompatible hardware and incompatible communications protocols. Figure 7.11 illustrates the technology time line for FM.

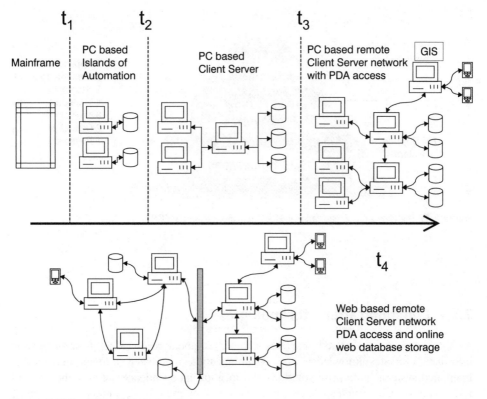

Figure 7.11 The FM technology time line.

The introduction of open source protocols based on the Open Source Interconnect (OSI) seven-layer protocol, and the drive by customers for greater transparency and interconnection supported by interoperability ushered in the integration of more widespread information technology infrastructure. That infrastructure now includes Ethernet based networking systems, digital communications infrastructures (cable and wireless based) that support remote access and the integration of multiple management sites across geographically isolated areas. The New Arup Organizational Model for Information (NAOMI, shown in Figure 7.12) is a model that has adopted the OSI seven-layer protocol and integrated it with a hierarchical model of building information systems to reflect FM.

The most dynamic technology factor to influence FM systems is the development of Internet technologies. This type of technology is well advanced and web-enabled FM (WFM) is transforming the way FM is carried out. Already systems developers such as *ArchibusFM* and *Aperture* provide fully integrated web based CAFM applications. All this may sound exciting, however, FM has been slow to adopt IT and communications technology, primarily because of the slow response times of the indoor environment.

In addition, the complexities alluded to earlier must not be understated: if it takes thirty minutes for a perceptible change in temperature to occur it is not critical that the system respond within milliseconds. This is changing as the scope of FM technology expands and hence there is a need to adopt current state of the art systems.

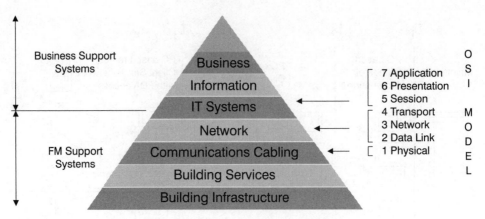

Figure 7.12 The New Arup Organizational Model for Information (NAOMI).

7.7.1 Communications technologies

At the outset it was stated that this chapter is about knowledge. Communicating information creates knowledge and communication is created through networking. Integrated systems, enterprise solutions and location independence are now the basis of FM technological systems. These systems are supported by a number of enabling technologies that include:

- distributed intelligence – rather than centralize control of processes and storage of information, different systems manage and influence the outcomes of applications in remote locations
- Virtual private networks (VPN) allow organizational information to move across public communications pathways while maintaining organizational security
- scalability of the technology to integrate new applications across the database platforms – the convergence of the Internet and operating systems, and the integration of OSI, has assisted the development of scalability
- portability provided by digital wireless communications using personal digital assistants (PDA) – these have changed the way FM is undertaken and has enabled FM professionals to cross the line of invisibility and engage with the customer, thus allowing new and changed information to be updated quickly and feedback supplied immediately
- integration across different communications platforms at various layers of the OSI model – integration at the physical and data levels of the OSI is well established, while integration at the application level is the goal
- networked building services, where the building is wired for work – rather than occupants entering an empty building they enter a building that is already completely wired (or wire-less, as the case may be) – the use of PDA's has already extended FM into the wireless environment.

7.8 Case study examples

This chapter has highlighted the breadth and depth of technology and its application to FM. In the same way that all organizations are both similar and different, so is the scope of innovation through technology. A consequence of this is that no one organization has applied all that technology has to offer and therefore the case studies presented here provide a snapshot of technology in FM rather than in-depth analysis of one particular organization.

The organizations listed in Table 7.2 have incorporated different aspects of FM technology. Primarily, the focus is people, place, process and planning. All have knowledge management as a secondary outcome.

Focus describes the intent of the technology. There are three types:

- strategic (S) – the focus is on improving strategic relationships between facility performance and organizational and business outcomes
- management (M) – the focus is on managing facility outcomes including management of contractors, management of the facility planning and procedures development process
- operational (O) – the focus is on operational aspects of service provision and implementation of planning and procedures.

In turn the '4 Ps' – people, process, place and planning – describe the main driver for the application of technology.

- people – change management, human resource development and training
- process – management practices, information management
- place – integration of value loops across different functional aspects of the organization
- planning – integration of CAFM, IT and process to enhance performance of the facility.

7.9 Futures

The future of FM is ultimately tied to the future of organizations and of business. Even the name, FM, is under pressure. Is it corporate real estate management or is it business infrastructure management? Or will it be something else? In a sense it is all of them. Professionals involved in the management and procurement of organizational space, both physical and virtual, must be cognizant of the relationships between space, business and organization from all viewpoints. A few trends and concepts worth noting are:

- workplaces are disaggregating – they are a mix of here, there and everywhere as they translate space and time
- FM technology will stratify to focus on different elements/aspects of the facility as technology converges
- web-based CAFM systems will dominate the FM infrastructure due to ease of access
- Internet technology will reduce the complexity of the system to the end user – all the technical activity (complexity) occurs elsewhere, i.e., in cyberspace – the major CAFM software suppliers offer this service: all data and information is delivered to the supplier over the Internet and all management of information is performed in cyberspace – while

Table 7.2 Case studies

Organization Name	Industry	Focus	Summary	FM Medium	People	Process	Place	Planning
University of Sydney	Education	S, M, O	Development of a new Information Model to support an integrated facility management practice.	CAFM	✓	✓	✓	✓
BP Amoco	Energy	M, O	Centralize facility and corporate service operations using a global shared services model.	CMMS		✓	✓	
Hewlett-Packard	Technology	S, M	Development of a comprehensive Facilities Information Management Strategy to optimize facilities IT.	CAFM	✓	✓		✓
Honda	Automotive	M, O	Certification for ISO14001.	EMS		✓	✓	
General Motors	Automotive	M, O	Implement increased efficiency, better communication, and measurable goals through a central facilities management process.	CMMS		✓		
Intel	Technology	S, M, O	Merger of the People Systems group and the Information Systems group to provide integrated service.	CAFM	✓	✓	✓	
Thomson Financial	Financial Services	M, O	Development of a web-based work request system to improve service delivery and better manage the performance of the facility portfolio.	CAFM		✓	✓	✓
Turner Broadcasting	Information Entertainment	M, O	Integration of two distinct FM systems using a project management interface tool.	CMMS		✓	✓	
IBM	Technology and Computer Services	S, M, O	An aggressive approach to reducing the impact on the environment. The major benefit areas were: recycling, energy use reduction, and water use reduction.	EMS	✓	✓	✓	

S, strategic; M, management; O, operational

this may offer flexibility of access it limits the opportunity to change systems, or manipulate data easily, thus 'locking' end users to systems

- simple intuitive interfaces will be commonplace (through web browser technology)
- meeting the demands of a diverse group of stakeholders for the same space as space becomes less defined will be the greatest challenge for FM.

The question is: can technology support this reality?

7.10 Conclusion

Facility management is a multidisciplinary and broad-spectrum profession that extends from the purely operational to the purely strategic. The technologies that are used at each of these extremes is influencing the ability of FM professionals to improve facility performance and add value to the organization.

This chapter has focused on the what, where, why and how of technology and its application to collection and management of information and the value added creation of knowledge for internal (FM professionals) and external (FM customers) use.

Financial management technology is an enabling tool. Like other information technology related industries it is heading toward convergence. Convergence will follow a similar path to that which is occurring in other information technology related professions and with it will come greater amounts of information. This may lead to facility managers becoming victims of information overload. Ongoing change, as a constant rather than a variable, in all aspects of life and business is a challenge that FM professionals will have to address if they are to be clear about what is important to know and what is not – in essence, to create knowledge. The opportunity is there for facility professionals to lead the future in how technology is applied to facilities management.

References and bibliography

Alexander, K. (1994) A strategy for facilities management. *Facilities*, **12** (11), 6–10.

Allen, J. (2002) *BP–Amoco Merger Spurs Process Refinement*. www.tradelineinc.com

Amabile, T.M. (1997) Motivating creativity in organisations [on doing what you love and loving what you do]. *California Management Review*, **40** (1), 39–57.

Anonymous (1996) Automating facilities management – buyers guide to computer aided facility management. *Buildings*, May, 38–48.

Anonymous (2000) *Case Study – Honda Transmission Manufacturing Plant*. www.p2pays.org/iso/casestudies.htm#case2

Anonymous (2002a) *Procter & Gamble Restructures to Function Globally*. www.tradelineinc.com

Anonymous (2002b) *Intel's Corporate Services Merges Information and People Systems Restructuring Facilities E-Learn Strategy*. www.tradelineinc.com

Aperture Technologies (1999) How technology is re-engineering FM processes. Special Supplement to *Facilities Design & Management*, September.

Aronoff, S. and Kaplan, A. (1995) *Total Workplace Performance – Rethinking the Office Environment* (Ottawa: WDL Publications).

Baird, G., Gray, J., Isaacs, N., Kernohan, D. and McIndoe, G. (1995) *Building Evaluation Techniques* (Sydney: McGraw-Hill).

Baldry, C. (1997) The social construction of office space. *International Labour Review*, **136** (3), 365–79.

Beech, N. (1997) Learning to build customers into facilities management: the case of Reuters. *International Journal of Facilities Management*, **1** (1), 51–8.

Brown, J.S. and Duguid, P. (2000) *The Social Life of Information* (Cambridge, MA: Harvard Business Press).

Bruton, K. (2002) Computerised Maintenance Management System (CMMS). *FM Magazine*, **10** (1), 40–3.

Burger, D.E. (1999) High performance communications systems in buildings. Presented at the *Joint IEEE/IREE and IEE Meeting*, March.

Calopio, J. (1998) *Implementation – Making Workplace Innovation and Technical Change Happen* (Sydney: McGraw-Hill).

Cook, R.J. (1993) New strategies and design methodologies for the virtual workplace. *Site Selection & Industrial Development*, December.

Drucker, P.F. (1998) The discipline of innovation. *Harvard Business Review*, November/December, 149–57.

Duffy, F. (2000) Design and facilities management in a time of change. *Facilities*, **18** (10/11/12), 371–5.

Dunphy, D. and Griffiths, A. (1998) *The Sustainable Organisation* (Sydney: Allen & Unwin).

Dunphy, D., Benveniste, J., Griffiths, A. and Sutton, P. (2000*) Sustainability – The Challenge of the 21st Century* (Sydney: Allen & Unwin).

Figueiredo, M. (1996) *Identification of Environmental Quality Costs and Technological and Environmental Risk Factors*. www.trst.com

Fisher, D. (2001) BACnet system specifying is on the rise. *Facilities Design & Management*, August. www.fmlink.com/ProfResources/Magazines/article.cgi?Facilities:fac0801b.htm

Ford, B. (1995) Integrating people, process and place – the workplace of the future. *Eighth National Quality Management Conference*, 21–37.

Gehl, P. (1999) Maintenance attitudes have evolved with technology. *AFE Facilities Engineering Journal*, **26** (3), May/June, 36–8.

Gratton, L. (2000) *Living Strategy – Putting People at the Heart of Corporate Purpose* (London: Prentice-Hall).

Greengard, S. (1998) How to make KM a reality. *Workforce*, **77** (10), 90–2.

Grimshaw, B. and Cairns, G. (2000) Chasing the miracle: managing facilities in a virtual world. *Facilities*, **18** (10/11/12), 392–401.

Guy, S. (1998) *Alternative Developments: the Social Construction of Green Buildings*. Research project of the Royal Institute of Chartered Surveyors (London: RICS).

Harnett, J. (1999) Digital economy. *Premises and Facilities Management*, October, 31–2.

Harrison, A. (1997) Converging technologies for virtual organisations. In: Worthington, J. (ed.) *Reinventing the Workplace* (Oxford: Architectural Press).

Hartkopf, V., Loftness, V. and Shankavaram, J. (1996) Facility managers as indispensable partners in corporate strategic planning. *Proceedings of the World Workplace IFMA Conference*, 727–42.

Horan, T.A. (2000) *Digital Places – Building Our City of Bits* (Washington: ULI).

Huang, J. (2001) Future space: a new blueprint for business architecture. *Harvard Business Review*, April, 149–57.

Huston, J. (1999) Building integration: mastering the facility. *Buildings*, **93** (12), 51–4.

Jensen, W.D. (2000) *Simplicity – The New Corporate Advantage* (New York: Perseus Books).

Jessup, L. (1997) A new workplace solution for Hoffmann-La Roche. *FM Data Monthly*, **16** (5), 8–11.

Jones, G. and Goffee, R. (1996) What holds the modern company together? *Harvard Business Review*, November/December, 133–48.

Jones, I. (1995) *Facilities Information Management Systems* (Manchester: CFM Publications).

Kanter, R.M. (2001) *Evolve – Succeeding in the Digital Culture of Tomorrow* (Cambridge, MA: Harvard Business Press).

Levine, N., Pitt, D., Beech, N. and Isaac, R. (1997) Organisations for the new millennium: facilities management and the new organisational environment. *International Journal of Facilities Management*, **1** (1), 11–20.

Madsen, J. (2001) GM shifts facilities management into high gear. *Buildings*, January. www.buildings.com/Buildingsmag/

McFadzean, E. (2001) Critical factors for enhancing creativity. *Strategic Change*, **10** (5), 267–83.

McGirr, K. and Stanley, D.E. (1998) Contextualising: technology, relationship and time in a financial services virtual organisation. *The Services Industry Journal*, July, 70–89.

Mintzberg, H. (1973) Strategy-making in three modes. *California Management Review*, **16** (2), 44–53.

Morgan, G. (2001) Thirteen 'must ask' questions about e-learning products and services. *The Learning Organization*, **8** (5), 203–10.

Myers, P. (1996) *Knowledge Management and Organisational Design: An Introduction* (Oxford: Butterworth-Heinemann).

Newman, H.M. (1994) *Direct Digital Control of Buildings: Theory and Practice* (New York: Wiley Science).

Porter, M. and Millar, V. (1985) How information gives you competitive advantage (transforming the value chain). *Harvard Business Review*, July/August, 149–60.

Porter, M., Pralahad, C.K and Hamel, G. (1997) Rethinking competition. In: Gibson, R. (ed.) *Rethinking the Future – Business, Principles, Competition, Control, Leadership, Markets and the World* (London: Nicholas Brealey).

Quinn, R., Faerman, S., Thompson, M. and McGrath, M. (1996) *Becoming a Master Manager*, 2nd Edition (New York: Wiley).

Rodgers, C. and Teicholz, E. (2001) *What does 'Workflow' Really Mean when Terminology is Constantly Misstated?* www.graphysys.com

Sheldon, C. (1997) *ISO4001 and Beyond: Environmental Management Systems in the Real World* (Sheffield: Greenleaf).

Smith, S. (1999a) The impact of communications technology on the office of the future. *FM Magazine*, April, 42.

Smith, S. (1999b) The use of quality functional deployment in the design of the office of the future. In: Karim, K. (ed.) *2nd International Conference on Construction Management*, Sydney, July, 146–55.

Smith-Bers, J. (1994) Just the facts about building systems integration. *Facilities Design and Management*, **13** (11), 50–3.

Southwood, B (1997) Can you dig it? The information challenge. *Building Services Journal*, March, 25–7.

Stahl, N. (1998) Genentech implements new CMMS. *FM Data Monthly*, **17** (3), March, 12–15.

Steelcase (1999) *Assessing Workplace Intangibles: Techniques for Understanding*. Steelcase Knowledge Paper. www.steelcase.com

Tatum, R. (1997) *CAFM brings more firepower to facility tasks*. www.facilitiesnet.com/fn/NS/NS3b7hh.html

Taylor, A.S. (2002) Using technology to integrate FM. *FM Magazine*, **10** (1), 35–9.

Turner, G. (1998) Organisational culture change and office environments. *Journal of Management Learning*, **29** (2).

Weitzman, L. (2002) *Thompson Financial Low Cost Web Based Request System*. www.tradelineinc.com

World Commission on Environment and Development (1987) *Our Common Future*. (Oxford: Oxford University Press).

8

Risk management

Deepak Bajaj*

Editorial comment

In every activity that involves decisions about future events, there is risk. However, as recent world history has shown, there is also risk of disaster that can impact on business operations and endanger facility occupants. Risk management, if it was not previously, has now become one of the most important competencies for the facility manager. While prediction of future catastrophes remains in the realm of prophets and witchcraft, planning for the possibility of such events and how to recover from them in minimum time is a necessary task and one that indirectly adds value.

Risk management is generally held to comprise the sequential processes of risk identification, risk analysis and risk response/treatment. In this way uncertainty can be objectively dealt with and converted into tangible plans to accept or reject proposals, to avoid high risk issues, to take out appropriate insurance cover, to transfer risk to other parties or to set aside some contingency. Although uncertainty might remain high, the risk to the organization can be limited through prudent management. To ignore risk is to invite disaster.

Individuals can display characteristics that identify them as generally risk-seeking, risk-neutral or risk-averse. The more individuals operate in collectives, the more likely it is that they will behave in a risk-averse manner. The ultimate collective, society, is highly risk-averse to risks that endanger safety, quality of life, economic viability, environmental resources, and so on. Businesses are generally risk-averse but are willing to accept increased risks as long as the anticipated returns offer sufficient compensation.

Continuity planning is a strategic activity that involves identifying risk centres and developing ways to minimize exposure or, when the event occurs, to recover from it as efficiently as possible. Uppermost in the mind of the facility manager is how to protect the organization's assets, i.e., its people, its data, its buildings and its inventory. Insurance is

* University of Technology Sydney, Australia

one form of response, but many assets are irreplaceable and therefore safeguarding them becomes a priority.

Risks can also come in the form of failure to comply with regulatory guidelines and standards. Such failure may result in economic loss such as fines or compensation, or reputation loss such as bad press or consumer anger. Managers must ensure that all reasonable steps are taken to reduce these types of risks. One approach is to conduct a due diligence review of operations to identify problems and to recommend remedial action.

Risk management is closely aligned to value, not so much in the sense of increasing it through better practice, but in protecting it from poor practice. Where risks are not properly identified or are ignored, the organization and its executives are exposed to potential litigation, share price drops and even criminal prosecution. Understanding risk is therefore of vital importance.

8.1 Introduction

Facility management (FM), which is sometimes misunderstood to be property maintenance or management, is much more than that. It is not just about the operation and maintenance of buildings, the provision of cleaning services or arranging furniture in offices. While FM is about all of the above at an operational level, it is also, at a strategic level, about planning for effective provision of services that can offer an organization opportunities for economy, efficiency, effectiveness and competitive advantage. Managing risks in FM is central to the delivery of value in all aspects of service provision.

8.1.1 How risks vary across facility types

FM is the practice, skill and science of ensuring that the workspace contributes positively to organizational performance. FM is a profession that requires of its practitioners proficiency in asset management, design and engineering, performance management, customer service delivery, and human behaviour (Arthur Andersen, 1999).

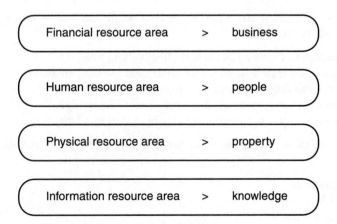

Figure 8.1 Areas of risk management.

Table 8.1 Areas of facility management

Facility type	Total building area (%)
Industrial	62
Educational	17
Office	7
Hotel	5
Retail	5
Health	4
Total	100% or 246 500 000 m²

The risks depend on the range of services offered at the facility. FM involves activities in the four areas shown in Figure 8.1.

Depending on the type of facility being managed, the direction and the focus of risk management changes with the mix of the four types of resources (i.e. business, people, property and knowledge) that are always central to the FM function and service provision. In effect, these four areas are the sources of risk for a facility manager.

The facility types are industrial, office, retail, hotel, educational and health facilities. The Australian stock data by facility type was published by the FMA (1999); a summary is provided in Table 8.1.

The facility service delivery data reveals significant differences in how facility services are delivered between facility market segments. This analysis highlights significant trends in terms of facility services delivered by in-house teams and outsourced providers. Some key trends are:

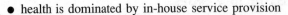

- health is dominated by in-house service provision
- offices exhibit the highest proportion of outsourcing
- education and industrial have a blend of in-house team and outsourcing.

8.1.2 FM processes and context

FM processes are cyclical and relate needs to a result that can be tested against user satisfaction with the service:

- space – adapted to changing needs and effectively utilized
- environment – to create a healthy and sustainable working environment
- information technology – to support effective communications
- support services – to provide quality services that satisfy users
- infrastructure – to provide appropriate capability and reliability.

In the context of these FM processes are the following categories of risk that need to be managed:

- health-related risk
- financial risk
- business-related risk
- marketing-related risk

- legal risk
- contractual risk
- management-related risk
- regulatory risk
- safety-related risk.

Generally the risks in FM are a combination of the above and as previously described the combination depends on the facility type.

8.1.3 Benefits of managing risks in FM

With proper risk management processes being followed, the facility manager could expect improvement in the following areas of the service provision:

- productivity
- trust
- quality
- environment
- image
- publicity
- business financial position
- control of facility services.

Additionally the facility manager can have more confidence in his/her decisions and actions due to the improved level of understanding and control of potential risks. All the above areas have a significant impact on the services provided and the satisfaction level of the clients who benefit from the service provided by the facility manager.

8.2 Broad principles of risk management and risk identification

8.2.1 The risk management process

Risk management is a process of formally identifying, analysing and responding to the uncertainties in a facility life cycle (see Figure 8.2). Risk responses should be such that they facilitate the satisfaction of facility objectives (Clark *et al.*, 1990). Furthermore, risk management should not be seen as a discrete activity, but as fundamental to the FM technique and the responsibility of the whole facility team.

Research and best practice demonstrate that maximum benefits can be derived from risk management if the process is applied continuously throughout the life cycle of the facility, from conception to completion. There are three steps in risk assessment:

- identifying and predicting
- assessing and consolidating into a manageable form
- managing and diffusing.

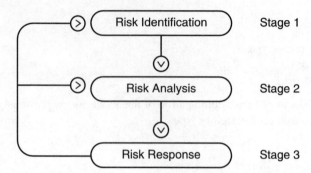

Figure 8.2 and stages layout:

Risk Identification	Stage 1
Risk Analysis	Stage 2
Risk Response	Stage 3

Figure 8.2 The risk management cycle.

A three step analytical approach based on Figure 8.2 involves the following steps:

- risk identification – developing an understanding of the nature and impact of risk on the current and potential future activities of the organization
- risk measurement – assessing and classifying the risk situations
- risk evaluation and re-evaluation – judging how to handle the risks and the possible need to re-evaluate the risk options.

This approach is further expanded into a process with sub-steps as shown in Figure 8.3.

Figure 8.3 illustrates the steps involved in risk identification, analysis and response. It is a logical process in which the source of the risk has to be established before the event itself. The effect of the event is established so the details that will help in the quantification of the risk can be isolated. Finally, a response to the risks based on the risk quantification and the probability of the risks eventuating is devised. This response will involve

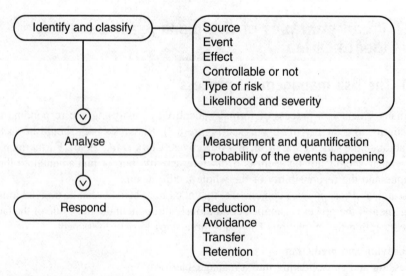

Figure 8.3 The risk management process.

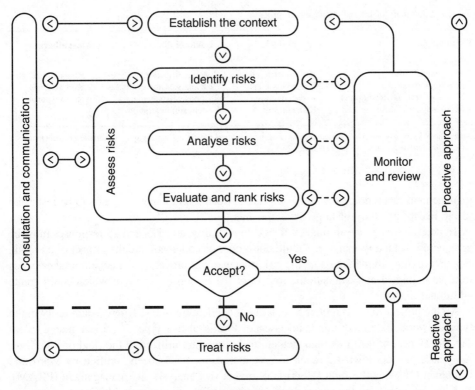

Figure 8.4 Risk management process (AS/NZS4360:1995).

reduction, avoidance, transfer or retention. Figure 8.4, which shows the AS/NZS 4360:1995 risk management process, illustrates this.

The Australian and New Zealand Standard emphasizes the pro-active approach to risk management and the need for consultation and communication amongst the parties involved. It also recommends that the decision-makers refrain from using a reactive approach. Yet, a study undertaken by Uher and Toakley (1997) supports the results of the pilot study by Bajaj *et al.* (1997) that the reactive approach is the most popular amongst many organizations.

Hertz and Thomas (1983) proposed a two-phase theory comprising a *Process of Risk Determination and Positioning Phase* and a *Facility Risk Evaluation Process.* They presume that risk management can be effective for risk/return trade-offs in strategic planning and management. The risk management process as discussed and presented by various authors and researchers is summarized in Table 8.2.

The first stage of risk management is identifying the risks and determining whether the risks pose an unacceptable level of uncertainty for the facility. Once the risks have been identified the next stage is to analyse them so they can be quantified according to their effects on the facility objectives. To ensure facility objectives are achieved, risk responses need to be developed. The decision makers now have the choice of risk removal, reduction, avoidance or transfer for each specific risk. However, this type of risk

Table 8.2 Summary of the risk management process

PMI (2000)	Turner (1993)	Al-Bahar (1988)	Most others
1 Risk identification	1 Risk identification	1 Risk identification	1 Risk identification
2 Risk quantification	2 Risk assessment	2 Risk analysis and evaluation	2 Risk analysis
3 Risk response and development	3 Risk reduction	3 Response management	3 Risk response
4 Risk response control	4 Risk control	4 System administration	

Note: The objectives of the risk assessment and management process remain the same despite the different nomenclature used by the various practitioners and experts.

management methodology, where each risk is treated on its merits, seems to be restricted to the feasibility study of large developments.

To determine the overall impact of risk on a facility, all risks and all responses must be considered. A facility manager should not ignore or underestimate the impact of any risk. For this reason, after Stage 3 of the risk management process, the decision-maker needs to review Stage 1 and evaluate the impact of the responses, some of which could create additional risks.

All the project risk techniques have a direct application in FM. There is similarity in the type of work undertaken, and, in essence, FM is also a type of project management service. Hence, project risk management techniques are utilized in the field of FM. Over the years, the top-down approach to risk management has been codified by Ward and Chapman (1995) and Simon (1995) into project risk analysis and management (PRAM). It is a derivative of the synergistic, combinatorial evaluation and review technique (SCERT) devised by Cooper and Chapman in 1987.

Turner (1999) has summarized PRAM into nine stages and SCERT into three stages, each with two phases. The SCERT technique involves the following risk management steps, summarized below, involving three stages, each having two phases:

- qualitative – scope and structure phase
- quantitative – individual risks and combined risks phase
- management – plan response and monitoring phase.

Risk management in high-quality facilities has the following pre-requisites:

- detailed and analytical specifications of the facility, and all associated risks
- clear perception of the risks being borne by each party
- sufficient capability, competence and experience to manage the risks
- motivation to manage risks that require a clear linkage between a party's management of risks and the party's receipt of reward.

It is generally accepted that the greater the detail provided by the owner in relation to the risks that are to be borne in whole or in part by the owner, the less the facility manager has to price for risk.

The benefits of risk management are summarized in the following six points:

- facility issues are clarified, understood and allowed for from the start
- decisions are supported by thorough analysis of the available data
- structure and definition of the facility are continually and objectively monitored

- contingency planning allows prompt, controlled and pre-evaluated responses to risks that occur
- definitions of the specific risks associated with a facility are clear
- statistical compilation of a historical risk profile will provide better modelling for the future.

8.2.2 Risk identification and measurement selection techniques

Risk identification approaches

Ashley (1984) suggests that there are two basic approaches to risk identification, the bottom-up approach and the top-down approach. The bottom-up approach works with the pieces and tries to link them together into more meaningful and controlled relationships. This is done at a more detailed scale in an attempt to build up an evaluation from these parts (Uher, 1993). There are various methods that utilize the bottom-up approach (Figure 8.5).

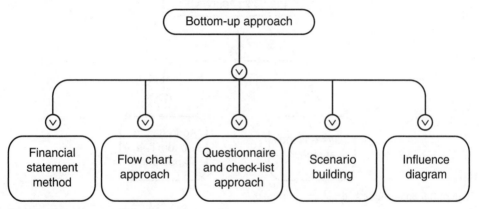

Figure 8.5 Bottom-up risk identification approach.

The various methods have been summarized by Uher (1993):

- *Financial statement method* – based on the assumption that financial statements serve as reminders of the various exposures to economic loss.
- *Flow chart approach* – attempts to identify risks by charting the company's operations.
- *Questionnaire and check-list approach* – based on producing a check-list of possible events by combining the experience of managers and historical records of the facilities involved to identify and record actual events.
- *Scenario building* – based on developing two extreme scenarios: in one scenario all goes as expected and in the second scenario everything goes wrong. In this approach an attempt is made to identify facility risks through a detailed analysis of both likely and negative factors influencing the facility.
- *Influence diagram* – an expansion of the flow chart approach. Shachter (1986) describes it as follows:

An influence diagram is a graphical structure for modelling uncertain variables and decisions and explicitly revealing probabilistic dependence and the flow of information. It is an intuitive framework in which to formulate problems as perceived by decision makers and to incorporate the knowledge of experts.

The four basic elements of the influence diagram are decision alternatives, risk factors or uncertainties, given quantities and calculated quantities or value models (Ashley and Avots, 1984). Influence diagrams can play an important role in the identification of risks, revealing the uncertainties through a structured approach to the flow of information, and identifying relevant variables and their relationships.

A top-down approach to risk identification involves a more holistic reasoning procedure (Uher, 1993) in which the facility is studied from an overall point of view and the risks identified from that perspective, without a detailed analysis of the facility.

Two methods with the top-down approach are shown in Figure 8.6.

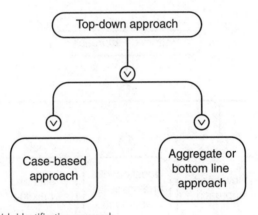

Figure 8.6 Bottom-up risk identification approach.

These two approaches can be summarized as follows (Uher, 1993):

- *Case-based approach* – uses the experience of similar facilities built in similar conditions as the starting point for identifying risk events. This approach gives the decision maker a starting point but fails to identify some unique risks related to the particular facility; also it useless if a particular facility is unique and there is no previous experience to draw on.
- *Aggregate or bottom line approach* – the most commonly used technique of risk identification in FM (Ashley, 1984). It is a very simple approach, as a certain percentage for contingencies is just added on to the basic costs. It does not provide any information on likely risk events and hence makes risk analysis and risk response very difficult.

None of the above approaches is ideal, and there is obviously a need for a combination of approaches. Al-Bahar and Crandall (1990) proposed a risk identification process with the following steps: preliminary check-list, identify risk events/consequence scenarios, risk mapping, risk classification and risk category summary sheet. For accurate analysis, the identification process is most important; in fact, it is believed that the principal benefits of risk management come from the identification rather than analysis stage.

There are two important considerations influencing the selection of a technique. One is the nature of the decision to be made, and the other the appropriateness of the technique to the decision context. In summary the two considerations are:

i The important characteristics of the decision are the:
   ● perceived seriousness of the risk consequences
   ● complexity of the decision context
   ● availability of the required analytical expertise.
ii The attributes of the risk measurement technique which require consideration are the:
   ● type of risk which is best modelled by the technique
   ● kind of output provided to the decision-maker
   ● nature of the available data
   ● ability of the technique to handle complex relationships and correlations between variables.

There are three types of technique used in risk analysis:

● qualitative – words, scales
● semi-quantitative – values
● quantitative – numeric values.

Comprehensive methods of risk analysis are still rather cumbersome and there is a need for a simpler approach. There are several other recognized processes for identifying risk and the most appropriate process for risk identification is to use a combination of techniques:

● 'riskstorming' – this is a structured 'brainstorming' session involving management representatives and staff
● employee consensus – managers and employees involved in the process are required to submit a list of the major risks that impact on their specific area of responsibility, as well as on the project as a whole; these risks are consolidated into a table, duplications are eliminated and the table used as the basis for more detailed analysis
● physical inspections – inspections of the facilities may help identify risks
● loss history review – a review of past incidents
● process mapping – this involves separating the activity, process or project into a set of elements and mapping these steps in sequence; by studying each element in turn and listing all influencing factors impacting on it, the major risks may be determined.

A trained facilitator should be appointed for the risk identification exercise. This may be an employee who has received appropriate in-house training or an external consultant.

Risk ranking

Although risk-ranking techniques are well established in the safety and reliability analysis of processes (Tweeddale, 1992), they have not been applied at the estimating stage of FM. In fact, with the exception of Al-Bahar and Crandall (1990) and a few other researchers, and some reference to it in the NSW Government's *Risk Management Guidelines*, there is little mention of risk ranking in the literature. One reason for this is that there is no published material/data on the probabilities of multifarious risks such as fire hazards (i.e., risks that affect people and property) occurring.

If risks are to be compared, a ranking mechanism must be established. For complex assessments, a formal structured approach is most suitable. Although there are many methods for doing this, the assessment of risk factors is straightforward. A risk factor represents the likelihood and severity of the impact of a risk event. If a risk is likely to occur *or* its impact is large, the risk factor will be high. It will be even higher if it is likely to occur *and* its impact is large (DPW&S, 1993, p. 16).

The components of the risk factor at the detailed level can be represented by a risk profile. This risk profile can be used to identify acceptable risks and allocate management priorities. A typical risk profile follows. The measurements for risk likelihood and risk impact can be from 0 to 1 and the risk factor value represents the overall importance of risk. The risk factor value then helps the facility manager in prioritizing the use of resources and energy needed to manage the major risks.

RF (Risk factor) = $(L + I) - (L \times I)$

where L = risk likelihood measure and I = impact measure

A graphical presentation of risk events according to their impact and severity (see Figure 8.7) can assist with reporting. Typically the three groups are classified as minor, moderate and major.

The three groups may be defined as (DPW&S, 1993):

- minor risks – can be accepted or ignored
- moderate risks – those that either have a high impact or a high likelihood but not both; management should specify measures for all moderate risks
- major risks – are likely to occur and have high impact; management should focus on preparing an adequate risk action schedule.

Because it is impossible to analyse all risks, short listing and ranking will help to allocate resources and focus attention on areas of greatest risk. As well as quantifying the probability of a risk occurring, ranking techniques also predict the severity of its consequences. In the industry these two variables are hard to quantify because the data, if

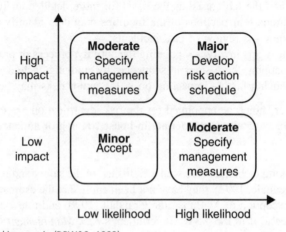

Figure 8.7 Risk ranking matrix (DPW&S, 1993).

kept at all, is not in this particular format. Relatively few organizations keep risk databases for future decision-making. This partly explains the industry's reluctance to adopt formal risk management techniques.

Risk mapping

After scenarios for the risk event and its consequences have been developed, a preliminary assessment of the significance and the relative importance of the risk can be made. This preliminary assessment is called *risk mapping*. The risk map is a two-dimensional graph in which one axis plots frequency and the other potential severity. Such a two-dimensional graph is very useful because it enables facility managers at an early stage to assess the significance of exposure to a potential risk.

However, contrary to general belief among contracting organizations, the significance of an exposure to a risk depends mostly upon the potential risk severity and not upon its frequency. Although risk frequency cannot be ignored, a potential risk with catastrophic consequences, though infrequent, is more serious than one expected to produce frequent small losses. Therefore, if two potential risks are characterized by the same loss severity, the one with greater frequency has the higher ranking. Most experts do emphasize, however, that both frequency and severity are important in evaluating the significance of a risk.

Figure 8.8 is a risk-mapping graph showing the magnitude of frequency and potential severity. The curves are called 'risk curves' and plot estimates of risk exposure. A family of these curves represents the full range of risks. The curve farthest away from the axis is the one that is higher risk and the one nearest to the axis is lower risk. For the facility manager, the priority should be to address the major risks first. In risk mapping, facility managers might define qualitative terms for frequency and potential severity.

Certainly some risks can be limited as to the extent of potential severity either by the nature of the risk, e.g., the complete loss of a cofferdam, or by contract or by law. Some others, however, are limited, e.g., pollution (risk 1) and environmental hazards (risk 2).

The four definitions below are qualitative categories used to describe the potential severity of a risk. If two potential risks are characterized by the same loss severity, the risk with greater frequency has to get higher ranking.

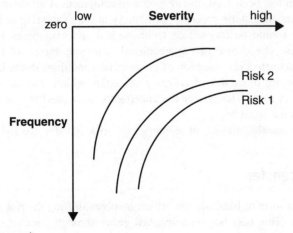

Figure 8.8 The risk mapping concept.

- catastrophic – risks having a massive and direct impact on the facility, causing severe restrictions on its use or even leading to its being abandoned altogether
- major – risks that, if they occurred, would result in considerable re-planning and major losses
- customary – risks that cause significant financial losses
- minor – risks that may be ignored due to their minimal affect on the facility; it is sufficient to recognize that they exist and usually no action is necessary.

On the other hand, there is another approach that focuses on frequency. This approach requires the risk manager to subjectively assign one of the four ratings listed below for each identified loss event.

- almost nil – in the opinion of the facility manager, the event will not occur
- slight – the event has not happened before, but could occur in the future
- moderate – the event occurs occasionally, and is expected to occur again in the future
- definite – the event has happened regularly, and is definitely expected to occur in the future.

These two qualitative approaches are sometimes considered inappropriate because they are too reliant on interpretation. Each word, therefore, has to have a specific definition, one agreed on by the entire facility team. The use of a quantitative scale for measuring and evaluating risk is more appropriate than a qualitative description. Most experts recommend that risk managers should, as far possible, assign numerical scales to subjective qualitative scales when using them in risk mapping. This will assure agreement within the facility team on the assessed level of risk exposure.

8.3 Risk response

Risk response is the final step in the risk management approach. It is an action or a series of actions on the part of the decision-maker in response to the presence of risks and their significance. Regardless of whether a 'gut feel' approach has been used or a scientific risk analysis approach has been used, risk response development is an essential ingredient in any business decision-making process, particularly at the estimating or tendering stage.

Risk response should be focused on technological and managerial issues as well as contractual issues. The choice of organizational structure, choice of type of contract, method of contract selection, selection of contractual conditions through which risks are allocated – these are examples of contract strategies which may play a vital part in effective control of risks. The greater the uncertainty associated with a facility the more flexible the response must be.

There are two possible choices of response, i.e., risk transfer and risk control.

8.3.1 Risk transfer

'Risk transfer is a form of handling risk which involves shifting the risk burden from one party to another.' This may be 'accomplished either through contract conditions or by insurance' (Uher, 1992).

Contractual transfer

Risk transferring contracts commonly exist between various parties concerned in the industry. The direction and intensity of the transfer are often governed by the contractual strength of individual parties. A client could, if necessary, place a greater burden of risk with a contractor, while a contractor, after securing a head contract, could transfer risks to a subcontractor.

Parties to whom risk has been transferred generally respond by including an appropriate risk allowance in cost estimates. The problem for the client is that she or he does not know the extent of risk allowances that reflect the value of risks transferred by the contractor and subcontractors. The greater the intensity of risk, the greater the amount of risk allowance. If the risk allowance is too high, she or he pays too much for the facility. If the risk allowance is too low, she or he runs the risk of either the contractor or subcontractors losing money. This could lead to more contractual claims, lowering of quality, or even bankruptcy of the contractor or subcontractors.

Sharing

The essential principle of transfer response is that risks, if they occur, should be equitably shared among the parties to a contract on the basis of their ability to control and their capacity to sustain such risks. Risks may be managed by segmentation, with each party being responsible for a defined portion of the activity concerned.

Additionally, 'risk allocation depends on the willingness of parties to take risks on' (Ward and Chapman, 1995).

Insurances

A common method of transferring risk is by way of insurance, where part or all of the risk is transferred to the insurance company. Under the terms of an insurance contract an insurance company agrees, for a monetary consideration, to assume the financial impact of a particular risk for a given time period. Insurance does not eliminate the risks involved in contracting, but it does shift most of the financial threat to a professional risk-bearer.

The purpose of insurance is to convert the risk (often expressed as contingency against a degree of uncertainty) into a fixed cost. As a result, the real cost to the insured party will be known. However, not all risk can be covered by insurance, and for those that are insurable, the cost of the premium may be considerable. The decision-maker should decide how much she or he is willing to pay in premiums for the insurance of risks, after taking into account the probability of such risks occurring.

Worker's compensation, public liability insurance and all risks insurance insure against accidental hazards.

8.3.2 Risk control

When risks cannot be transferred, for whatever reason, management action is required to reduce, avoid or retain the risk.

Risk avoidance or reduction

Where the level of risk is such that it is deemed unacceptable for the risk to be retained, the best method of dealing with the risk is to avoid the activity with which the risk is associated. A contract that places an excessive burden of risk upon one party is likely to turn many bidders away since they choose to avoid the risks. Contracting organizations cannot, however, avoid risks indefinitely if they are to remain in business. They must accept some level of risk and ensure that the benefits to be gained by accepting the risk are greater than potential losses. This is a difficult, but not impossible, task.

In summary, elimination is seldom an option and reduction of risks is more appropriate for most organizations. If risk cannot be fully avoided 'its impact could be reduced by reassessing strategy, developing alternative solutions or even redesigning the facility' (Uher, 1990). The redesign of a facility is most effective when the risk involves excessive uncertainty and it is impractical for any party to control it.

Risk retention

If risk cannot be avoided, reduced, financed or transferred, risk must be retained. If risk is being retained, good management and business judgement must be applied to decide how to control residual risks. Management may have some degree of control over residual risks; the control may be exerted to reduce the likelihood of occurrence of a risk event and also to minimize the impact if the event occurs. More specific information about an activity or activities affected by a risk event may reduce the degree of uncertainty.

For those risks that cannot be controlled, a contingency allowance is commonly determined. The contingency sums may be applied in the form of a single figure based on the best estimate of final cost, or as a range of figures to provide different probabilities of protection against risk (Perry, 1986).

The single figure approach has several weaknesses:

- it is most likely that the percentage figure has been arrived at arbitrarily and is not appropriate for the specific facility
- there is a tendency to double-count risk as some estimators are inclined to include contingencies in their best estimates of individual cost items
- a percentage addition results in a single figure prediction of the estimated cost, implying a degree of certainty that is simply not justified
- it reflects only the potential for detrimental or downside risk – the approach does not highlight any potential for cost reduction, and may therefore be used to hide poor management performance
- because it captures all risk in terms of a cost contingency, it tends to direct attention away from time and performance or quality risks.

8.4 Contingency planning

Contingency covers possible or unforeseen occurrences; in estimating, the word contingency is used for two types of estimates: the first is the expected value of a possible identified event, and the second is the possible cost of unforeseen events. It is the second that needs closer attention, because it carries a greater margin for error (Carr, 1989).

If previous experiences are recorded as a database, they will be of immense value for future reference. If suitable historical data is available, the estimator can have a reasonable degree of confidence in the accuracy of the estimate of the possible costs associated with a risk event. The first type of contingency can be made more accurate with formal risk identification, monitoring and recording events. This will lead to the establishment of a database that will inform an estimator of the different components of risk as discussed earlier, from similar types of facilities managed in the past, and finally lead to an expected value for that identified event.

One method of risk identification, within the top-down approach, is the aggregate or bottom line approach. It is the easiest and most readily used approach by which allowance can be made for the uncertainties of a facility at the estimating stage, however, the approach is unscientific and does not make the estimator plan or think of the facility in depth. That is where the drawback of this method lies: either the allowance will be high or low and it is more likely to be based on the intuition of the decision-maker(s). The shortcomings of this approach include:

● the arbitrary nature of the percentage figure
● potential double counting of contingencies
● it highlights 'downside' risk and does not address any potential cost reductions (Groves, 1990).

Apart from monetary contingency amounts, there is a need for contingency planning for disaster situations. Such contingency plans are called disaster recovery plans (DRP). It is beneficial for the facility manager to have one for each facility. A DRP goes through the same process as that of a risk management plan, as discussed below.

8.5 Developing and documenting a risk management plan

The risk management plan is developed for the risks identified and ranked with the help of one of the methods of ranking, whether qualitative, semi-qualitative or quantitative. Regardless of the method of ranking, the risks would be categorized as major, moderate and minor risks. Some facilities might require a different categorization to include catastrophic, major, moderate, minor and insignificant. The plan of action for major and catastrophic would require a risk action schedule as shown in Figure 8.9. Each individual major risk, due to both the severity and the frequency, would need a detailed action schedule of its own. Taking the example of Figure 8.7, for all moderate risks, the facility manager needs to specify management measures to be taken to mitigate, reduce or manage the risk. Minor risks are generally accepted and noted. The action plan for the major risks goes into much greater detail such as who is the responsible person and what resources are required to manage the risk.

8.6 Conclusion

Risk management in the current world climate is an important activity for developing confidence and adding value to FM services. Effective communication of risk management plans by the facility manager is an important step towards ensuring that strategic decisions are properly informed and well balanced.

A facility manager should be comfortable using simple techniques of risk ranking and identification. It is no good trying to adopt sophisticated techniques and then using 'gut feeling' to quantify and rank the risks. It is far more meaningful and informative if

Activity / Risk:	Risk Number:
Risk Description (causes, consequences):	
Current Controls and Plans:	
Additional Actions Recommended:	
Responsibility:	
Resources Required:	
Timing (key milestones, closure):	
Reporting (to whom, when, in what form):	
References (to other documents or plans as appropriate):	

Figure 8.9 Template for a risk action schedule.

potential risks are clearly understood and form part of facility decisions. Risk management is therefore an important protection device that ensures that anticipated value gain is fully realized.

References and bibliography

Al-Bahar, J.F. (1988) Risk Management in Construction Projects: A Systematic Analytical Approach for Contractors. PhD Dissertation for University of California, USA.

Al-Bahar J.F. and Crandall, K.C. (1990) Systematic risk management approach for construction projects. *Journal of Construction Engineering and Management*, **116** (3) September, 533–46.

Arthur Anderson (1999) *Industry Research into Facility Management in Australia* (Facility Management Association of Australia).

AS/NZS 4360 (1995) *Risk Management Guidelines* (Standards Australia and Standards New Zealand).

Ashley, D.B. and Avots, I. (1984) Influence diagramming for analysis of project risks. *Project Management Journal*, March, 56–62.

Ashley, D.B. (1989) Project Risk Identification using Inference Subjective Expert Assessment and Historical Data in Transactions of the Internet International Expert Seminar. *State of the Art in Project Risk Management*, Institute of Technology, Atlanta, Oct 12–13, pp. 9–28.

Bajaj, D., Oluwoye, J. and Lenard, D. (1997) An analysis of contractors' approaches to risk identification in New South Wales, Australia. *Construction Management and Economics Journal*, **15** (4) July, 363–9.

Carr, R.I. (1989) Cost-estimating principles. *Journal of Construction Engineering and Management*, **115** (4) December, 545–51.

Clark, R.C., Pledger, M. and Needler, H.M. (1990) Risk analysis in the evaluation of non-aerospace projects. *International Journal of Project Management*, **8** (1) February, 17–24.

Cooper, D.F. and Chapman, C.B. (1987) *Risk Analysis for Large Projects: Models, Methods and Cases* (New York: John Wiley).

Cotts, D.G. (1999) *The Facility Management Handbook*, 2nd Edition (New York: AMACOM).

Cross, J. (1997) *Risk Management* (Sydney: Department of Safety Science, University of NSW).

DPW&S (1993) *Total Asset Management – Risk Management Guidelines* (Sydney: NSW Department of Public Works and Services).

Groves, L. (1990) Project risk management from concept to design to tender – new initiatives at BHP engineering. In: *Proceedings of 1990 National Engineering Project Management Conference and Forum*, Sydney, September (IES).

Hertz, D.B. and Thomas, H. (1983) Risk analysis approaches and strategic management. *Journal of Advances in Strategic Management*, **1**, 145–58.

Marshall, V. (1997) *Design/Build: The Project Delivery System of Choice*. Graduate Report, University of Florida.

Nutt, B. and McLennan, P. (2000) *Facility Management: Risks and Opportunities* (Oxford: Blackwell Science).

Perry, J.G. (1986) Risk management – an approach for project managers. *Project Management*, **4** (4) November, 211–16.

Petrocelly, K.L. and Thumann, A. (2000) *Facilities Evaluation Handbook: Safety, Fire Protection and Environmental Compliance*, 2nd Edition (Englewood Cliffs, NJ: Prentice-Hall).

PMI (2000) *A Guide to the Project Management Body of Knowledge (PMBOK)*, 3rd edition, Project Management Institute, 2000.

Shachter, R.D. (1986) Evaluating influence diagrams. *Operations Research*, **34** (6) November/December, 871–82.

Simon, P. (1995) Project risk analysis and management. *Australian Project Manager*, **14**, 10–15.

Toakley, A.R. (1991) The nature of risk and uncertainty in the building procurement process. *The Australian Institute of Building Papers*, **4**, 171–81.

Turner, J.R. and Cochrane, R.A. (1993) Goals-and-Methods Matrix: Coping with Projects with Ill-Defined Goals and/or Methods of Achieving Them. *International Journal on Project Management*, **11** (2), 93–102.

Turner, R. (1999) *The Handbook of Project-based Management – Improving the Processes for Achieving Strategic Objectives*, 2nd Edition (London: McGraw-Hill).

Tweeddale, C.S. (1992) Some experiences in risk ranking and short listing. *Journal of Loss Prevention in Process Industries*, **5** (5) May, 279–88.

Uher, T.E. (1990) *The effect of risks and uncertainties on subcontract bid prices*. PhD Thesis, University of NSW.

Uher, T.E. (1992) *Risk Classification*. Seminar notes, unpublished.

Uher, T.E. (1993) Risk management in the building industry. In: *Proceedings of AIPM Conference*, Coolum, Australia, March.

Uher, T.E. and Toakley, A.R. (1997) *Risk Management and the Conceptual Phase of the Project Development Cycle* (Risk Management Research Unit, School of Building, University of New South Wales).

Ward, S.C. and Chapman, C.B. (1995) Risk-management perspective on the project lifecycle. *International Journal of Project Management*, **13** (3), 145–9.

9

Human resource management

Suzanne Wilkinson*
David Leifer†

Editorial comment

Human resource management is a necessary competency for facility managers, as they are likely to find themselves leading a multidisciplinary team of in-house people as well as outsourcing work to external consultants and suppliers. Adding value to an organization usually means using its people effectively. Recruiting, training and staff development procedures are important for getting the right people and keeping them motivated and fully utilized.

The decision as to whether to use in-house or outsourced expertise is not simple, as there are advantages and disadvantages that must be carefully weighed. Insofar as permanent staff are concerned, it is vital to ensure that job satisfaction is high through articulation of a clear vision, team involvement, effective working environment, the right tools for the job and granting appropriate responsibility. As facility managers are becoming increasingly occupied with strategic functions, the work setting must provide opportunities for others to gain knowledge and build expertise.

On a much broader level, human resource management is a key support function for all organizations. Traditionally it has been managed as a separate entity, with little interaction between recruitment and facilities, or between staff development and information technology (IT) usage. However, some organizations have chosen to merge two or more of these areas under a single senior manager's direction. Human resources, facility management and IT can be envisaged as the legs on a three-legged stool. They have increased strength and value in enabling the primary function when they are connected together at the top; but if one leg is removed or isolated then the entire stool will collapse and be of little use. Assisting this approach are new reporting strategies known as

* The University of Auckland, New Zealand
† University of Sydney, Australia

enterprise resource planning (ERP) systems, which collect and integrate key performance data and make it accessible across the organization.

Some facilities can be designed to minimize the number of staff, or at least to minimize time lost through unproductive travel to and from various parts of the building, unloading and transporting materials, and communicating with colleagues. An understanding of these issues can be useful when directing refurbishment work or when moving into new premises. In this context a strong link can be forged between human resource management and space management, particularly at a strategic level.

People add value. The way people are treated, engaged, managed, rewarded and developed dictates the success of the entire organization and all the subprojects and activities that it undertakes.

9.1 Introduction

Human resource management is the management of people and its purpose is to get the best effort from the labour resource of an organization. This includes getting and retaining the best people for the job, and providing and maintaining an environment that will assist people to give their best.

Human resource management as a discipline has developed into a diverse set of theories and practices ranging from motivation and efficiency studies, on the one hand, with practical aspects such as training strategies and recruitment, to employment contracts, on the other. The challenge is to examine these aspects of human resource management and put them into the context of facilities management. Clearly human resource considerations in the 'facilities' discipline will include not only the recruitment, retention and performance of staff, but the nature of the environment in which the organization's staff work. This is a fundamental aspect of facility management's core business. It is in this area that human resource management and facilities management overlap and are interwoven.

9.2 What is human resource management?

People are an important resource for any organization, because it is through them that organizations get their work completed and make a profit. Moreover, salaries and wages are a major operating cost for an organization. The discipline of human resource management recognizes this and assists organizations to achieve success. They do this by providing a nurturing environment in which people can develop their skills and contribute effectively to the growth of organizations and also by making the most effective use of the money expended on salaries, e.g., by developing good, cost-effective training strategies. An understanding of human behaviour is required if this it be done successfully. How people react to various situations, what motivates people and how organizations can provide the most comfortable environment for people to work are all questions which require answers if organizations are to get the best from their employees. Human resource management, however, does not just deal with individuals, but with teams and how teams can work together to produce the best team output. Human resource management incorporates the management of individuals and groups, the management of tasks for ensuring a successful and motivated workforce, and it deals with the workplace environment, ensuring that people are able to perform their allocated duties.

9.3 Management theories and human resource management strategies

The discipline of human resource management has developed in part from applying theoretical ideas in practice. Some of the more influential management theorists and their contribution to human resource management and its impact on facilities management are discussed here. Their ideas provide some of the guiding principles for human resource managers and make them more effective. These theorists are grouped into four main streams: people are efficient machines, people work better in a structured environment, people need to take control of their own tasks and people need basic comforts to perform efficiently.

9.3.1 People are efficient machines

Fredrick Taylor, who developed his ideas in the late nineteenth century, is one of the most commonly mentioned theorists who subscribed to the idea of people as 'efficient machines' (Taylor, 1967). Taylor, or more specifically, Taylorism, has become a common by-word in human resource management circles because Taylor tried to develop strategies that would allow workers to perform their tasks in the best way possible, and so improve worker efficiency and output. What Taylor tried to achieve was a system that would maximize the amount of work a person could do. To do this he observed how people performed tasks, measured how often they did the task or how much they produced and then tried to provide a rationalized system for their respective tasks that would optimize the amount they could produce. This scientific approach to the way people do their work is still evident today, especially in construction projects where durations for construction tasks are estimated using man-hours, i.e., the time it takes a person to complete a particular task. These times are based on how long it would take a competent and trained person to complete the task. Today this approach is commonly called work-study.

Taylor was not the only theorist promoting the idea that people are efficient machines: Frank and Lillian Gilbreth were also working around the same time and they, too, were concerned with promoting efficient working practices based on human performance. They recognized that many jobs, e.g., bricklaying, could be standardized. By examining the pattern of repetitive work, the Gilbreths discovered that the time taken for many tasks could be reduced by developing more efficient systems of work. The impact of the Gilbreths' theories today is mainly seen in the development of the field of ergonomics which has had an influence on the way individuals in the construction industry perform their work. Whilst Taylor and the Gilbreths examined the activities of people, Henri Fayol examined the activities of managers and of management in general. He believed that there were patterns of management practice that could also be systematized and developed methods of management that could be learnt by managers, making them more efficient. He summarized management into various functions such as planning, organizing, commanding, co-ordinating and controlling (Thomason, 1988). He believed that if managers were to be efficient they needed to understand these basic management functions. Today, managers such as facilities managers still use these basic functional descriptions in describing their managerial roles.

9.3.2 People work better in a structured environment

Whilst Taylor and the Gilbreths were working directly on making people more efficient, some other theorists were pursuing the same goal by altering the administrative environment in which they worked.

Barnard and Weber, for instance, examined the management of organizations (Harrison, 1993). Chester Barnard was what is known today as a systems theorist – what he tried to do was to define systems for the promotion of an efficient organization through the co-ordination of the activities of the workforce. One of the ways of doing this was through communication, using supervisors in an organization as messengers between the various levels of the hierarchy. Barnard believed that communication in the organization between all levels was essential and would promote efficiency. His ideas were crucial for facilities managers, as without them communication systems would not have developed into the systems available today.

Max Weber also believed in defining organizational systems that would promote efficiency (Inkson and Kolb, 1998). He suggested that a bureaucratic system would promote the best worker efficiency as it produced an organization with clear lines of communication and responsibility, where people were given explicit tasks and responsibilities. Weber's ideas are useful for facilities managers in the understanding of structured approaches to management of the workplace, although today organizations take on many forms other than those modelled on Weber's bureaucratic systems.

9.3.3 People need to take control of their own tasks

Two theorists who have influenced human resource management, in that they produced theories that support the idea of worker autonomy leading to better worker efficiency, are Mary Parker Follett and Elton Mayo (Graham and Bennett, 1995). Follett advocated a system in which managers should decide, in collaboration with the workforce of an organization, on the best approach for the individual or team to achieve the best output. Any decision-making or policy formulation would then be done through a system of communication between the various levels of the organization and not by managers alone. This, she believed, would give the workers a feeling of control over their tasks and their working environment.

Mayo also believed that people should be given some autonomy over their working practices. He suggested that when workers are given freedom to control their own working conditions and tasks, efficiency increases. His work involved examining both group behaviour and individual behaviour, and suggested that where groups existed they influenced efficiency by exerting group control over the individual (Inkson and Kolb, 1998).

9.3.4 People need basic comforts to perform efficiently

One of the main areas of human resource management concerns the analysis of the behaviour of people and their requirements in order for them to succeed. Abraham Maslow (1987), Douglas McGregor (1960) and Frederick Hertzburg *et al.* (1959) all produced theories to help with the management of people based on the idea that people need basic comforts to perform efficiently. Maslow and Hertzberg *et al.* provided frameworks for the analysis of human behaviour and suggested that if people were provided with basic needs then they would work better. Maslow's framework dealt with motivation based on a hierarchy of human needs. He identified five levels that he believed should be met, in ascending order, for a person to work effectively. He believed that at the most basic level a person requires that their physiological needs, such as hunger and thirst, be met. Furthermore they require a safe environment, and then the satisfaction of social needs such as friendship, social interaction and a sense of belonging. Finally, the hierarchy led to esteem, translated as confidence, self-respect and recognition, with the last level termed *self-actualization* or *the realization of one's full potential*.

Human resource managers can use the hierarchy to assist with the way in which their organizations provide for their employees' needs, how they motivate their staff, and what kind of environment they provide. However, human resource managers are just one set of managers among several. They do not usually decide on issues such as the pace of automation but will be fitting the employees to jobs that will usually be set by other factors. Maslow's hierarchy helps the human resource manager by providing a framework to assist with altering the working environment and not with altering the actual job function.

Hertzberg provided a similar set of factors that can be used as motivators for employees (Payne, 1996). He suggested that there are two main factors affecting motivation at work: on the one hand, hygiene factors and, on the other hand, motivators. The hygiene factors concern the working environment and need to be managed in order for the employees to feel satisfied. These factors might include working conditions, salary and leadership. The motivators are concerned with the nature of the work and need to be carefully managed so that employees are motivated to work; these factors include achievement, recognition and responsibility. Human resource managers still use these hygiene factors and motivators in order to achieve a satisfied workforce.

Douglas McGregor also considered the idea that people need basic comforts and suggested that managers can have an impact on the satisfaction of their employees and that their style would lead to employees reacting in different ways. He classified managers into Theory X or Theory Y categories, which basically meant either dictatorial and adversarial (Theory X) or inclusive and collaborative (Theory Y). He suggested that Theory X managers believed that most people need to be carefully controlled whilst performing their tasks, whereas Theory Y managers believed people should be allowed to exercise autonomy and responsibility over their work. McGregor believed, as do most human resource managers today, that Theory Y is the better route to establishing and maintaining an efficient workforce.

9.4 Human resource management functions

In addition to the four main areas discussed above, the human resource manager is concerned with how employees are recruited and nurtured throughout their time with a company. The human resource management process consists of managing the people in an organization, looking after their needs from before they enter an organization until they leave. Human resource managers therefore need to have an overall understanding of human behaviour, i.e., how people react in certain situations, what their requirements are, how they are motivated and what personality type they are, coupled with an understanding of the organization's needs and requirements. This section examines the main functions of human resource management from job analysis and job specifications, recruitment and selection strategies, through to training, promotion, employee relations, job satisfaction and motivation.

Inkson and Kolb (1998) suggested that there are eight basic functions in the human resource process. These are:

- human resource planning
- recruitment
- selection
- orientation
- training
- development
- performance
- rewards.

To this can be added three other functions which are traditionally performed by human resource managers: external mobility, employee relations and, important to facilities managers, the workplace environment.

9.4.1 Human resource planning

In most organizations the human resource manager is responsible for the overall strategic planning for the employees. They are required to have an understanding of current personnel shortages and where they are going to be in the future. For the facilities manager a key planning consideration is assessing what work will be handled internally and what ought to be outsourced. This impacts on the work of the human resource manager, who needs to work closely with the facilities manager when designing staffing projections. Experience suggests that one should outsource wherever expensive labour, equipment or tools are under-utilized within the organization. Outsourcing saves on depreciation of equipment and labour overheads on the basis that the outside contractor, by fully deploying their resources, has lower unit costs. Personal experience in providing out-sourced facility management services to the leisure industry indicates that 15% savings can be made by putting the labour intensive functions such as cleaning and security out to tender (Leifer, pers. com., 1998). It is prudent, however, for supervision and policing of the out-sourced work to be retained in-house. The 'institutional knowledge' of why things have been done needs to be documented and kept current as the loss of this knowledge can prove a great (and expensive) impediment further down the track. This is particularly true of details of power and communication cabling.

Another option that facilities managers should consider is multiskilling of existing employees if this will defer the need for new staff. Again, the facilities manager would need to work along side the human resource manager to discuss multiskilling and training issues when undertaking planning exercises.

Where in-house staffing is necessary, planning is complex. Most building services systems within the facility manager's remit run throughout the premises, need to be maintained to legislated standards, are fundamental to the operation of the business, and require specialists skills for maintenance. Maintaining a full spectrum of skills on call through direct employment is not within the capacity of most organizations. If an organization decides to have specialists on the staff, the facility manager must first set up a system that ensures that these staff have a full schedule of preventative maintenance to carry out so that they are not left idle at any time.

Where larger companies have the capacity to recruit a number of staff regularly, for instance in an annual graduate intake, smaller companies tend to react to staff turnover as and when required. The human resource manager must be aware of the likelihood of members of staff leaving the organization and plan for the impact that this might have – if unusually high numbers of staff in a particular area leave in a short period of time, this is obviously of concern and needs to be investigated. It is the human resource manager who will make such an investigation. In addition to this, human resource planning requires an analysis of future needs so that the organization is in a strong position to react to any changes in market conditions, such as shortages of skills in specific areas, e.g., information technology or engineering.

9.4.2 Recruitment

One of the main tasks of a human resource manager is the recruitment of staff. This not only requires an analysis of the job that the potential employee is required to do, but also an understanding of the availability of people, locally, nationally or even internationally, with the required skills. Many of the skills needed in the facility management area are technical and these are the areas in which there is a high demand for staff due to a shortage of trained personnel. Filling these places is therefore a challenge and word-of-mouth and headhunting are fairly commonly used methods of recruitment in these areas.

Many companies find that they need to use different recruitment strategies depending on a range of factors such as whether the job requires skills which can be found in house, whether there are people in other companies who are mobile, what the current market conditions are for recruiting certain skills, and what level of skill is required. The human resource manager will research and report to senior managers on these issues.

Part of the recruitment process also involves analysis of the type of job and writing a job specification so that potential candidates can be adequately screened. For instance, in recruiting a construction project manager, a requirement might be that all potential candidates have a technical background, be computer literate or have project accounting experience. These requirements are made explicit in the job specification.

There are various ways in which companies can recruit, with a popular mode being the use of Internet advertising companies. However, the mode of recruitment needs to be considered since the target market needs to have access to the chosen mode of advertisement. For instance, a non-technical person may be more likely to look for employment through traditional modes of advertisement, such as newspapers or word-of-mouth, whereas a technical person may approach recruitment consultants who could specialize in finding technical staff or use Internet searches.

Recruitment strategies also vary depending on the size of the company: larger companies may make regular recruitment rounds to universities and select a certain number of candidates each year, whereas smaller companies will generally recruit when a staff member leaves or when there is an increased workload due to company growth. As with most management recruitment, when recruiting facilities managers, human resource managers will usually use a combination of approaches in order to secure the best possible candidate.

9.4.3 Selection

The human resource manager is responsible for selection processes used by the company. Since the aim of the company is to ensure that the most suitable candidate for the job is selected, various methods of selection have been developed. Most companies will pre-screen candidates through an application form. What the company is looking for here is usually whether candidates meet the general requirements for the job. For example, in recruiting a facilities manager, a requirement might be education to degree level in certain disciplines; if candidates do not have this qualification then they may be rejected at application level. Once candidates pass the pre-screening stage, referees nominated by the candidates may be asked to provide references. In this case, a selection panel may be looking for some indication of the ability of the candidate to do a job, plus some indication of their personal characteristics. When asking for references, the human resource manager needs to supply a list of areas that referees should comment on so that valid comparisons can be made between candidates. Candidates who are short-listed at this level will usually move onto the next stage of selection. This next stage in the process may vary from company to company with some using telephone or tele-conference interviews while others use face-to-face interviews. Some companies may ask for a presentation of work or request to see previous evidence of the applicant's work, while others may use this in combination with various tests, such as mathematical or problem solving tests or team tests, as a means of selection.

Interviews, either informal, with key members of the company, or formal, with a panel of people, still appear to be the most popular mode of selection and often provide the interview panel with the greatest amount of information on the candidate. Langford *et al.* (1995) discuss the interview process and refer to an important factor in the process, that of prejudice of the interviewers, and how this can be avoided. They point out that personal attitudes towards a person's appearance and, to some extent, their beliefs should not prejudice the process or detract from the key issue of whether a person has the ability to do the job.

The selection of the most suitable candidate will be made using some combination of methods and will often involve a rating of candidates from acceptable or appointable to not acceptable, for whatever reasons. The most suitable candidate will then be offered the employment or be asked to negotiate conditions for the position. If applicable, the human resource manager will form part of the negotiating team. Acceptance of the candidate for the position may be conditional, depending on satisfaction of some specified requirements.

It is the role of the human resource manager to have an understanding of the types of employment benefits that may be acceptable to various candidates. Research by Wilkinson (1996) into the factors that motivate civil engineering graduates to accept employment identified six key factors that can be used by companies to attract civil engineering graduates and found that these factors were not gender specific. The top six factors, in descending ranked order of importance, were opportunity to do interesting work, opportunity to do varied work, the organization's training programmes for graduates, involvement in new developments, opportunity for quick advancement and salary. Bennet (1996) also found that these factors are important as motivators to senior engineering students.

The human resource manager needs to be aware of the factors that motivate people when accepting employment and also be able to match these with the company's ability to offer any or all of the factors. Certain benefits are more important to different individuals – pension scheme, company car, workplace facilities (such as health and fitness club membership), child care, and so on, can be used by the human resource manager as enticements for a preferred candidate so that the company gets the best candidate for the job.

9.4.4 Orientation

Once a candidate has been selected they will need to undergo some form of initiation or orientation into the company. How this is done varies from company to company. For small companies this may consist of meeting the team that the selected candidate will be working with, and meeting any of the clients that the candidate will have immediate communication with, finding out about company processes and procedures, and basically becoming familiar with the working environment and what is offered in terms of space and technology. In larger organizations a series of orientation events may be used, such as weekend team-building exercises for new graduates, a team dinner, informal lunches or centralized orientation days that provide information about administrative and other procedures.

For facilities staff, orientation should include a tour around the facility to be managed, an exposition of the in-house safety procedures and systems (on which the inductee will sign-off for record purposes), instruction on where manuals and key documentation are kept, and instruction in work-order and business procedures. A human resource manager will be instrumental in setting up the orientation events and making sure that the new employee attends the most appropriate events.

9.4.5 Training

Part of the facility manager's human resource role is to be responsible for assessing and meeting the training requirements of employees. To this end they need to be aware of the various training options. Bennet (1996) summarizes training requirements for engineers that can be generically applied to facilities managers. He suggests that there are six ways in which an engineer can keep up to date with their profession and latest developments. These are on-the-job training, formalized graduate education, professional registration, non-credit short courses and workshops, professional society activity and reading. Training can also be classified as internal and external to a company (Inkson and Kolb, 1998).

Companies can sometimes provide all the training needs of an employee internally by providing adequate supervision by experienced personnel whose role it is to pass down required skills and techniques – to this end some companies use mentoring as a training strategy, others use a 'train yourself' approach. External training, however, has the added benefit of providing professional employees with contact with other professionals in their field and this cross-fertilization of ideas often has added benefits for the company. For instance, external training on a new piece of software for managing a facility may produce a trained employee who has knowledge of current technology and also an understanding of how it is being used in other companies. The human resource manager must be able to look at each employee's requirements and decide the best option, within budget constraints, for each individual and then assess whether the training undertaken by the employee is meeting the requirements of the company.

For many facility tasks trade certified staff are essential; this is particularly true for staff maintaining fire, electrical, and mechanical systems. Mention has been made of multiskilling and it is usually sensible to merge roles where possible, e.g., plumbing and gas fitting. It is standard practice in top facility management organizations to maintain a register of all skills and interests of their staff. This can be mutually beneficial: a common example is the training and certification of hobbyist welders on staff.

9.4.6 Development

Once an employee is part of an organization one of the roles of the human resource manager is to ensure that they are adequately catered for in terms of professional and personal development. This includes making sure the employee is satisfied with his/her job. This can be done in a variety of ways including interviews, informal feedback or questionnaires that gather information about job satisfaction. Inkson and Kolb (1998) define development as 'activities designed to develop the person's longer-term potential in an organization, including development for positions beyond the present job'. To this end the human resource manager must be able to identify, with the employee, a potential career structure and then provide the advice required for the person to be able to advance along the defined career path. Advancing employees has a roll-on effect and the organization must ensure that staff promotions do not leave voids elsewhere.

9.4.7 Performance

In order for an employee to stay employed, there must be some form of evaluation, whereby the employee and the employer fully understand the requirements and ambitions of the individual and are sure that the employee is doing the work to the required standard and beyond. Performance is directly related to promotion and demotion. For promotion, various companies operate different schemes. The most common way appears to be promotion schemes based on merit that is assessed through an annual review. The review may recognize the employee's contribution to the company and make adjustments to the salary of the individual, or may redeploy the individual to other work of a higher status, or may provide the individual with more responsibility. Promotion needs to be seen to be fair and equitable for all employees and, therefore part of the human resource manager's role is to assist with devising promotion criteria that are applicable to the employees. Bennet (1996) suggested a link between the career development programme and the performance appraisal system where each performance session is designed to set future objectives and appraise past objectives.

9.4.8 Rewards

At the most basic level reward for employment is in the form of a salary. When an employee has exceeded their requirements or produced extra benefits to the company then they may be rewarded for their achievements. The human resource manager has to be able to provide incentives for employees to achieve and ensure that rewards are fairly distributed. Rewards can vary from extra salary payments through to bonus payments through to internal recognition using certificates, prizes and promotion or additional bonuses such as a company car, hotel vouchers, and similar incentives.

9.4.9 External mobility

One of the areas that concern the human resource manager is the mobility of the workforce in a company. These include issues such as dismissal, redundancy, transfer, retirement and resignation.

Where an employee fails to perform adequately, or has acted in a capacity that contravenes their employment contract, they may be warned to improve their performance or face dismissal. Subsequent warnings and failure to conform may result in dismissal. The human resource manager needs to be careful to act fairly and guide the process in these cases, since their failure could lead to unfair dismissal claims in the court.

Where a company faces loss of profits and is struggling to maintain their position in the marketplace, or is undergoing a process of restructuring, the company may look to downsize the workforce by natural attrition (i.e., people leaving and not being replaced), or by redundancy. Redundancy clauses are usually found in individual employment contracts and provide the mechanism by which an employee may be made redundant.

Occasionally employees are transferred from one part of employment or location to another. There are many reasons why an employee may be transferred, such as promotion, work more common in another location (as is often the case with construction projects), or the person's skills are required elsewhere. For the human resource manager, transfers can cause problems due to the employee being unwilling to move, usually because of external factors such as family and mortgage commitments.

Retirement of individuals has to be managed in order for the human resource manager to plan adequately for replacement staff. Finally, the human resource manager must keep track of the resignation situation of staff. In particular they need to find out why a person is leaving the organization. One way of doing this is through an exit interview, where the human resource and other managers discuss with the person why they are leaving and, if applicable, use the information gathered to improve the current status of employees.

9.4.10 Employee relations

Employer/employee relations are one of the cornerstones of human resource management. Contracts need to be managed and negotiated; conditions of employment need to be regularly updated as individuals are transferred or promoted. In order to do this a thorough understanding of employment law is required. The human resource manager would usually work in conjunction with an employment lawyer to draft and refine contracts. Finally, the human resource manager may have to negotiate with trade unions for conditions of employment and salary increases. Trade unions may be collectively representing the employees or may be called in to assist with the mediation of grievances that an employee may bring against the company.

9.4.11 Environment

Human resource managers need to be aware of the environment in which employees work. This has the potential to impact on the efficiency of an organization and to affect profits. Concern for the work environment is the area where the role of the facilities and human resource managers become interwoven. The work environment is of concern to the human resource manager since there is a perceived link between productivity and the workplace. This assumed relation becomes increasingly important as in many service sector industries salaries and wages constitute some 80% of the costs of doing business. If the workplace is restricting productivity, then this translates into significant inefficiency.

Inefficient premises lead to inefficient organizational performance. An example may be seen in a resort hotel where gold plated tapware was installed in their 400 bedrooms rather than the brushed stainless steel as recommended by the designers. Unlike polished stainless steel the gold plate required daily polishing to remove fingermarks. Assuming that each set of taps required 20 seconds to polish, with three sets of taps in each room, the result is that one extra employee is required just to clean taps!

The link between the workplace and productivity is, however, difficult to demonstrate outside of a manufacturing context. When productivity can be measured by the number of items produced, or key strokes on a computer in a given time, changes in facility variables can be compared with their effects on output. With 'knowledge work', however, there is no output measure; the loss of one good idea could be literally 'immeasurable'. In the absence of a more definitive understanding of these interrelationships a more 'econometric' approach can be taken.

The Marans and Sprekelmeyer (1982) model depicted in Figure 9.1 may be used to explain the interaction between the physical workplace and the managerial environment. At the centre of the model the building user is deemed to have individual characteristics that are located in an organizational context, i.e., the employer's workplace. Both the physical and managerial environments of this workplace have measurable characteristics.

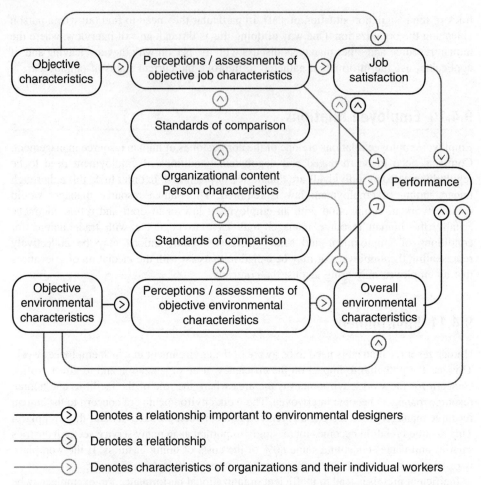

Figure 9.1 The Marans and Sprekelmeyer model of workplace environment (1982).

However, the way that these objective characteristics are perceived is tempered by the user's standards of comparison, which are further tempered by their psychological state. Thus it is not unusual to find highly motivated staff working well in a poor quality environment without complaint, and poorly motivated staff complaining about conditions in an excellent building. In one case complaints of a sick building coincided with the 'downsizing' of the staff. The complaints could not be substantiated by direct measure, but stopped once the downsizing had been completed.

It is the role of the facility manager, in conjunction with the human resource manager to ascertain whether the employees' job satisfaction and productivity are being affected by external factors to do with the environment. One tool that is available for relating workplace factors with productivity is the Dillon and Visher (1987) User Satisfaction Survey Instrument. This survey links user satisfaction and (self-assessed) productivity to the environmental factors of thermal comfort, acoustic privacy and other issues. This survey was expanded by Crosby (1996) to include building health elements.

The Crosby survey questionnaire is shown in the Appendix, along with an illustrative chart on which the results are plotted. Whilst the survey is necessarily simplistic, it is intended that it be used longitudinally over time. It is the deviations of the most recent results from the benchmarks built up from previous studies that are of importance. The average values for the environmental issues are plotted on the chart that also shows the 95% and 99% confidence zones (calculated from the standard deviation). This means that if a surveyed average falls outside the confidence zones the facility manager can assume that there is a problem rather than just a quirk in the sampling.

The survey asks respondents to rate their satisfaction with a number of issues by entering scores of between 1 and 5 – generally, a score of 3 indicates no opinion one way or the other. The questions are analysed in groups, producing a score in 11 areas:

- temperature comfort
- ventilation comfort
- internal noise
- spatial comfort
- privacy
- lighting
- external noise
- neuro-specific health symptoms
- allergic reaction symptoms
- overall satisfaction with the workspace
- ability to do one's work (self-assessed).

One further question identifies asthmatics to ensure that differences in the sample do not skew the results. Managers need to monitor these results for any significant changes, allowing pro-active measures to be instituted before problems surface. Sick building syndrome, for example, is a crucial issue.

Where an untoward change in satisfaction or ability to work becomes apparent, the facility manager can then return to the primary data to see which of the issues, and the factors behind them, are perceived by the staff as the underlying problem. Facility managers can then allocate scare resources to tackling the problem. Moreover, the facility manager will have objective arguments that can be put to resource allocaters to leverage the necessary funding for remedial action. Human resource managers can work with facilities managers in this case to assist with improving the workplace environment.

9.5 Conclusion

Human resource management is the management of people. It is specifically aimed at getting the best effort from people in an organization, getting the best people for the job initially and providing the environment required that will enable people to give their best. Theoretical approaches to human resource management have assisted the development of human resource practices, particularly in relation to motivation, administration and rewards. Human resource functions are complex and varied, and include such factors as devising training strategies, planning recruitment, monitoring performance, preparing employment contracts and ensuring that employees are satisfied with their jobs. An important part of the human resource manager's role is dealing with the environment in which employees work as this has the potential to impact on the efficiency of the employees. Human resource managers can assess the workplace environment in a variety of ways and it is at this point that the role of the human resource manager and the facilities manager become interwoven. Both play a crucial role in ensuring job satisfaction and employee efficiency, key issues of concern in human resource management.

References and bibliography

Bennet, F.L. (1996) *The Management of Engineering* (New York: John Wiley).

Crosby, A. (1996) *A Tool for Measuring Sick Building Syndrome: Development & Application.* Final Report to Health Research Council (University of Auckland).

Dillon, R. and Visher, J. (1987) User Manual Tenant Questionnaire Survey, Public Works Canada. *Architectural & Engineering Services*, AES/SAG 1–4 November, 87–8.

Environmental Protection Agency (1991) *Building Air Quality: A Guide for Building Owners and Managers.* Environmental Protection Agency, EPA/4001–91, DHSS (NIOSH).

Graham, H. and Bennett, R. (1995) *Human Resource Management* (London: Pitman Publishing).

Harrison, R. (1993) *Human Resource Management* (Reading, MA: Addison-Wesley).

Herzberg, F., Mausner, B. and Snyderman, B.B. (1959) *The Motivation to Work* (New York: John Wiley).

Inkson, K. and Kolb, D. (1998) *Management – Perspectives for New Zealand* (New Zealand: Addison-Wesley Longman).

Kraatz, J.A. (1985) *Air Quality, Air Conditioning and Health.* Scholarship Report (University of Queensland).

Langford, D., Hancock, M.R., Fellows, R. and Gale, A.W. (1995) *Human Resources Management in Construction* (London: Longman Scientific & Technical).

Leifer, D. (1997) Monitoring Sick Building Syndrome in Auckland University Buildings, In: *Proceedings of the 31st Annual Conference of the Australia & New Zealand Architectural Science Association*, Brisbane, pp. 153–6 (Pictorial Press).

Marans, R.W. and Sprekelmeyer, K. F. (1982) Measuring overall architectural quality: a component of building evaluation. *Environment and Behaviour*, **14** (6) November, 652–70.

Maslow, A.H. (1987) *Motivation and Personality* (New York: Harper & Row).

McGregor, D. (1960) *The Human Side of Enterprise* (New York: McGraw-Hill).

Occupational Health Working Environment Series No. 15 (1981) *Clean Air at Work* (Australian Government Publishing Service).

Payne, A. (1996) *Management for Engineers* (New York: John Wiley).

Standards Australia (1991) *Australian Standard 1668 Part 2: Mechanical ventilation for acceptable indoor air quality* (Standards Australia).

Taylor, F.W. (1967) *Principles of Scientific Management* (New York: Norton).

Thomason, G. (1988) *A Textbook of Human Resource Management* (London: Institute of Personnel Management).

Wilkinson, S.J. (1996) The factors affecting the career choice of male and female civil engineering students in the UK. *Career Development International*, **1** (5), 45–50.

Appendix

APPENDIX The Survey Questionnaire
(based on Dillon and Visher, 1987, modified by Crosby, 1996)

Please rate the following attributes of your particular work location in this building by circling the number between 1 and 5 that best summarizes your experience in working here.

		1	2	3	4	5
1.	temperature comfort	bad				good
2.	how hot it gets	too hot				comfortable
3.	how cold it gets	too cold				comfortable
4.	temperature shifts	too frequent				constant
5.	ventilation comfort	bad				good
6.	air freshness	stale air				fresh air
7.	air movement	stuffy				circulating
8.	noise distractions	bad				good
9.	background office noise level	too noisy				comfortable
10.	specific office noises (voices and equipment)	disturbing				no problem
11.	furniture arrangement in your work space	bad				good
12.	amount of space you have in your workplace	bad				good
13.	work storage	insufficient				adequate
14.	personal storage	insufficient				adequate
15.	visual privacy at your desk	bad				good
16.	voice privacy at your desk	bad				good
17.	telephone privacy at your desk	bad				good
18.	electric lighting	bad				good

	APPENDIX The Survey Questionnaire (continued)					
19.	how bright the lights are	**1** too bright	**2**	**3**	**4**	**5** too dull
20.	glare from the lights	**1** high glare	**2**	**3**	**4**	**5** no glare
21.	noise from air systems	**1** disturbing	**2**	**3**	**4**	**5** no problem
22.	noise from office lighting	**1** buzz / noisy	**2**	**3**	**4**	**5** no problem
23.	noise from outside the bulding	**1** disturbing	**2**	**3**	**4**	**5** no problem

Please rate the following conditions of your health while you are at your workstation. (Normal would refer to what you would consider your usual state of health when not at work)

24.	skin (poor indicating what you consider to be excessive levels of irritation, itchiness, dryness, reddening, etc)	**1** poor	**2**	**3**	**4**	**5** normal
25.	eyes (poor indicating what you consider to be excessive levels of irritation, itchiness, dryness, watering, etc)	**1** poor	**2**	**3**	**4**	**5** normal
26.	nose (poor indicating what you consider to be excessive levels of irritation, sneezing, congestion, runniness, etc	**1** poor	**2**	**3**	**4**	**5** normal
27.	throat (poor indicating what you consider to be excessive levels of dryness, irritation, coughing, etc	**1** poor	**2**	**3**	**4**	**5** normal
28.	chest (poor indicating what you consider to be excessive levels of breathing difficulty, wheezing, tightness, etc	**1** poor	**2**	**3**	**4**	**5** normal

To what extent do you experience the following whilst at your work station?

29.	headaches	**1** regularly	**2**	**3**	**4**	**5** never
30.	lethargy	**1** regularly	**2**	**3**	**4**	**5** never
31.	tiredness	**1** regularly	**2**	**3**	**4**	**5** never
32.	do you suffer asthma?	**1** yes	**2**	**3**	**4**	**5** no
33.	how would you rate your overall satisfaction with your workplace?	**1** dissatisfied	**2**	**3**	**4**	**5** very satisfied
34.	please rate how this space affects your ability to do your work	**1** makes it difficult	**2**	**3**	**4**	**5** makes it easy

Please return this form to _____   **Thank you**

Figure 9.A1 The survey questionnaire.

Environmental issue	1	2	3	4	5
thermal comfort			▭●▭		
air quality		▭●▭			
noise control			▭●▭		
spatial comfort			▭●▭		
privacy			▭●▭		
lighting				▭●▭	
building noise control				▭●▭	
specific symptoms				▭●▭	
neurotoxic symptoms			▭●▭		
Overall satisfaction			▭●▭		
ability to work			▭●▭		

Generic benchmark for comparing building user satisfaction survey results.

This chart is based on illustrative results from several hundred Australasian respondents (Leifer, 1997).

thermal comfort	=	average responses of Q1 + Q2 + Q3 + Q4
air quality	=	average responses of Q5 + Q6 + Q7
noise control	=	average responses of Q8 + Q9 + Q10
spatial comfort	=	average responses of Q11 + Q12 + Q13 + Q14
privacy	=	average responses of Q15 + Q16 + Q17
lighting	=	average responses of Q18 + Q19 + Q20
building noise control	=	average responses of Q21 + Q22 + Q23
specific symptoms	=	average responses of Q24 + Q25 + Q26 + Q27 + Q28
neurotoxic symptoms	=	average responses of Q29 + Q30 + Q31
overall satisfaction	=	average responses of Q33
ability to work	=	average responses of Q34

Key

averaged value

95% confidence (1.6 SD) N > 50 respondents

99% confidence (2.3 SD) N > 50 respondents

Figrue 9.A2 Survey results and benchmarks.

10

Financial management

Craig Langston*

Editorial comment

Effective financial management lies at the heart of achieving outcomes that represent value for money. It pervades all aspects of facility management at all stages of the property life cycle, and therefore is a fundamental competency. Financial criteria are normally used to measure business success and to benchmark performance. In modern times when managers are required to continually do more with less and budget cuts are sadly routine, it is more important than ever to identify areas of potential saving that can reduce cost without lowering quality standards or increasing risk exposure.

Financial management is a complex balancing of costs and benefits in order to maximize return to society, investors and other stakeholders. It involves estimating, economic forecasting, quantification, adjustment, accounting, reporting, monitoring and benchmarking activities as well as an understanding of the broader business context. Investment analysis, project cost control and operational reporting underpin financial decisions and ultimate success indicators. Feasibility studies, budgets and life-cost studies are key techniques for facility managers.

The interesting thing about facility management is that it involves long-term financial decisions that are both complex and illusive. The effect of time is therefore an important factor that cannot be overlooked. A discounting methodology is employed to disadvantage costs and benefits in future time periods so that they can be equated with present-day values. A life-cost methodology is used to help make decisions on available alternatives and to plan for operational resources and expenditure timing. Cost centres are valuable in the management of actual expenditure and to compare against budget targets. Collectively these tools enable proper financial management and apply equally to equity or debt funding sources.

Taxation considerations are part of the overall financial picture and comprise issues such as deductions for operating expenditure and depreciation of capital works. They

* Deakin University, Geelong, Australia

affect decisions by enabling advantage through realization of government incentives, rebates and allowances. Capital gains tax is also important, particularly when developing and selling property at a profit. Facility managers, while not being taxation experts, must have an appreciation of these matters if they are to make informed decisions regarding acquisition, leasing and disposal of assets.

Benchmarking is a well-recognized technique for comparing performance against best practice within a specific industry. Facility managers routinely use benchmarking to highlight areas of potential improvement and to demonstrate successful achievements. Comparing against the 'best of breed' is not dissimilar to setting hurdles to drive better performance outcomes, but the biggest advantage is being able to identify areas of poor performance, and then, by devoting additional resources to those areas, restructuring them and increasing overall value.

10.1 Introduction

Financial management is a critical ingredient in the effective deployment and operation of facilities. Few managers are able to divorce themselves from either financial constraints or the overarching aim of delivering value for money to their organization or to the community to which their organization contributes. Whether public or private sector, financial imperatives exist and must be respected. Not only is it important to make correct decisions about new initiatives, but issues of budgeting and controlling cash outflows, and ensuring that limited funds are used wisely, are of equal concern.

This chapter deals with a very complex topic by examining the principles of financial management for both new initiatives and routine activities, and perhaps more importantly, the link between them. There is a wide range of tools that can be applied at various stages of a facility's life cycle, applicable for various purposes, yet there are principles which pervade the entire process and tie everything back to a 'bottom line' focus. However, rather than financial management being seen as a constraint or an onerous obligation, it can also be viewed more positively as an opportunity to explore options, to improve performance output per unit of resource input, and to introduce rigour into the decision-making process that can highlight both competitive advantage and risk exposure.

10.2 Strategic planning

All financial decisions start and end at a strategic level. New initiatives are born out of a strategic need for progress, measured in the context of profitability, productivity, sustainability and/or quality enhancement objectives. Strategic planning is an idea generation process, but nevertheless relies on a thorough understanding of the issues, constraints and externalities that apply. It is therefore part creative and part rational, requiring imagination, vision, market knowledge and risk assessment. Financial judgement underpins all deliberations.

It is common to separate the creative and the rational, so that the latter does not dominate and lead to stereotyped solutions and a lack of innovation. Idea generation can be assisted by tools like brainstorming, which are purposely applied so as not to introduce criticism, judgement or participation barriers in the early stages. The outcome of a strategic planning process is a series of options that have some likelihood of satisfying identified objectives.

Profitability is a common objective. This may be translated more specifically into goals such as expanding market share for an existing product, opening branch offices in other countries, increasing shareholder wealth through new investments and the like. Other generic objectives such as productivity, sustainability and quality enhancement also usually have a profitability aspect, therefore many of the evaluative tools that are later applied to potential new initiatives, with a view to ranking and prioritizing actions, are monetary based.

At the other end of the sequence, past initiatives are subject to review, perhaps ultimately resulting in corrective action being taken. Actual performance thus leads to strategic decisions about continuity, replication, expansion or transfer as may be considered appropriate, not only now, but at future times when the environment within which the performance is being measured is itself subject to change. Performance assessment is often monetary-based, or might become so when conditions dictate.

In between this notion of conception and review lies a vast range of routine activities, involving service provision, fine-tuning, realignment, adaptation and renewal. Often less strategic in nature, although perhaps tactical, and certainly operational, these activities are continual and integrated with other organizational functions. Some measure of financial control is provided through budgets, cash flows, cost plans, expenditure audits and general accounting instruments to ensure that value for money (or financial efficiency) is an ongoing commitment.

As one moves from creativity to rationalization, issues like financial return, benefit-cost ratio, payback period and investment risk take on increased importance. Judgement is required to assess the results of objective analyses and weigh up the advantages and disadvantages before taking the next step. Objective considerations are often financial and expressed in money, and in some cases could include social and environmental attributes. The combination of objective and subjective aspects may be quite formidable and require considerable wisdom, as well as experience and methodological understanding.

But ultimately the process gives rise to a short list of options from which one (or more) is selected and implemented. Before any such conclusions can be drawn, however, an analysis of customer requirements must be made. This may be a market analysis to measure demand for a new product or service, or to identify organizational needs that are to be addressed. Either way, failing to assign this issue adequate investigation time will undermine the credibility of key assumptions used in the financial analysis.

10.3 Market analysis

Marketplace demand and customer needs are vital inputs to any financial study. They are expectations of the future climate within which new initiatives will operate. Growth in demand will create confidence in capital investment and the likelihood of good returns, whereas falling demand will add to risk and uncertainty. The built environment is renowned for being sensitive to situations of economic growth and slowdown or recession, affecting interest rates, investment, rent and other facility demand indicators. Predicting these cycles is, of course, vital.

Demand assumptions are usually formulated through market research. This approach may comprise surveys of customers or facility users about their views, needs, expectations and wants. While this may be done either before or after the generation of ideas for new initiatives, it is typical that supporting evidence is obtained prior to carrying out a feasibility study on identified options.

If the underlying assumptions for demand or need are unreliable, the feasibility study will be of little use. However, this can be obviated to some extent by a thorough risk assessment of the assumptions and the implications of change to base values, whether positive or negative. Following this approach, the analyst can identify a feasible range for each key assumption and make recommendations on the basis of the resultant variances. Given the probabilities of low and high returns, judgement can be exercised as to whether the new initiative is likely to be worthwhile.

In addition to questions of demand or need, there are other assumptions that must be made and tested before a feasibility study can be prepared. These include time horizon decisions, economic forecasts, discount rate selection and taxation impacts. Normally these are reasonable starting points against which variations can be assessed.

Once the market analysis is complete and options have been both conceived and ranked, the task often becomes one of financial evaluation. The generic tool that is used to perform this process is the feasibility study.

10.4 Feasibility studies

There are many types of feasibility studies, but they are all characterized by their methodical assessment of benefits (inflows) and costs (outflows). Benefits are calculated as the dollar worth of the project upon completion and becoming fully operational as a 'going concern'. Costs include all the expenditure that is required to bring the project to completion, including finance and holding charges. The bottom line is usually referred to as the investment profit or development margin.

Feasibility studies are a primary financial management tool. They are used to identify options that are worth implementing from a profitability perspective. Depending on the situation, they have either short-term or long-term time horizons, and quite different approaches are used in either case. This section will discuss short-term studies, while the next section will focus on studies with a longer time frame (normally of more than three years duration).

The aim of any new initiative, from a financial perspective, is to maximize returns by delivering a project with high market value while keeping development costs as low as possible. Projects are feasible if they generate benefits in excess of costs, and the resultant profit is more than what would be realized via other investment opportunities. However, high profit may also entail high risk, so a balanced decision is necessary to ensure that the project matches the attributes of the organization and fulfils intended requirements.

Short-term investments are typically judged on both investment profit and yield. However, risk analysis is necessary to gauge the likelihood of the estimates being realized. The more uncertain the estimate, the higher the profit or yield should be to provide compensation for the risk of loss. Risk management is a key component of feasibility studies and critical where the process includes forecasting future events.

Profit is the amount of money left over after the investment is sold and all development costs have been paid. It is normally expressed as a dollar amount (development margin) and as a percentage (profit) of total development costs. Profit is relevant to short-term investments and must compare favourably with alternative uses of capital.

Investment yield is also used where profit is the basis for decisions and is defined as generated net income divided by investment value. For property transactions, value is interpreted as the sum of land, building, finance and profit; investment yield is, therefore, simply the ratio of net income to total development costs and can be alternatively described as return on investment.

There are two main ways in which a short-term feasibility study is set out. The first, known as the profitability approach, commences with the expected value of the completed project, and deducts the costs of development (including finance costs) needed to achieve it. The remainder is profit and indicates the balance of benefits over costs. Obviously a positive answer is required if the proposed project is to be of interest to an investor. Often such studies include a range of options, and the higher the anticipated profit compared to initial development costs the better the project looks as an opportunity for financial return.

The second method, called the residual approach, is used specifically in facility acquisition scenarios. It calculates the land value based on expected market value less construction, finance and profit. If the result is less than the advertised price for the land, then the potential investor would be interested, and vice versa.

Short-term feasibility studies include specific treatment of interest received and paid, and express all benefits and costs in current dollars. Inflation is generally ignored, although a cost escalation allowance could easily be included, since the time horizon is quite short. As will be seen in the next section, this approach is inappropriate for longer periods, and a discounted cash flow (DCF) approach is then required.

10.4.1 Discounted cash flow (DCF) analysis

DCF is used for long-term investment decisions. The DCF approach is quite different from the methods described previously because it includes the consideration of the impact of time value. It is also generally regarded as a superior method for longer-term investments than those that ignore time value (such as simple payback and accounting rate of return methods). Nevertheless, DCF can be used for short-term investments by creating monthly cash flows rather than the following the typical yearly approach, and applying a monthly discount factor.

In any case, costs are deducted from benefits to calculate the net benefit per period which is then discounted. The sum of the discounted net benefits indicates whether a project is financially feasible or not. The capitalized value of the property, if relevant, is included as a theoretical sale in the final period of the study. It is a more complex method to use, but it is more relevant to ongoing facility management than the short-term options commonly used by developers and speculators.

Under a DCF approach, investment profit and yield give way to net present value (NPV). Projects are favourable where NPV is positive, and in cases of mutually exclusive investments, the higher the NPV the more attractive the option. Projects are chosen provided they are expected to provide sufficient return to cover expected profit and risk allowances, and thus make them worthwhile. Note that the use of a discounted cash flow approach builds in interest paid and received (or debt and equity commitments) as part of the discount rate and therefore interest should not be separately included in yearly cash flows.

NPV is defined as the sum of discounted benefits less the sum of discounted costs over a given time horizon. The further in the future that costs or benefits are anticipated, the more they are reduced (discounted) so that their impact is effectively lessened. This calculation accounts for the fact that money has investment potential and therefore the same sum of money is more valuable today than tomorrow. At discount rates of 5% or more, costs and benefits incurred or received after 20–25 years are effectively rendered negligible.

Internal rate of return (IRR) is another useful assessment criterion. IRR determines the level of profitability and risk contingency implicit in a project. At a given real discount rate (based on the true time value of money), the difference between the IRR and the discount rate can be considered as equal to the profit and risk allowance. The larger the difference, the more attractive the investment.

Investment ratios such as benefit–cost ratio can also be used to help judge return. They generally apply to DCF applications, but some ratios can also be used in non-DCF applications. Benefit–cost ratio is defined as the sum of the discounted benefits divided by the sum of the discounted costs. A ratio greater than 1 indicates a financial gain; a ratio of 2, for example, indicates that benefits outweigh costs by a factor of two. The ratio is useful because it places the expected return in the context of the cost necessary to generate it and in this regard is similar to the investment yield discussed previously.

A DCF approach by its nature involves a comparison between two or more alternative investments. Most feasibility studies involve some form of comparison, but where no explicit alternative is identified, an implicit comparison still occurs. The discount rate reflects the opportunity cost of investing in the financial marketplace at a secure rate of interest and thus equates to a theoretical 'do nothing' option. A positive NPV is better than doing nothing and a negative NPV is worse.

The literature frequently describes the composition of the discount rate in terms of investment return, inflation and (where relevant) taxation. Formulae are available that allow calculation of the discount rate to any number of decimal places, yet in practice it is common to assess the discount rate rather than to calculate it. This assessment is often based on personal judgement, intuition and previous experience. It is therefore difficult under these circumstances to know exactly what is incorporated in the discount rate and this may have contributed over time to the general uneasiness that many people have about the discounting process.

Government authorities increasingly set discount rates for economic appraisal based on Treasury advice. Three rates are used, called 'test' discount rates, and are based on likely, pessimistic and optimistic scenarios. For example, the advocated discount rate may be 7% but a risk analysis is required using both 4% and 10% to highlight the sensitivity of the investment. If the project is viable at all discount rates then there is some degree of confidence in the proposal.

The discounting process is better served by removing those factors that are not well represented by the compound interest approach. The discount rate will therefore be based on investment return, inflation and perhaps taxation. Profit expectations and risk contingencies are dealt with separately. Such a strategy, apart from being conceptually correct, will demystify much of the theory surrounding the discounting technique and will lead to greater understanding and confidence.

It is recommended that the discount rate should be equal to the after-tax weighted average of equity and borrowed capital applicable to the project being appraised where the weighted average is calculated as the equity rate of interest multiplied by the proportion of equity funds used in the project, plus the borrowing rate of interest multiplied by the proportion of borrowed funds used in the project, e.g., 5% equity for 80% of needs and 10% borrowing for 20% of needs gives a weighted average of 6%. The individual cash flows should include income tax deductions, depreciation and other taxation concessions or liabilities but must exclude all interest received or payable. Individual cash flows should preferably be expressed in present-day terms (real terms) and if so the discount rate must be exclusive of inflation. Profit and risk contingencies are assessed external to the discounting process.

The weighted cost of capital, after inflation and taxation have been removed, is unlikely to be high. Under certain conditions the resultant discount rate may even be negative. Although it may seem incongruous that any discount rate can be less than zero, in reality this situation highlights that the after tax investment return is offset by inflation to the point where money is devaluing over time despite its productive use. Future cash flows therefore increase in comparative value.

It must be remembered that the discounting process is merely a technique available for comparative purposes that disadvantages future costs and benefits relative to present values. Discounting is based on the compound interest principle in reverse and thus can be described as a negative exponential. Over long time periods the effect, particularly at high rates of discount, can be such as to make future costs and benefits irrelevant.

The following are recommendations about the factors that influence the choice of discount rate and how they should be treated:

- the discount rate should be based on forecasts of investment return, inflation and taxation appropriate to the selected study period
- interest payable on borrowings or interest lost on use of equity should not be explicitly included in the cash flow forecasts but dealt with by the discount rate
- discount rates should be real (i.e., inflation should be removed) and the cash flow forecasts should be expressed in present-day terms
- taxation deductions, depreciation and liabilities, where relevant, should be explicitly included in the cash flow forecasts and the discount rate should be net after tax
- profit should not be built into the discount rate but its adequacy should be judged by the difference between the discount rate and the internal rate of return
- risk should not be built into the discount rate but should be assessed separately using one or more specialist risk analysis techniques
- the discount rate should be calculated as the after-tax weighted average of equity and borrowed capital applicable to the individual projects being appraised.

The discount rate is project-related, not purely market-related as is generally understood in the literature, and therefore may differ between competing investment opportunities. Furthermore, the use of artificially high rates of discount unreasonably distorts the equitable balance of initial and subsequent cash flows.

The above approach to feasibility studies assumes that the benefits and costs that are estimated in each year reflect the tangible inflow and outflow of cash for the project. The study is therefore investor-centred and calculates the financial impact of the project on the investor. Where a broader social impact is to be evaluated, the concept is expanded, and the feasibility study is usually known as a cost–benefit analysis (CBA).

10.4.2 Cost-benefit analysis (CBA)

An investor-centred feasibility study can be called an (economic) CBA. However, the inclusion of social and environmental issues that are external to the investor or providing authority give rise to what is commonly called a social CBA.

The only real difference between the two types is the nature of the costs and benefits themselves. A social CBA aims to identify net social gain, regardless of who are the winners and who are the losers, so that if a positive NPV results then the project is seen as potentially worthwhile. Benefits therefore are not cash inflows to an investor or providing authority, but represent advantageous outcomes that may or may not be tangible cash flows. A similar case applies to costs, but in reverse. Examples of social benefits may be time savings from shorter travel journeys for commuters or improvements in patient health care from a new hospital, while examples of social costs include environmental pollution arising from a power plant or traffic congestion from a new road diversion.

Social CBA uses the concept of collective utility to measure the effect of an investment project on the community. Intangibles and externalities are assessed in monetary terms and included in the cash flow forecasts, even though in many cases there is no market valuation available as a source for the estimates. Nevertheless the technique is useful in that it attempts to take both financial and welfare issues into account so that projects delivering the maximum benefit can be identified and selected.

The technique has come under attack from conservationists who claim it frequently estimates environmental goods and services at a level below their true social value. It is used to assess sustainable development and to comment on the effective deployment of scarce resources, but it can be manipulated and used for political justification.

Despite its shortcomings, social CBA is still widely used in both developed and developing countries. Other techniques, such as multicriteria analysis, offer advantages to some of the issues of objective and subjective assessment, but may introduce different problems. Much research is being conducted in this area with a view to linking issues of profitability, productivity, sustainability and quality enhancement into a common methodology with a single decision criterion.

10.5 Budgeting

The outcome of a feasibility study, by whatever name, is the recommendation of a course of action to be pursued. This is not only the result of numeric calculations involving costs and benefits, but also the investigation of risk exposure and probability of the risk occurring. Once a recommendation is made and accepted, however, the focus of financial management shifts from evaluation to control processes and the first of the new tools to be used is budgeting.

Budgeting is the disciplined pre-determination of cost, and as such is a significant element of planning and control for any business or project. It can deal with benefits or costs with equal capability, and portrays the likely consequences of associated design, management or operational decisions in terms of dollars. It also identifies major cost factors and areas of risk. It is the first step in the overall planning and control process for any activity.

Accountants readily recognize the importance of budgeting in business concerns. In these cases the budget provides management with the right tool to make effective use of the capital at its command. The basic budget is seen by management as both an early warning system and a frame of reference for evaluating the financial consequences of operational decisions. The degree of sophistication in a company's budgeting system will usually depend on its size, but even small businesses will find that an elementary level of budgeting will assist in the control of the business.

The rising costs of facilities in recent years has emphasized the need for careful control of cost. The budget is the mechanism by which control can be achieved. It provides a standard for comparison, without which there can be no benchmark of financial performance.

The budget becomes the mechanism within which the facility management team must work to achieve their objectives. It provides a considered measure for comparison and judgement of actual results and so is useful in controlling all aspects of cost both now and in the future. Budgets should, however, be realistic – the setting of unrealistically low budgets may initiate a large amount of investigation into new methods or ways of doing things, but where this is too difficult it may lead, in the end, to an overrun of cost or a reduction in quality. Setting excessively low budgets to force change can be counter-productive.

The process of budgeting should be understood within the organization at all levels. An ideal arrangement is that the process should support the following principles:

- transparency of information
- dissemination of divisional budgets throughout the organization
- consistency of budgeting process from one year to the next
- accountability of performance
- safety net mechanisms for divisions when things go wrong
- reward for efficiency
- equity
- reward for entrepreneurialism or innovation
- collective decision making.

A culture needs to be developed where allocated funding does not have to be spent or lost, but in which divisions can save for future years to cover major items of expenditure. This is particularly important for the facility management division.

Budgets therefore can apply to new initiatives (strategic) and to routine activities (tactical and operational). They share a common methodology and purpose. They not only establish a framework within which subsequent performance can be judged, but they drive the search for new solutions that can deliver required outcomes at lower cost with equal or higher quality. Financial management is, therefore, once again concerned with evaluation and idea generation, but at a more detailed level, and within the parameters established by earlier feasibility studies.

Budgets should not be routinely changed – once set, they should be used as a benchmark for measuring actual performance. It is pointless to continually revise budgets so that they reflect the latest thinking and directions, for at the end of the day there is nothing to conclude other than that the budget was satisfied, which in such cases is a hollow victory. Instead organizations can learn from historical experiences of being under or over budget after a period of time or after a project is completed, and perform better next time round.

10.6 Cost planning

If budgets are the benchmarks, then cost plans are the action statements that guide decisions to achieve agreed targets. A cost plan can really take any form: for facility construction it may be based on elements of the building such as upper floors external walls, roof, and so on, while for customer service it may be based on human resources such as personnel requirements, workload and duties. For occupancy considerations, it may include recurrent costs such as energy, cleaning, supplies, maintenance and other commitments. It can apply to a fixed-term project, to a year of operations or to many years of facility ownership.

Cost plans are used to guide and to inform. As the name suggests, they are about costs and therefore do not normally address issues of income in broader feasibility or budgetary processes. They list items, usually quantified in some way and priced to reflect the likely total cost. Sometimes this can be translated into cash flows so that the timing of expenditure is also identified. Cost plans can be issued, revised, reformulated and reissued as necessary to reach solutions that fall within previously set budget parameters.

For built facilities, cost plans should include both capital and operating costs. The total is usually described as life cost (or life cycle cost) and illustrates the cost consequences of construction, cleaning, energy, maintenance and replacement activities over the life of the facility. Life-cost planning is emerging as a popular tool in overall financial management and offers the opportunity to optimize total cost by perhaps increasing initial expenditure with a view to lowering annual expenditure, which over a number of years may produce a significant saving.

Value management (VM) is a useful technique when the purpose is to find better ways of achieving a particular goal (function). It is a structured yet otherwise commonsense approach to systematically investigating a problem, generating ideas about possible solutions, evaluating them and making recommendations. The intended outcome is to increase value for money, or in other words, improve function and/or lower cost so that the ratio of function to cost is increased.

VM is typically a qualitative technique that uses weighted scores to rank alternatives based on predetermined performance criteria. The technique can, however, be enhanced by separating the subjective (functional) issues from the objective (financial) ones. In this case the former is expressed as a value score and the latter as a total cost or cost per unit. By dividing the value score by the total cost or cost per unit, a value for money index is created. The option that has the highest index represents the solution that has the best ratio of benefit to cost. In these types of analyses, cost should be inclusive of both capital and operating expenditure.

VM is typically a team-based exercise. It draws on a wide range of experiences and knowledge and integrates them so that there is potential for innovative or 'lateral' solutions to be created. The recommendations that flow from such a study can inform decisions that are ultimately recorded in cost plans and other planning instruments.

10.7 Cost control

Budgets and/or cost plans essentially act as targets that guide and help control activity costs. While the overall budget may be considered as a maximum limit, the distribution of funds within the various divisions or categories can be considerably more flexible. Therefore an unexpected overrun in one area can be funded by finding a saving in another area, so that the net result is no change.

Actual performance therefore has to be closely monitored and continually compared to budget forecasts. This process is called cost control. Differences between actual and planned expenditure require review and perhaps corrective action. Monitoring also enables future budgets to be improved.

Regular (typically monthly) comparisons of performance are vital. Performance indicators can assist in highlighting those areas that are going well and those that are not. The financial planner or facility manager needs to continually act on this information and fine tune operations so that final targets are achieved. If no action is taken, small problems can turn into big ones and the potential for recovery is diminished.

Senior management may also be interested in monthly comparisons of performance. Where overruns are incurred, plans for redressing the situation may require formal submission and endorsement. In some organizations, cost information is online for all managers to access, giving an up-to-date picture of financial performance and status.

Routine data collection concerning actual costs and the reasons for variations to original targets is an important activity that assists future budget preparation, cost estimation and risk assessment. It is important that it be collected in a form that can be properly interpreted and adjusted to different time periods and qualitative comments about special events that gave rise to unusual costs should be noted.

Monitoring costs without follow-up action is unlikely to have significant benefit. The key is being able to identify a problem, formulate a solution and implement it, and so minimize the impact of the problem. Action plans, which may in fact incorporate comparisons across a range of possible options, are developed to resolve problems and to manage implementation.

10.8 Feedback and performance assessment

The final step in the process of financial management is both reflective and informative. It is important that lessons are learnt from past decisions and actions so that future performance will be improved. A loop must be formed that enables the benefit of hindsight to feed back into the early stages of decision-making on other projects and activities in succeeding years. This must be a formalized loop, and methods for accessing and incorporating such information are critical.

Feedback is useful for a whole range of purposes, such as generating ideas for strategic planning, market analysis, feasibility studies, discounted cash flow analysis, CBA, budgeting and cost planning. The data arises as part of the process of cost control, but has numerous applications. Feedback may be quantitative (monetary based) or qualitative (performance based) as both are relevant to effective decision-making and the pursuit of value for money goals.

In the case of built facilities, post-occupancy evaluation (POE) is an effective feedback device. Through the investigation of actual performance, perhaps through structured interviews with staff, customers and other users, the dissemination of questionnaires about such things as workplace design, service quality, comfort or even physical measurements of indoor air quality, temperature, lighting and the like, useful knowledge can be gathered. It can be used to improve performance for the facility from which it was collected, as well as to inform designers and decision-makers about new facilities yet to be procured.

10.9 Conclusion

Financial management applied to facilities is a complex process. It demonstrates the need for integrated systems to assist managers and the acquisition of sophisticated competencies and skills. The processes discussed in this chapter are specific to facilities and should be viewed in the context of conventional accounting procedures that attend to overall business management.

The linkage between the various financial tools and procedures follows a time line from the conception of new initiatives to the control of operational consequences. Investment analysis, procurement and occupancy reflect the generic life cycle of facilities, yet there is actually a continual micro-cycle of renewal within this framework that financial processes must address. Facility management in this context is a broad and all-encompassing discipline.

It is recommended that facility managers possess appropriate financial capabilities to perform the full range of tasks outlined previously, or at least to effectively participate in and understand the output provided by specialist consultants. In many parts of the world, quantity surveyors are taking a leading role in the financial management of facilities and the economic consequences of their construction and operation as applied to society at large.

References and bibliography

Alexander, K. (1996) *Facilities Management: Theory and Practice* (E. & F.N. Spon).

Ashworth, A. (1999) *Cost Studies of Buildings*, 3rd Edition (Longman).

Atkin, B. and Brooks, A. (2000) *Total Facilities Management* (Blackwell Science).

Barrett, G.V. and Blair, J.P. (1982) *How to Conduct and Analyze Real Estate Market and Feasibility Studies* (Van Nostrand Reinhold).

Barrett, P. (1995) *Facilities Management: Towards Best Practice* (Blackwell Science).

Bernard Williams Associates (1999) *Facilities Economics* (Building Economics Bureau Limited).

Boardman, A.E. (1996) *Cost–Benefit Analysis: Concepts and Practice* (Prentice-Hall).

Bowers, J.K. (1997) *Sustainability and Environmental Economics: An Alternative Text* (Longman).

Bull, J.W. (1992) *Life Cycle Costing for Construction* (Thomson Science and Professional).

Clifton, D.S. and Fyffe, D.E. (1977) *Project Feasibility and Analysis: A Guide for Profitable New Ventures* (Wiley).

Cotts, D.G. (1999) *The Facility Management Handbook*, 2nd Edition (AMACOM).

Daly, H.E. (1997) *Beyond Growth: The Economics of Sustainable Development* (Beacon Printing).

Damiani, A.S. (1998) *Moving up the Organization in Facilities Management: Proven Strategies to Increase Productivity in your Workforce* (Scitech Publishing).

Dell'Isola, A.J. and Kirk, S.J. (1995a) *Life Cycle Costing for Design Professionals*, 2nd Edition (McGraw-Hill).

Dell'Isola, A.J. and Kirk, S.J. (1995b) *Life Cycle Cost Data*, 2nd Edition (McGraw-Hill).

Department of Finance (1991) *Handbook of CBA* (Australian Government Publishing Service).

Diesendorf, M. and Hamilton, C. (1997) *Human Ecology, Human Economy* (Allen & Unwin).

Fabrycky, W.J. and Blanchard, B.S. (1991) *Life-Cycle Cost and Economic Analysis* (Prentice-Hall).

Ferry, D.J., Brandon, P.S. and Ferry, J.D. (1999) *Cost Planning of Buildings*, 7th Edition (Blackwell Science).

Field, B.C. (1997) *Environmental Economics: An Introduction*, 2nd Edition (McGraw-Hill).

Flanagan, R. and Norman, G. (1983) *Life Cycle Costing for Construction* (Surveyors Publications).

Flanagan, R. and Tate, B. (1997) *Cost Control in Building Design: An Interactive Learning Tool* (Blackwell Science).

Flanagan, R., Norman, G., Meadows, J. and Robinson, G. (1989) *Life Cycle Costing: Theory and Practice* (BSP Professional Books).

Gilbert, B. and Yates, A. (1989) *The Appraisal of Capital Investment in Property* (Surveyor Publications).

Gilpin, A. (2000) *Environmental Economics: A Critical Overview* (Wiley).

JLW Research and Consultancy (1992) *Capitalisation and Discounted Cash Flow Valuation: Bridging the gap* (JLW).

Kolstad, C.D. (2000) *Environmental Economics* (Oxford University Press).

Langston, C. (1991a) *The Measurement of Life-Costs* (NSW Department of Public Works).

Langston, C. (1991b) *Guidelines for Life-Cost Planning and Analysis of Buildings* (NSW Department of Public Works).

Langston, C. (1996) Life-cost studies. In: *Environment Design Guide (GEN10)* (Building Design Professions, Australia).

Langston, C. and Ding, G. (2001) *Sustainable Practices in the Built Environment*, 2nd Edition (Butterworth-Heinemann).

Langston, C. and Lauge-Kristensen, R. (2002) *Strategic Management of Built Facilities* (Butterworth-Heinemann).

Layard, R. and Glaister, S. (eds) (1994) *Cost–Benefit Analysis*, 2nd Edition (Cambridge University Press).

McGregor, W. and Then, D. (1999) *Facilities Management and the Business of Space* (Arnold).

Nagle, G. and Spencer, K. (1997) *Sustainable Development* (Hodder & Stoughton).

Park, A. (1994) *Facilities Management: An explanation* (Macmillan).

Pearce, D.W. (1988) *Cost–Benefit Analysis* (Macmillan).

Price, C. (1993) *Time, Discounting and Value* (Blackwell Publishers).

Robinson, J.R.W. (1989) *Property Valuation and Investment Analysis: A Cash Flow Approach* (Law Book Company).

Roddewig, R.J. and Shlaes, J. (1983) *Analyzing the Economic Feasibility of a Development Project: A guide for planners* (American Planning Association).

Rondeau, E.P., Brown, R.K. and Lapides, P.D. (1995) *Facility Management* (Wiley).

Seeley, I.H. (1997) *Building Economics* (Macmillan Press).

Stevens, D. (1997) *Strategic Thinking: Success Secrets of Big Business Projects* (McGraw-Hill).

Tompkins, J.A. (1996) *Facilities Planning*, 2nd Edition (John Wiley).

van Pelt, M.J.F. (1993) *Ecological Sustainability and Project Appraisal* (Avebury).

Wright, M.G. (1990) *Using Discounted Cash Flow in Investment Appraisal* (McGraw-Hill).

11

Operations and maintenance management

Mohammad A. Hassanain* and Thomas M. Froese*
Dana J. Vanier†

Editorial comment

Another important competency that provides value-adding to built facilities is the co-ordination of operations and maintenance management. Included within this category are energy efficiency, maintenance work and non-core support services such as cleaning, catering, security, fleet management, childcare, library facilities and landscaping. Many non-core services are increasingly subjected to outsourcing decisions to enable core business to remain in focus. Facility managers are often judged on their performance in this area, regardless of whether or not it is undertaken by in-house staff or external contractors, as it is both fundamental and highly visible.

One of the more controversial aspects of operations management is energy planning. Commercial buildings alone are responsible for a significant proportion of total energy demand in developed countries. The cost of energy is borne by the organization from its gross profit, and by society through the provision of electricity generating infrastructure, greenhouse gas emissions, non-renewable resource consumption and polluting waste products. Energy-saving strategies can release funds for more productive purposes without noticeable reduction in the comfort of working environments. As limits to the world's resources draw closer, it is expected that governments will introduce higher compliance standards for energy performance.

Energy efficiency must also consider upstream processes embodied in finished products. While largely a design consideration, embodied energy is relevant to maintenance and refurbishment work undertaken after original construction is completed. Maintenance is a significant activity. It is usually labour intense and, due to the confined nature of the work and its low economy of scale, it is relatively expensive. Initial design choices that lead to low maintenance requirements often have high embodied energy and refurbishment inflexibility.

* University of British Columbia, Canada
† National Research Council, Canada

Maintenance management demands prioritizing required work within available funding constraints. No matter whether a condition-based or preventative approach is taken, maintenance needs to be carefully scheduled so as not to interfere with normal business operations. Data concerning the frequency and cause of necessary repairs can assist greatly with future maintenance planning and cash flow forecasts.

Modern facilities have a number of supporting services that are either essential or desirable to productive workplaces. Security is a good example. This is a non-core service and is often outsourced to external agencies with specific expertise and knowledge. Facility managers are nevertheless responsible for the performance of these support services, which have the potential to add considerable value to an organization if they are properly co-ordinated. Because these types of services are visible to most facility occupants and customers, they take on an operational profile that belies their real strategic importance.

The actual value of service contracts is not reflected by their invoiced cost but by their customer focus. Where control over this aspect is lost, a low negotiated payment is of little consolation. Outsourcing that does not raise the quality of customer service is not worth doing.

11.1 Introduction

A building can be considered as an asset or an investment that needs to be maintained so that its optimal value is retained over its life cycle. Building systems, such as roofing, mechanical or electrical, usually have a much shorter life span than their supporting structures. These services are in constant need of regular maintenance to ensure that they continue to function properly and that they retain their value and good appearance. Maintenance, as per British Standard 3811, may be defined as 'the combination of all technical and administrative actions intended to retain an item in, or restore it to, a state in which it can perform its required function' (BSI, 1984).

Magee (1988) defines facility maintenance as 'the set of ordered activities which, when properly managed, allow for the continual operation of a facility'. While the wording of the definition of the term 'maintenance' may differ from one reference to another, the objectives of carrying out maintenance are universal. Maintenance management objectives may include: extending the useful life of assets, assuring the optimum availability of installed equipment for production and/or services, ensuring readiness of equipment needed for emergency use at all times, ensuring the safety of personnel using facilities, and guaranteeing customer satisfaction (Korka et al., 1997).

In the past, the significance of maintenance was not fully recognized, mainly because nearly all early structures were massive, over-designed, made from natural materials, existed in an unpolluted atmosphere and wore out very slowly (Allen, 1993). Nowadays, buildings have become more complex and the cost of their services forms a major portion of the initial cost, thus affecting the cost of maintenance. The field of facility management has evolved to deal with the complexities of buildings, and their continual use and environments. It is now accepted that the process of facility management commences upon the completion and commissioning of the building, and terminates when the building ceases to be useful and is abandoned or demolished.

11.2 Maintenance information management

Information plays an important role in properly managing any facility. Mottonen and Niskala (1992) reported on the categories of information needed in maintenance management from the findings of research aimed at creating an information system for housing maintenance and management in Finland. These categories include:

- information on the building and the whole property – this information is created at the planning, design and construction stages of the project; in the case of new buildings, it is practical to transfer these data directly from the information system maintained by the planner to the system that would be employed in facility management, including information on the site and buildings, spatial system (room, floor, etc.) and structural and technical systems
- information on the operation and maintenance of the technical systems such as the estimated service life of items of equipment – this information is obtainable from designers, manufacturers and the literature in general; some of the information is created or supplemented in the course of maintenance, e.g., information concerned with the operation and condition of devices and structural elements, user experiences or operating and maintenance costs
- information on the maintenance organization, staff, goals and principles of the organization, maintenance routines and monitoring procedures.

11.2.1 Making decisions in maintenance organizations

With all the information needed at hand, maintenance management decisions are taken at three levels (Aikivuori and Mottonen, 1992; Gordon and Shore, 1998). It can be seen that the need for information is different at different levels. These levels are as follows:

- strategic level – involves the long term planning of facilities, usually beyond a five-year horizon, where decisions are concerned with the operational goals of the organization, their changes and the resources available for achieving them
- tactical (managerial) level – involves the tactical positioning of the organization in the two- to five-year time frame, where decisions are concerned with the acquisition of resources and their effective and productive use
- operational level – concerns itself with the implementation of the strategic and tactical levels in day-to-day operations, normally within the current budget year.

This information is presented in a form of a hierarchical pyramid. The principle behind the pyramid is that information is transferred from the operational level to the strategic level through the tactical level. The decision levels are not necessarily separated within the maintenance organization by distinct boundaries, but can be examined in the form of a continuum. Figure 11.1 is a schematic description of management and information usage on different levels within a maintenance management organization.

Levels at which decisions are made   Information requirements

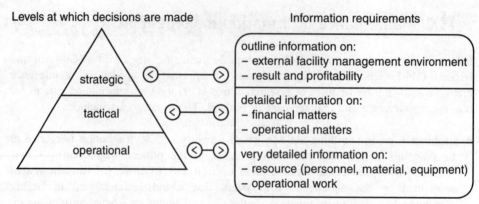

Figure 11.1 The decision pyramid in a maintenance organization (Svensson, 1998).

11.2.2 Classification of maintenance operations

Vanier and Lacasse (1996) reported on the various types of maintenance actions, as shown in Figure 11.2. Preventive maintenance inspection carried out annually or cyclically is termed condition-based inspection, whereas condition-independent inspection is under-taken in accordance with manufacturer's recommendations, for example, or in relation to known deterioration profiles. The inspection may be initiated by a reported failure (response) or by the two preventative maintenance inspection types described earlier. These inspections will necessitate carrying out either normal or emergency action. Emergency maintenance occurs in the event that inaction would create serious safety or health problems. Programmed (or phased) corrective maintenance is needed for failures requiring extensive repairs or where budgets do not permit immediate remedial action.

Then (1982) presented a broad classification of the types of maintenance operations carried out in constructed facilities. Two main generic groups of maintenance works were identified: preventative or planned maintenance and corrective maintenance. Further, he identified separate work categories within each of the two groups. Figure 11.3 illustrates Then's classification of maintenance operations.

11.2.3 Milestones in maintenance management projects

As an example of the typical milestones involved in maintenance of projects, Korka *et al.* (1997) reported on the ten basic steps of the maintenance management programme in the Naval Facilities Engineering Command within the Department of Navy. These steps are as follows:

Preventive maintenance cycle   Inspection type   Action   Maintenance

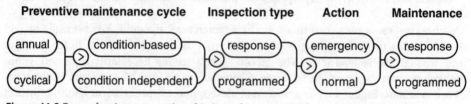

Figure 11.2 Types of maintenance actions (Vanier and Lacasse, 1996).

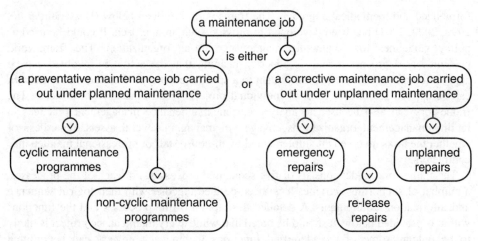

Figure 11.3 Classification of maintenance operations (Then, 1982).

- request – requests to perform maintenance work may be transmitted in different ways, including verbally, via telephone call or by written request
- approve – simple jobs are handled by a maintenance supervisor; when large expenditures are involved, several levels of management approvals are required
- plan – involves planning and estimating a specific work request, ensuring an adequate take-off has been performed
- schedule – involves scheduling the work based on available resources, priorities, and job assignment
- perform work – involves craftsmen executing the work order
- record data – data recording varies from simply listing the actual hours to keeping comprehensive records of material charges, equipment identification, work assigned and performed, and other pertinent data
- account for costs – knowing where and for what purpose the money is being allocated
- develop management information – providing facts on current work, including costs, accumulated data, productivity, equipment identification, job assignments and scheduling
- update equipment history – history records may vary from little or no data to on-line upgrading of all equipment, showing use, downtime, and maintenance labour and material costs expended on pieces of equipment
- management control reports – as management information is developed, control reports covering expenditures, performance, backlog, equipment data and work data can be generated regularly.

11.3 Maintenance management operations process model

Given the complexity of modern buildings and their importance as investments and factors of production for their owners, managing the maintenance of built assets requires a formalized and methodical approach. The framework described below (Hassanain *et al.*, 1999, 2000, 2001) has been developed to provide such an approach. It could be used as policy guidelines for conducting maintenance in an organization. The framework, presented as a process model, is generic, meaning that the activities involved can be applied to non-specific assets, rather than to a specific asset type. Further, the framework can be applied at both the level of individual projects or on a network of projects. The framework can also be used to analyse current maintenance management practices in facility management organizations engaged in managing several assets, regardless of whether the tasks involved are implemented by in-house staff or professional maintenance contractors.

The framework model consists of five sequential processes. For each of the processes, a number of supporting activities have been defined together with their logical sequence and information requirements. A detailed description of the processes and the functions within is provided below. It should be noted that while every maintenance project is likely to be unique, some of the identified functions within each process can be omitted depending on the characteristics of the asset being examined. The five processes forming the framework are as follows:

- Identify Assets (referred to as node A)
- Identify Performance Requirements (node R)
- Assess Performance (node P)
- Plan Maintenance (node M)
- Manage Maintenance Operations (node O).

The generic framework is described schematically as an $IDEF_0$ process model diagram, as shown in Figure 11.4. A process model describes the activities that exist within a business process. It defines the tasks that need to be undertaken within each process, and illustrates how and what information needs to be communicated between tasks (NIST, 1993). A series of interrelated diagrams illustrating information flow from one activity to another, at different level of details, is presented throughout this chapter. A description of the process modelling methodology used is provided in the Appendix.

11.3.1 The 'Identify Assets' model

The 'Identify Assets' process (node A in Figure 11.4) involves carrying out an inventory activity to identify the assets that may require maintenance operations within their service life. An asset may be defined as a uniquely identifiable element or group of elements which has a financial value and against which maintenance actions are recorded (IAI, 1999). Service life may be defined as the actual period of time during which the asset, or any of its components, performs without unforeseen costs of disruption due to maintenance and/or repair (CSA, 1995). Asset data is obtained from asset registers. An asset register is a table that records the assets of an organization, within which the value of all assets can be determined (IAI, 1999). The inputs necessary to carry out the 'Identify Asset' process are an existing facility and a set of resources. The output is a list of assets requiring maintenance. It is envisaged that the

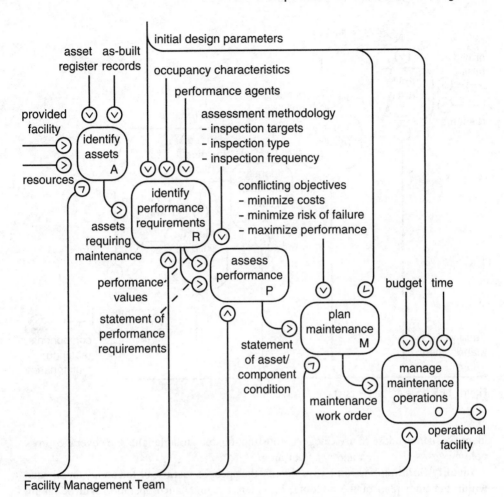

Figure 11.4 General processes involved in maintenance management model.

execution of this process would be optimal in Design–Build–Operate projects, where information on physical assets may be captured and recorded during both design and construction phases – such projects are becoming increasingly common in the construction industry in the UK (Wix *et al.*, 1999) and consequently the opportunities for following this course should also increase.

In identifying a particular asset (e.g., a technical system in a building located within a campus), this process is broken down into five functions as shown in the IDEF$_0$ diagram in Figure 11.5.

The functions involved are:

Identify Facility (A1): identify the facility and/or campus (stock of buildings) in which a specific building is located.

Identify Building (A2): identify the specific building within the facility in which a maintenance operation will be taking place. Data identified may include building

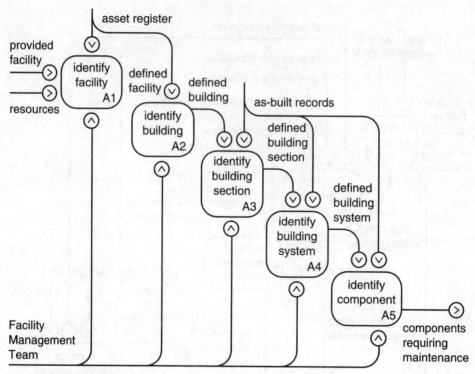

Figure 11.5 Node A: identify assets.

identifier, name, address, use category, construction date, total height, site coverage, gross volume, gross floor area and net floor area.

Identify Building Section (A3): identify the specific section, or the architectural zone within the floor plan of the building, in which a maintenance operation will be taking place. Managing buildings on the building-section level provides a more precise means of evaluating condition and determining maintenance requirements. Data identified may include building section identifier, name, and net floor area.

Identify Building System (A4): identify the systems that may require a maintenance action to restore them to their original condition. Data identified may include building system identifier, type, installation data, and projected service life.

Identify Component (A5): identify the particular components of the system that may require condition assessment, and hence a maintenance action to restore them to their original condition. Data identified may include component identifier, installation date, manufacturer and supplier.

11.3.2 The 'Identify Performance Requirements' model

The 'Identify Performance Requirements' process (node 'R' in Figure 11.4) includes functions required to identify categories of performance requirements of an asset as a unified entity (e.g., a building), as well as the components that make up the assembly of

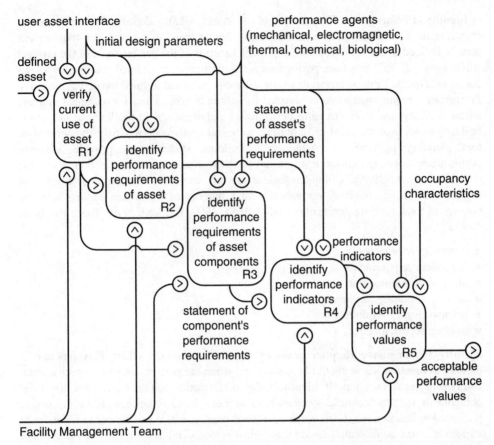

user asset interface

initial design parameters

performance agents
(mechanical, electromagnetic,
thermal, chemical, biological)

defined
asset

verify
current
use of
asset
R1

identify
performance
requirements
of asset
R2

statement
of asset's
performance
requirements

identify
performance
requirements
of asset
components
R3

statement of
component's
performance
requirements

performance
indicators

identify
performance
indicators
R4

occupancy
characteristics

identify
performance
values
R5

acceptable
performance
values

Facility Management Team

Figure 11.6 Node R: identify performance requirements.

the asset (e.g., technical building systems). Performance may be defined as the behaviour of a product related to use (ISO, 1984). The scope of this process extends to identifying performance indicators and their means of expression within each category of performance requirements. The input to this process is a list of assets (e.g., buildings and/ or technical building systems and system components) requiring maintenance which is obtained from asset registers. The outputs are statements of the performance requirements as well as a range of acceptable performance values. This process is broken down into five functions as shown in the IDEF$_0$ diagram in Figure 11.6.

The functions involved are:

Verify Current Use of Asset (R1): verify and/or analyse the current use of the asset and the occupancy conditions against those specified at the beginning of the commissioning phase. At this stage, factors impacting the performance of the asset are identified according to their nature and origin (ISO, 1984). These factors may be mechanical, electromagnetic, thermal, chemical or biological factors. Their origin may be either external to the building envelope, and caused by the atmosphere around the building, or internal to the building envelope, caused by the building occupants.

Identify Performance Requirements of Asset (R2): identify the performance requirements that the asset, as a unified entity, has to meet. A performance requirement may be defined as user requirements expressed in terms of the performance of the product (ISO, 1984). In this function, performance requirements are defined without imposing constraints on the form or materials of the solutions proposed to fulfil these requirements. Performance requirements can be defined for whole buildings and for technical systems within buildings, however, treating performance requirements on the level of the whole building rather than the level of individual technical systems does satisfy the concept of total building performance. Although a building system may provide adequate performance in one dimension it might fail in other areas due to specification or context (Hartkopf *et al.*, 1986). Compiled below are the most representative categories of performance requirements in agreement with a number of references dealing with the concept of total building performance (ISO, 1984, 1992; Hartkopf *et al.*, 1986; Blachere, 1993):

- durability requirements
- fire safety requirements
- air quality requirements
- acoustical quality requirements
- thermal quality requirements
- lighting requirements.

Identify Performance Requirements of Asset Components (R3): This function is parallel to function R2. It should be considered when the performance of a specific asset component is in question. It identifies the performance requirements that the asset components such as technical systems have to meet. In accordance with the discussion presented in function R2, Hartkopf *et al.* (1986) suggest that it is wise not to identify the various measurements with individual technical systems and assemblies, such as roofs or walls, in isolation since it is usually the interfaces of these systems or assemblies which fail as technical systems are designed to their component performance requirements, resulting in the inability of two systems (component-to-component interfaces) to satisfy all the performance requirements specified.

The following example lists the performance requirements for flat, or low-slope, conventional roofing systems as described by Lounis *et al.* (1998):

- water tightness – prevention of water leakage into the building, a requirement ensured by the waterproofing membrane and flashings
- energy control – prevention or minimization of heat (or cooling) exchange between the interior and exterior, a requirement ensured by the thermal insulation
- condensation control – prevention of water vapour condensation within the roofing system using the vapour barrier
- air leakage control – minimization of air leakage through the roof system using the air barrier
- load accommodation – ability to sustain dead and live loads by the structural deck
- maintainability – capability of economic repair.

Identify Performance Indicators (R4): identify the parameters adequate to measure all aspects of performance in a performance category. The diversity of performance requirement categories of assets and/or components defies the definition of a single parameter which is adequate to measure all aspects of performance, so to judge performance effectively each category of performance is considered separately. For example, while some of indicators in the durability requirement category include the existence of deflections, cracks and corrosion, some of those in the fire safety requirement category include duration of evacuation time, survival time, provisions of smoke detectors and exit signs.

Identify Performance Values (R5): states the upper and lower limits of acceptable performance values thus providing a range of acceptable solutions to fulfil the performance requirements. While international standards may specify performance categories for particular assets and components, specification of performance values is the task of building designers (ISO, 1986). Each performance requirement has a 'comfort zone' that establishes the limits of acceptability for the type of occupancy in a building. These limits in turn are translated into regional standards and codes. Such limits are established by the physiological, psychological, sociological and economic requirements of the occupancy (Hartkopf *et al.*, 1986).

11.3.3 The 'Assess Performance' model

The 'Assess Performance' process (node 'P' in Figure 11.4) includes functions required to assess the condition of an asset and to determine the deviation in the performance which has occurred through its service life. It involves identifying the performance assessment method(s) and their pre-set frequencies, depending on the configuration of the asset being examined. The objective of this process is to catalogue assets and/or components that have ceased to meet the performance requirements specified in process R and hence require a maintenance, repair, renewal or 'do nothing' action. Maintenance includes general activities such as cleaning drains, removing obstructions in roofs and replenishing depleted protection fluids in mechanical equipment. Repair includes unplanned intervention activities performed to rectify distresses found. Renewal includes activities such as installing a new asset and/or component to replace the existing one for economic, obsolescence, modernization or compatibility reasons (Vanier, 2000). One example would be the installation of a new roofing system either above the existing system, or in place of the old roofing system. 'Do nothing' includes postponing or ignoring maintenance, repair or renewal. The inputs to this function are statements of acceptable performance values from process R. The outputs are statements of the asset condition, and a range of management options that objectively specify a set of actions that should be taken when a specific set of conditions occurs. This process is broken down into four functions as shown in the IDEF$_0$ in Figure 11.7.

The following functions are involved:

Identify Condition Assessment Technique (P1): this function identifies the condition assessment technique that would be followed to assess the performance of an asset and/or its components, and indicate the necessity for required corrections. Condition Assessment Surveys (CAS) may vary from being a simple, visual walk-through to a thorough analysis that may include an in-depth review of background documentation, in-situ and laboratory testing, and disassembly of selected components (Cole and Waltz, 1995).

Assess Asset Condition (P2): This function is the core function of the 'Assess Performance' process. All other functions within this process exist to support this primary function.

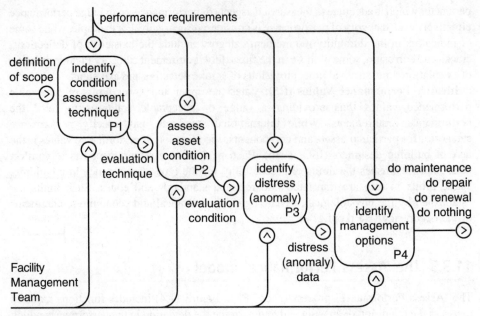

Figure 11.7 Node P: assess performance.

Identify Distress (Anomaly) (P3): identify the distress or anomaly found in the asset through the CAS. Identifying distresses is achieved through carrying out particular functions, depending on the type of asset being assessed. This function is broken down into four subfunctions as shown in the IDEF$_0$ in Figure 11.8.

The subfunctions involved are:

Identify Distress (Anomaly) Type (P31): describe the type of the distress found.

Identify Distress Severity Level (P32): describe the severity level of the distress found – these might range from low to medium to high.

Determine Distress Quantity (P33): quantities are measured as number of units, or combined length, or areas of distress, depending on the type of distress found.

Document Distress Cause(s) (P34): causes might include an aggressive environment, inadequate design, poor workmanship and/or lack of maintenance.

Identify Management Options (P4): describe the range of management options available when specific sets of conditions occur or are imminent. These include carrying out maintenance, repair or renewal or some combination of these, or doing nothing. An input to this function is a statement of the condition of the asset being examined. Asset condition may be expressed quantitatively as a numerical rating (i.e., a condition index) or qualitatively as a categorical rating.

Table 11.1 illustrates a range of maintenance, repair and renewal options corresponding to condition values for roofing systems. Implementing a particular action is dependent on the value of the roofing condition index obtained from MicroRoofer, an engineered management system (Bailey *et al.*, 1990).

The qualitative or subjective approach to assigning condition ratings, and hence implementing a particular maintenance, repair and renewal option, is illustrated in

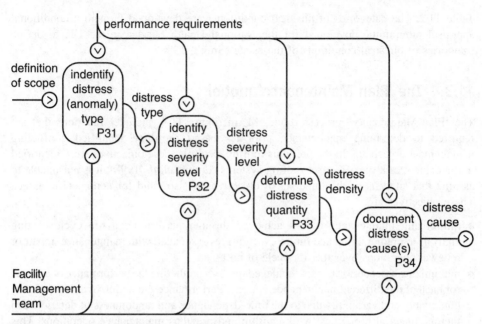

Figure 11.8 Node P3: identify distress (anomaly).

Table 11.1 Options corresponding to values of roofing condition index

Roofing condition index	Maintenance, repair and renewal options
86–100	Routine maintenance
71–85	Minor repairs needed
56–70	Moderate repairs needed
41–55	Major repairs needed
26–40	Replacement probable
11–25	Replacement needed
00–10	Replacement critical

Table 11.2 Subjective listing of condition rating categories

Condition	Description
A	The element is as new and can be expected to perform adequately for its full normal life.
B	The element is sound, operationally safe and exhibits only minor deterioration, which can be corrected by routine maintenance.
C	The element is operational but major repair or replacement will be needed soon. Usually a time frame is established.
D	This category would cover those elements where there is a serious risk or imminent breakdown or of them being unacceptable on health and safety grounds.

Table 11.2. The categories of physical condition were developed through a conditional appraisal programme developed for the National Health Service in the UK to define categories of physical conditions of the estate (Smith, 1988).

11.3.4 The 'Plan Maintenance' model

The 'Plan Maintenance' process (node 'M' in Figure 11.4) includes functions that are required to determine maintenance priorities based on three identified conflicting management objectives. In the previous roofing example these objectives were identified in the context of a specific asset roofing systems (Lounis *et al.*, 1998); it is reasonable to assume that analysis of the same set of objectives is also valid for non-specific assets. These objectives are:

- minimizing maintenance cost – achieved through performing a life cycle costing analysis to predict initial and future expenditures associated with maintenance, repair or renewal operations over the life cycle of an asset
- maximizing asset performance – achieved through predicting the performance of an asset for each of the different maintenance options. Performance prediction of assets presents uncertainty and variability due to the time-dependence and randomness of deterioration factors, material properties, workmanship and previous maintenance operations. This calls for the use of a probabilistic method where modelling of performance requires only limited data on the condition of the asset at two or more points in time in order to derive the probabilities of transition from one state to another having lower condition. One method of performance prediction is based on the principles of the Markov chain, which determines the deterioration in condition through a series of algorithms using Markovian probability matrices and condition states (Lounis *et al.*, 1998)
- minimizing risk of failure – achieved by considering the probability of failure and the consequences of failure concurrently. One method of calculating the probability of failure is obtained using the Markovian model. A consequence of failure is a statement of costs associated with loss of productive time and damage to surroundings (Lounis *et al.*, 1998) or damage to other systems that could, in turn, exacerbate damage.

The inputs to this process are statements of the asset condition and/or the condition of its components, as well as a set of management options to be implemented when a specific set of conditions occurs or is about to occur. The output is an optimal decision or a strategy based on the result of the analyses carried out within this process. This optimal decision is translated into identifying maintenance workload to proceed and, as a result, a maintenance work request is issued so that maintenance jobs are implemented. Another output of this process is a list of deferred maintenance jobs, which are of secondary priority and are awaiting completion. This process is broken down into five functions as shown in the IDEF$_0$ in Figure 11.9.

The functions involved are:

Predict Remaining Service Life (M1): perform an analysis to predict the performance and the service life of an asset.

Estimate Cost of Maintenance (M2): estimate the resources required to carry out the maintenance work requested. This function is broken down into four subfunctions as shown in the IDEF$_0$ in Figure 11.10.

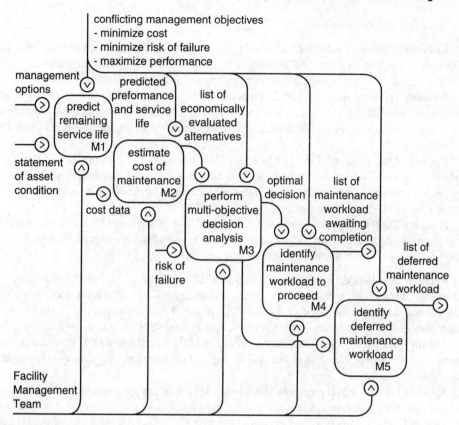

Figure 11.9 Node M: plan maintenance.

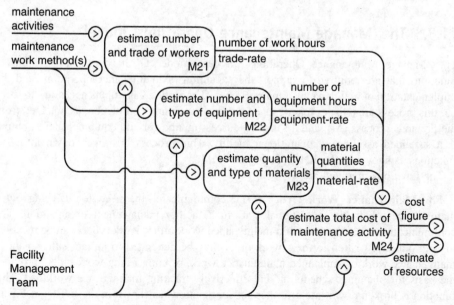

Figure 11.10 Node M2: estimate cost of maintenance.

The subfunctions involved are:

Estimate Number and Trade of Workers (M21): estimate the number and trade of workers needed to perform the maintenance work requested. Some maintenance jobs require a crew of a single trade. Some jobs require multiple crews of multiple trades.

Estimate Number and Type of Equipment (M22): estimate the number and type of equipment needed to execute the maintenance work. The number of hours the equipment is going to be used can be estimated. The hourly rate for using the equipment can be determined.

Estimate Quantity and Type of Materials (M23): estimate the quantities and the type of materials needed to perform the requested maintenance work. Some maintenance jobs are simple and require only one type of material. Some jobs are complex and require the combination of several materials.

Estimate Total Cost of Maintenance Activity (M24): estimate the total cost of carrying out the requested maintenance job. The estimated cost would be the sum of the man-hours, equipment/tools and materials needed to perform the work.

Perform Multiobjective Decision Analysis (M3): perform a risk-based multiobjective decision analysis to recommend a decision taking into consideration the following conflicting management objectives: minimization of maintenance and repair costs, maximization of the building system performance, and minimization of risk of failure.

Identify Maintenance Workload to Proceed (M4): identify first priority maintenance jobs to be carried out based on the results obtained from the above mentioned analyses.

Identify Deferred Maintenance Workload (M5): identify the remaining maintenance, repair or renewal jobs, i.e., that are of lower priority and have therefore been deferred due to lack of funds in annual budget cycles and should be carried out after the completion of the first priority jobs.

11.3.5 The 'Manage Maintenance Operations' model

The 'Manage Maintenance Operations' process (node 'O' in Figure 11.4) includes functions that are required to support the execution of maintenance operations and the implementation of maintenance, repair or renewal activities. The inputs necessary to carry out this process are a list of maintenance workload awaiting completion, as obtained from the previous process (M), and a set of resources (manpower and equipment). The output is a sustained asset in an operational facility. This process is broken down into five functions as shown in the $IDEF_0$ in Figure 11.11.

The functions involved are:

File Maintenance Work Order (O1): communicates the need for carrying out a maintenance job to the operational staff in a facility management organization. The communication takes the form of a maintenance work order. Work orders can be received in both oral and written form. A work order may be generated to initiate either planned maintenance work or unplanned maintenance work, or both. Filing work orders provides the basis for planning, scheduling and effectively tracking maintenance workload. This function is broken down into a further six subfunctions as shown in the $IDEF_0$ in Figure 11.12.

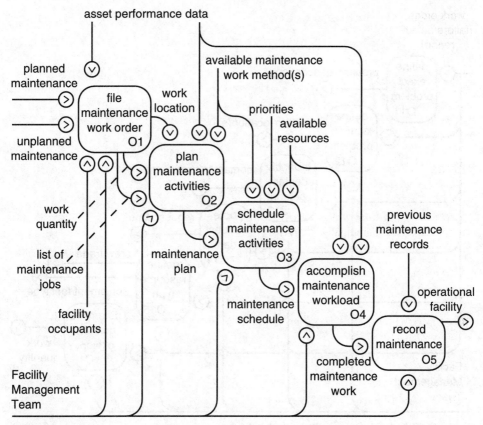

Figure 11.11 Node O: manage maintenance operations.

The typical tasks associated with filing and processing a maintenance work order are:

Define Exact Problem (O11): describe the failure or defect in an asset (e.g., a roofing system) that calls for a maintenance action to restore it to its original condition.

Define Location of Problem (O12): specify the location of the problem.

Identify Contact Person (O13): record contact details of the person requesting the maintenance action.

Note Request Date (O14): record the date of the request for a maintenance action.

Define Resource Type (O15): describe the type of resources needed to perform the maintenance work, i.e., manpower, equipment and/or special tools required to perform the work.

Determine Quantity of Work (O16): determine the quantity of maintenance work, based on the extent of the defect described in the work order.

Plan Maintenance Activities (O2): includes functions required to plan maintenance activities, and the method to be followed to achieve them. In essence this involves setting up a work plan, for which various questions must be answered, such as what, who, where, when and how the operational staff in a facility management organization will respond to a filed work order. The input to this function is a list of maintenance jobs awaiting

work order
(failure/defect
report)

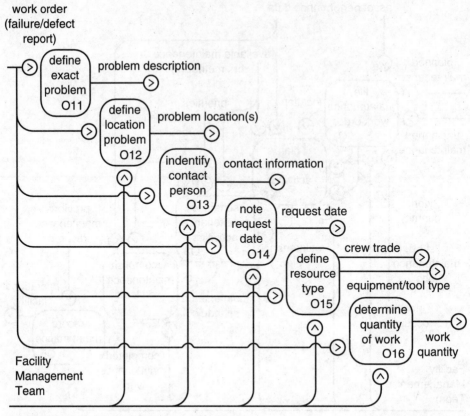

Figure 11.12 Node O1: file maintenance work order.

completion. The output is a maintenance work plan. It should be noted that every maintenance project is likely to be unique, meaning that some of the functions within this process can be overlooked depending on the characteristics of the asset undergoing maintenance. This function is broken down into four subfunctions as shown in the IDEF$_0$ in Figure 11.13.

The subfunctions involved are:

Choose Maintenance Work Method (O21): identify the method to be followed in performing the maintenance work. The choice of a specific maintenance work method over another will directly influence the cost and the duration of the maintenance work.

Define Maintenance Activities (O22): define a series of maintenance activities through the description of the given problem in the work order, and the work method chosen. This function is broken down into three subfunctions as shown in the IDEF$_0$ in Figure 11.14.

The functions involved are:

- *Define Repetitive Activities (O221):* list/outline the activities that would be carried out more than once as a result of processing a single work order, e.g., replacing window glazing to several floors in one building.

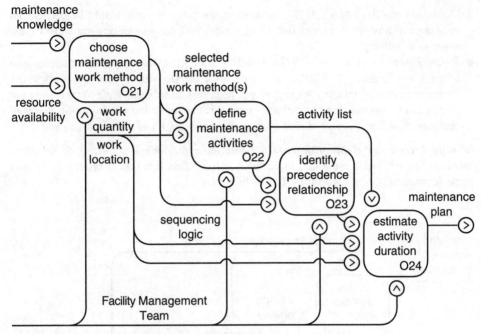

Figure 11.13 Node O2: plan maintenance activities.

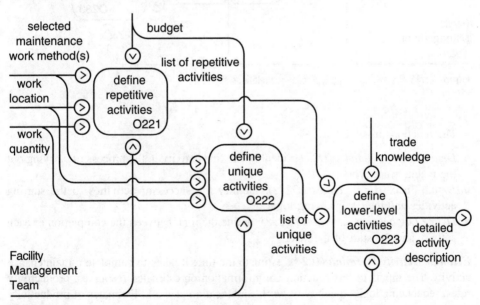

Figure 11.14 Node O22: define maintenance activities.

- *Define Unique Activities (O222):* list/outline the activities that would be carried out only once as a result of processing a single work order, e.g., painting the walls of one room in a building.
- *Define Lower-Level Activities (O223):* define implicit activities within a maintenance activity, e.g., painting would be the general description in a requested work order (either repetitive or unique). As part of this task, certain subtasks have to be carried out as part of the process (lower-level activities), such as removing old paint, sanding wall surfaces, applying a sealer, a first coat of paint and then a second coat of paint.

Identify Precedence Relationship (O23): determine the logical sequence of the steps involved to carry out a requested maintenance work. This function is broken down into three subfunctions as shown in the $IDEF_0$ in Figure 11.15.

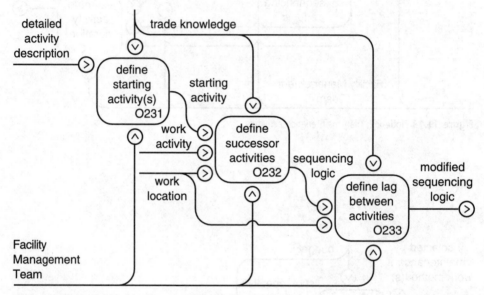

Figure 11.15 Node O23: identify precedence relationship.

The functions involved are:

- *Define Starting Activity (O231):* identify the first activity in the process of carrying out the maintenance work.
- *Define Successor Activities (O232):* identify the successor activities to the starting activities and their sequencing logic.
- *Define Lag between Activities (O233):* define the time between the completion of each activity and the start of the next.

Estimate Activity Duration (O.24): estimate the time it takes to complete a maintenance activity. The inputs to this function are information on estimated resources, productivity rates, sequencing logic, quantity of work, and location of work. Estimates of productivity rates for various facilities maintenance and repair tasks can be obtained from published cost data such as R.S. Means Facilities Cost Data (R.S. Means, 1997). The output of this task is a maintenance plan.

Schedule Maintenance Activities (O3): includes functions required to schedule maintenance activities. The inputs to this function are a list of maintenance activities, activity durations, estimated resources and sequencing logic. The output is a maintenance schedule. This function is broken down into three subfunctions as shown in the IDEF$_0$ in Figure 11.16.

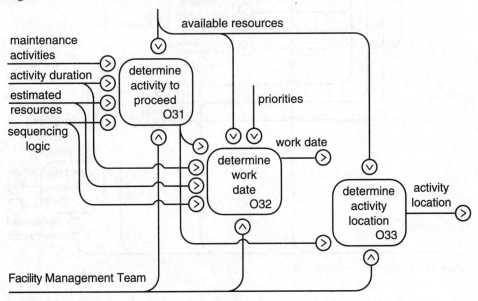

Figure 11.16 Node O3: schedule maintenance activities.

The subfunctions involved are:

Determine Activity to Proceed (O31): serves to denote which maintenance work to proceed with.

Determine Work Date (O32): indicates the date for the commencement of the work.

Determine Activity Location (O33): gives the location of the work.

Accomplish Maintenance Workload (O4): The inputs to this function are sequencing logic, activity location, the activity to proceed with, and activity duration. The output is a completed maintenance workload. This function is broken down into four subfunctions as shown in the IDEF$_0$ in Figure 11.17.

They are:

Set Up Work Area (O41): establish and organize the work-space depending on the complexity of the maintenance work requested.

Prepare Resources (O42): mobilize and co-ordinate the resources (manpower and equipment) needed to perform the work.

Perform Work (O43): this is the core function of the 'Accomplish Maintenance Workload' model. All other functions within this process exist to support this primary function.

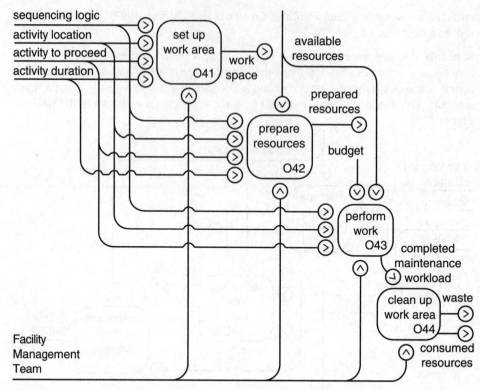

Figure 11.17 Node O4: accomplish maintenance workload.

Clean Up Work Area (O44): separate waste products and remove them from the work space after the maintenance work is carried out. Partially consumed resources would also be salvaged for future use if needed.

Record Maintenance (O5): includes functions required to record accomplished maintenance work. These functions are usually noted by the crew that has performed the work. The input to this function is a completed work unit. The output is an operational facility. This function is broken down into four subfunctions as shown in the $IDEF_0$ in Figure 11.18.

They are:

Report Completed Work (O51): report the completion of the work requested. In some circumstances, with the start of a maintenance operation, the crew might discover a much larger problem, necessitating an increase in both man-hours and material required. Such a change of work scope prompts a revision of estimates of man-hours and materials needed to reflect actual crew productivity and actual consumption of resources. Reporting completed work also serves to measure, for quality control purposes, how well the maintenance work has been performed.

Report Resources Consumed (O52): report the actual amount of resources consumed to carry out a maintenance work order. This monitoring helps to determine whether the work has been accomplished at the lowest cost by examining manpower utilization, material usage and costs. Manpower utilization can be measured by analysing data

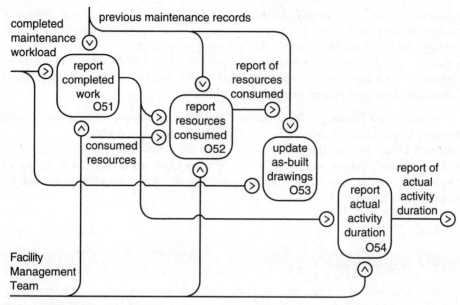

Figure 11.18 Node O5: record maintenance.

relating to the number of man-hours and the quantity of material that each trade used to complete work collected from all completed maintenance work orders. Comparisons can then be made to determine the relative efficiency of current operations (Magee, 1988). This function is broken down into three subfunctions as shown in the IDEF$_0$ in Figure 11.19.

- *Report Number and Trade of Workers (O521):* record the actual number and trade of the people who performed the work. This also allows comparison of actual man-hours spent against estimated man-hours.

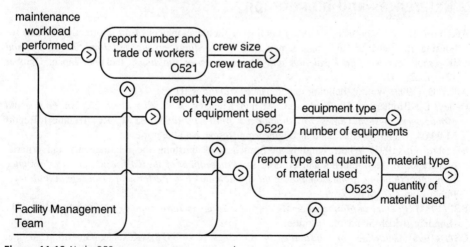

Figure 11.19 Node O52: report resources consumed.

- *Report Type and Number of Equipment Used (O522):* record the type and number of items of equipment used to perform the work. This allows comparison of actual equipment-hours needed against estimated equipment-hours.
- *Report Type and Quantity of Material(s) Used (O523):* record the type and quantity of material(s) needed to perform the work. This allows comparison of actual quantities of materials consumed against estimated quantities.

Update As-built Drawings (O53): instruct facility management staff to update as-built drawings to reflect the changes, if any to the configuration of the asset.

Report Actual Activity Duration (O54): record the actual time taken to perform the work. This allows comparison of the actual duration of an activity against its estimated duration. Such monitoring verifies that the assigned work crew is appropriate and that the productivity rate is acceptable.

11.4 Conclusion

The scheme presented is a generic framework for asset maintenance management. It acts as policy guidelines for the conduct of maintenance in an organization and provides a way of bridging the gaps in the practice of maintenance of built assets by asset managers. The framework is schematically described as an $IDEF_0$ process model.

The main advantages gained from using the $IDEF_0$ process models are clearly defined boundaries and responsibilities of functions within management processes, as well as improved levels of communication between participants (Sanvido, 1990). The model does, however, have the limitation of having no values or priorities assigned to functions; as a result all functions appear to have equal impact on the maintenance project rather than the scheme reflecting the actual impact of each function.

References and bibliography

Aikivuori, H. and Mottonen, V. (1992) Outline of the development of an information system for housing maintenance and management. In: *Proceedings of the 1st Joint Japan–Finland Workshop, Service Life Prediction and Maintenance of Buildings*, Tsukuba, Japan, October 12–16, 220–31.

Allen, D. (1993) What is building maintenance? *Facilities*, **11** (3), 7–12.

Bailey, D., Brotherson, D., Tobiasson, W. and Knehans, A. (1990) *ROOFER: An Engineered Management System (EMS) for Bituminous Built-Up Roofs.* USACERL Technical Report M-90/04 (US Army Construction Engineering Research Laboratories).

Blachere, G. (1993) Whole building performance, establishing the requirements and criteria, preparation of requirements and criteria. In: *Proceedings of the CIB W60: The Performance Concept in Building, Some Examples of the Application of the Performance Concept in Buildings*, Publication 157, May, 33–9.

BSI (1984) *Glossary of Maintenance Management Terms in Terotechnology.* British Standard 3811 (London: British Standards Institute).

CSA (1995) *Guidelines on Durability in Buildings (CSA-S478)* (Rexdale, Ontario: Canadian Standards Association).

Cole, G. and Waltz, M. (1995) How long will it last? Condition assessment of the building envelope. In: *Proceedings of the 13th ASCE Structures Congress, Restructuring: America and Beyond*, Boston, MA, April 2–5, 642–57.

Gordon, A. and Shore, K. (1998) Life cycle renewal as a business process. In: *Proceedings of APWA Congress – Innovation in Urban Infrastructure*, Las Vegas, NV, 41–53.

Hartkopf, V., Loftness, V. and Mill, P. (1986) The concept of total building performance and building diagnostics. In: Davis, G. (ed.) *Building Performance: Function, Preservation, and Rehabilitation*, ASTM STP 901, 5–22 (Philadelphia, PA: American Society for Testing and Materials).

Hassanain, M.A., Froese, T.M. and Vanier, D.J. (1999) Information analysis for roofing systems maintenance management integrated system. In: *Proceedings of the 8th International Conference on Durability of Building Components and Materials*, Vancouver, May 30–June 3, **4**, 2677–87.

Hassanain, M.A., Froese, T.M. and Vanier, D.J. (2000). IFC-based data model for integrated maintenance management. In: *Proceedings of the 8th International Conference on Computing and Building Engineering*, ASCE, Stanford University, CA, August 14–17, **1**, 796–803.

Hassanain, M., Froese, T. and Vanier, D. (2001) Development of a maintenance management model based on IAI Standards. *Artificial Intelligence in Engineering*, **15** (2), 177–93.

IAI (1999) *Specifications Development Guide. Industry Foundation Classes – Release 2.0* (International Alliance of Interoperability).

ISO (1984) *Performance Standards in Buildings – Principles and their Preparation and Factors to be Considered (ISO-6241)* (Geneva: International Organization for Standardization).

ISO (1986) *Performance Standards in Buildings – Presentation of Performance Levels of Façades Made of Same-Source Components (ISO-7361)* (Geneva: International Organization for Standardization).

ISO (1992) *Building Construction – Expression of User's Requirements, Part 2: Air Purity Requirements (ISO-6242-2)* (Geneva: International Organization for Standardization).

Korka, J.W., Oloufa, A.A. and Thomas, H.R. (1997) Facilities computerized maintenance management systems. *Journal of Architectural Engineering*, **3** (3), 118–23.

Lounis, Z., Lacasse, M. A., Vanier, D. J. and Kyle, B.R. (1998) Towards standardization of service life prediction of roofing membranes. *Roofing Research and Standards Developments*, **4**, ASTM-STP-1349, Nashville, TN, 3–18.

Magee, G.H. (1988). *Facilities Maintenance Management* (Kingston, MA: R.S. Means Co. Inc.).

Mottonen, V. and Niskala, M. (1992). Hyperdocuments in maintenance management of buildings. In: *Proceedings of CIB W70: Innovations in Management, Maintenance and Modernization of Buildings*, Rotterdam, October 28–30, **9**, Paper 12.1.3.

NIST (1993) *Federal Information Processing Standards 183 – Integration Definition for Function Modeling (IDEF$_0$)*. NIST Computer Systems Laboratory (Gaithersburg, MD: National Institute of Standards and Technology).

R.S. Means Co. Inc. (1997) *Facilities Maintenance and Repair Cost Data* (Kingston, MA: R.S. Means Co. Inc.).

Sanvido, V.E. (1990) *An Integrated Building Process Model*, Technical Report 1, Computer Integrated Construction Research Program, Department of Architectural Engineering, Pennsylvania State University.

Smith, R. (1988) Estate maintenance monitoring and appraisal. In: *Proceedings of the 9th National and the 2nd European Building Maintenance Conference*, London, November 4–5, 1.4.1–26.

Svensson, K. (1998) *Integrated Facilities Management Information: A Process and Product Model Approach*. PhD thesis, Royal Institute of Technology, Construction Management and Economics, Stockholm, Sweden.

Then, D.S. (1982) Computerizing the maintenance operations – a review of information needs for public sector housing maintenance. In: *Proceedings of the 7th National Building Maintenance Conference*, London, November 2–5, BM2-4-1–13.

Vanier, D.J. (2000). *Municipal Infrastructure Investment Planning (MIIP)*. Project: Statement of Work, Institute for Research in Construction, National Research Council of Canada, January.

Vanier, D.J. and Lacasse, M.A. (1996) BELCAM project: service life, durability and asset management research. In: *Proceedings of the 7th International Conference on the Durability of Building Materials and Components*, Stockholm, Sweden, 848–856.

Wix, J., Yu, K. and Ottosen, P.S. (1999) The development of industry foundation classes for facilities management. In: *Proceedings of the 8th International Conference on Durability of Building Components and Materials*, Vancouver, May 30–June 3, **4**, 2724–35.

Appendix – IDEF$_0$ process modelling notation guide

This Appendix describes the notation for IDEF$_0$ process modelling language (semantics and syntax) used to present graphically the proposed framework of maintenance management. It is not meant to provide a full description of the IDEF$_0$ notation, but to provide a description of the techniques used to develop the process models presented earlier in this chapter. IDEF stands for Integration Definition for Function Modelling. Use of IDEF$_0$ modelling language permits the construction of models comprising system functions (activities, actions, processes, operations), functional relationships, and data (information or objects) that support systems integration (NIST, 1993).

A.1 Background

The IDEF$_0$ notation was one of a series of techniques developed during the 1970s by the US Air Force Program for Integrated Computer Aided Manufacturing (ICAM), to improve manufacturing productivity through systematic application of computer technology. These techniques are as follows:

1 IDEF$_0$, used to produce a 'function model', which is a structured representation of functions, activities or processes within the modelled system or subject area

2 IDEF$_1$, used to produce an 'information model', which represents the structure and semantics of information within the modelled system or subject area

3 IDEF$_2$, used to produce a 'dynamics model', which represents the time-varying behavioural characteristics of the modelled system or subject area.

In 1993, the U.S. National Institute of Standards Technology (NIST) documented the IDEF$_0$ notation as Federal Information Processing Standard (FIPS) 183.

A.2 Structure of IDEF$_0$

A system is represented through IDEF$_0$ by means of a model composed of diagrams with supportive documentation that break a complex subject into its component parts. Each diagram consists of boxes and arrows representing the functional activities, data and function/data interfaces. Text accompanies each of the diagrams describing the activities in that diagram (NIST, 1993).

A.3 Schematic presentation

Boxes in the diagram represent functions, corresponding to activities, actions, processes, operations, or transformations. One or more inputs to the function are transformed into one or more outputs using the mechanism provided. The transformation process for the data is controlled by one or more controls. The data entities are illustrated schematically in Fig. 11.A1 and Table 11.A1 (NIST, 1993).

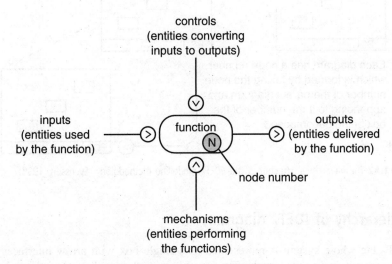

Figure 11.A1 Schematic presentation of the function box (NIST, 1993).

Table 11.A1 Data entity descriptions

Entity	Description
Function	An activity, action, process, operation, or transformation, which is described by an active verb. A function is shown as a box
Input	An entity, which undergoes a process or operation, and is typically transformed. It enters the left of the box, and may be any information or material resource
Output	An entity which results from a process or objects which are created by a function. An output is shown exiting the right side of the box
Control	An entity which influences or determines the process of converting inputs to outputs. A control is shown entering the top side of the box
Mechanism	An entity such as a person or a machine, which perform a process or operations. A mechanism is shown entering the bottom side of the box
Node	A unique identifier to every function, which is shown in the bottom right-hand corner of the box

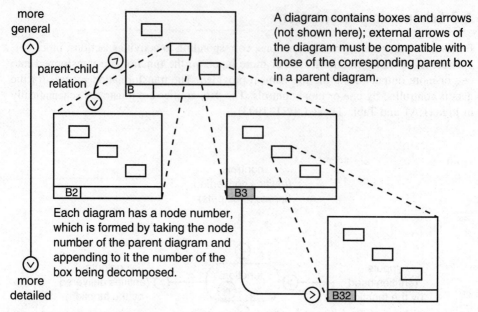

more
general

parent-child
relation

A diagram contains boxes and arrows
(not shown here); external arrows of
the diagram must be compatible with
those of the corresponding parent box
in a parent diagram.

B

B2

B3

Each diagram has a node number,
which is formed by taking the node
number of the parent diagram and
appending to it the number of the
box being decomposed.

more
detailed

B32

Figure 11.A2 The hierarchical structure of the IDEF$_0$ modelling methodology (Svensson, 1998).

A.4 Hierarchy of IDEF$_0$ diagrams

In IDEF$_0$ the whole system is represented as a single box with arrow interfaces to the environment external to the system. The function in the box is decomposed into between three to six functions, each of which may be further decomposed into subfunctions. This top-down decomposition may be continued, resulting in three to six 'child' or detailed diagrams for each function on any given level (NIST, 1993). A schematic view of the hierarchy of the diagrams is illustrated in Fig. 11.A2.

A.5 Decomposition

The number of functions within each diagram is limited to a minimum of three and maximum of six. These constraints limit the level of detail and complexity in any diagram, yet prevent the diagram from being trivial. Further, the level of detail is controlled by the position of the diagram in the hierarchy of diagrams. The amount of detail increases through each level of decomposition. This results in gradual exposition of detail. In this chapter each level of decomposition is illustrated as a separate process model, with each having a different figure number.

A.6 Modularity of IDEF$_0$ diagrams

When a function in a box is decomposed the scope of the function and its interface arrows create a bounded context for the subfunctions. The scope of the detail or 'child' diagram fits completely inside its 'parent' function. The interface arrows of the 'parent' box match the external arrows of the detail or 'child' diagram. Thus, arrows entering and exiting the detail diagram must be the same arrows interacting with the parent diagram (NIST, 1993).

A.7 Numbering the IDEF$_0$ diagrams

Each major function in the system is assigned a node number, which acts as a unique identifier for the function in the process model. The node number is shown in the bottom right-hand corner of the box. Decomposition of each box leads to a number of diagrams. Each diagram has a node number, which is formed by taking the node number of the parent diagram and appending to it the number of the box being decomposed (Svensson, 1998; NIST, 1993). The node numbering system allows the reader to trace the steps of decomposition back through the parent function of each diagram.

12

Portfolio management

Geert Dewulf*
Lydia Depuyt
Virginia Gibson‡

Editorial comment

Facilities are most often interpreted as the physical space owned or leased by an organization, and in this regard, therefore, translate to buildings and the land upon which they reside. An ability to strategically manage real estate is a fundamental competency for the facility manager. The emphasis placed on strategy is due to the influence that investment return holds in the decision-making process for property transactions and portfolio management. Real estate is a strategic resource and a significant asset for most organizations.

Not all facilities rely on real estate in the conventional sense; infrastructure can take many forms. There are, however, generic principles that apply to the management of infrastructure, and maximizing investment yield is clearly one of them. Businesses invest in their own facilities so it is important that this investment is effectively managed. Decisions need to be made as to whether to own or lease, when and how to expand, decentralize, dispose or refurbish in accordance with business directions but always within the context of investment return.

Large organizations typically have a diverse portfolio of real estate assets. These assets need constant attention, not only in terms of operation and maintenance, but also in terms of fitness for purpose and economic viability. Buildings naturally tend towards obsolescence in the long term, but can become obsolete within a shorter time frame due to changes in regulations, market perception, locality, functional use and environmental impact. Building owners will look to refurbishment strategies, and even adaptive reuse, to prolong economic life, and will consider relocation and leasing where financially feasible. Image and position also play a significant part in real estate decisions.

* University of Twente, The Netherlands
† Fortis Real Estate Development, The Netherlands
‡ University of Reading, UK

Relocation can be a complex process, and like many other non-core activities is often outsourced. The goal for any relocation exercise is to minimize disruption to business operations, to ensure safe transfer of assets (including data) and to keep all stakeholders fully informed. Relocations can result from a desire to centralize or decentralize business operations, downsize, expand into new markets or to pursue globalization strategies, although the most common reasons are to maximize investment yield or minimize leasing costs. The co-ordination of such moves is the responsibility of the facility manager.

Clearly real estate decisions can contribute to increased value for an organization, and therefore it is important that expertise is available in all aspects of property management. This extends to development initiatives, design and construction, and can encompass other key competencies such as financial management, risk management and project management.

12.1 Introduction

Managing a real estate portfolio within a corporation or a public organization from an occupational perspective differs from the way an investor manages his/her portfolio. Where an investor sees property as an asset offering a high return on investments, a corporate real estate (CRE) manager sees a real estate portfolio as a resource with the main goal of supporting the core business of a corporation. Managing a CRE portfolio therefore is about finding the right balance between adding value to the core business process and optimizing the (financial) performance of the real estate portfolio. CRE managers are therefore both *facilitators* and *investors*: they facilitate the core business by providing appropriate space in which to house the functions and activities of their organization, and they invest in physical assets, many of which are reported on the balance sheet. Consequently, CRE portfolio decisions have to be based on both future business requirements and trends in the real estate market.

Business change as well as real estate market movements are, however, hard to predict. This uncertainty and lack of ability to foretell the future is why a portfolio approach is a useful tool for the CRE manager. In many ways this is the essence of a portfolio management approach, i.e., managing a group of assets under uncertainty. In the context of a group of CRE assets, a portfolio approach attempts to find an appropriate mix of assets within the portfolio as a whole in order to balance risk and return. The risks are associated with change in both the business and real estate markets, while the returns are related to business performance and the financial impact of an asset. In this chapter, two ways of dealing with these uncertainties within a portfolio context are considered: scenario planning and creating a flexible portfolio.

In order to first establish a context for discussing a portfolio approach, the circumstances within which CRE decisions are taken are reviewed. One of the greatest challenges for CRE professionals is matching the changing needs of their organization as it identifies opportunities and threats within its own sector and business environment with emerging opportunities and new products in the real estate market. Both of these factors are considered here.

Two complementary approaches to CRE portfolio management are then examined. First, the application of scenario planning is discussed demonstrating how it is possible to develop more robust real estate solutions by exploring different feasible futures. In particular, the identification of the risks and returns associated with any solution make the CRE manager more mindful of where management effort needs to be directed. Secondly, the paper demonstrates how a CRE portfolio can be divided into segments that require different types and degrees of flexibility. This process allows a CRE manager again to target his/her effort so that risks and return are balanced in different ways for different parts of the portfolio.

12.2 The dynamic context of the portfolio

A CRE portfolio is constantly subject to pressures from both inside and outside the organization that makes change necessary. The task of the CRE manager is to attune the portfolio to the internal organizational requirements and, at the same time, to developments in the real estate market. Organizational change and market developments require creativity and a proactive approach to the management of the CRE in order to align the real estate portfolio with these changes and to optimize its performance.

12.2.1 Organizational change

Organizations are faced with an ever-changing environment. Improvements in information and communications technology (ICT) have had the potential to transform work processes in many organizations. Social trends have altered the structure of the workforce leading to changes such an increase in the proportion of women in employment. This has led, in turn, to the introduction of a broader range of employment contracts, with many individuals employed under part-time or other flexible arrangements. The increased concern for environmental issues has forced organizations to consider how and where they employ their staff and distribute their goods. The deregulation of some industries and the increased involvement of the private sector in public sector activities have created both opportunities and threats for many organizations. Taken together, these changes in the wider environment have posed real strategic challenges for organizations.

Organizations have therefore developed a range of strategies and tactics to deal with these influences. For instance, business process re-engineering (BPR) has been seen as a way of harnessing ICT in order to drive down costs and become more competitive. New working practices have been developed to increase the productivity of the workforce by allowing the staff to work in the most appropriate location and at the most suitable time. Activities that are not core to a business have been outsourced so that management energy can be diverted to those functions that really add value. Together these management initiatives have the ability to transform the way in which organizations operate.

Each of these strategies and tactics potentially could have an impact on the CRE portfolio. BPR might lead to the organization requiring less space but possibly requiring it in different locations. Flexible working practices may require new infrastructure and alternative office layouts. The transfer of some activities to a contractor through outsourcing may lead to a reduction in overall space requirements (Gibson and Lizieri, 1999). It is not yet clear, however, just how different organizations respond to each of these drivers of change – the possible may not be the probable.

There is little academic evidence of how organizational change actually takes place, or of how it impacts on a CRE portfolio. One study of large organizations within the UK suggests that change is often more evolutionary than might be predicted (Lizieri *et al.,* 1997). Bringing about major structural and cultural change is not a fast process; organizations take time to plan and implement these changes. Consequently, although the impact on the CRE portfolio may initially be minimal, the longer-term impact could be significant. A CRE manager therefore needs to be involved in any change management programme from the start.

Furthermore, the assets in a CRE portfolio are, by nature, long lasting and static, so any real estate decision made today is likely to have far-reaching consequences. Too often real estate decisions are made without understanding the future risks and rewards associated with those assets. There are costs associated with acquisition and disposal that need to be accounted for in any robust decision-making process. In practice, CRE managers often find themselves with a portfolio that does not even meet their current business requirements let alone their future ones (Gibson, 2000). In part this may be due to near-sighted decision-making processes.

The challenge therefore for a CRE manager is (1) to understand and track the external change drivers of their own organization, and (2) to find out how their organization is likely to react to these factors. Organizations are increasingly changing both their form and function yet the remit of the CRE manager is to find an optimal mix in the portfolio that best supports the business (Figure 12.1); however, it is almost impossible to define the optimal fit between organization and portfolio. For decades, architects, developers and consultants have looked for the optimal match between organization(s) and building(s).

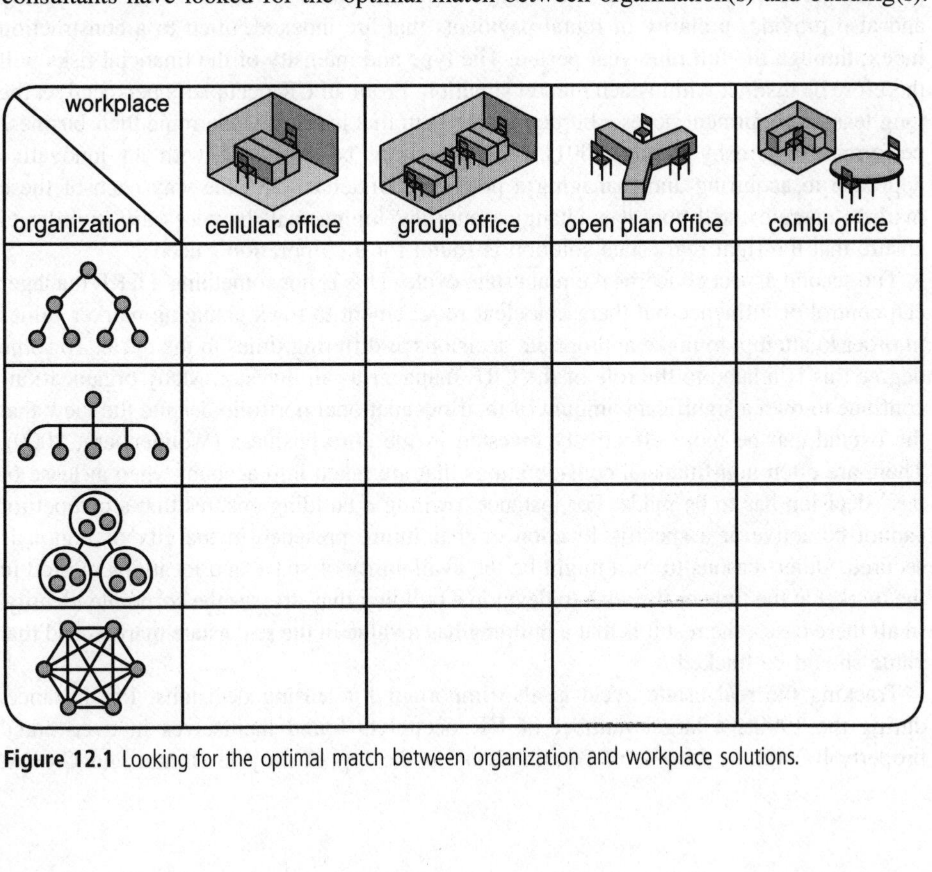

Figure 12.1 Looking for the optimal match between organization and workplace solutions.

For instance, many workplace gurus proclaim that flat organizations require open plans and hierarchical organizations require cellular offices. In practice, however, a generic workplace solution does not exist, and what has worked for one organization may not work for another.

Finally, the real estate that organizations occupy has a value both for the business and within the real estate market. It is this latter aspect that adds another layer of complexity; not only are CRE managers required to track their organization's business environments, they also need to understand changes within the real estate markets in which they operate.

12.2.2 Real estate market changes

There are at least three dimensions to understanding how changes in the real estate market can benefit or detract from an organization's performance. First, most CRE managers have portfolios that are geographically dispersed, often extending over a number of countries, all of which have their own unique real estate markets. These markets are subject to different influences and constraints and are at different points of market maturity (Keogh and D'Arcy, 1994). Any real estate decision is therefore grounded in a market and needs to be understood within that context. For instance, the traditional lease structure in the UK of a 15–25 year term with upward only rental adjustments can be contrasted with the 3–6–9 indexed lease structure. This latter structure, adopted in France and some other parts of Europe, allows tenants to terminate a lease at either the three- or six-year period and also provides a clarity of rental payments that are indexed, often to a construction index, through the full nine-year period. The type and intensity of the financial risks will therefore be distinct within each market situation. From an UK occupier's perspective, the long lease commitment poses a burden on the firm that is felt to undermine their business performance (Crosby *et al.*, 2001). This can only be mitigated with an innovative approach to acquiring and managing a portfolio. Understanding the way each of these markets operates, and how it is changing and developing, will be necessary in order to ensure that the right real estate solution is found for a corporation's needs.

The second aspect concerns the real estate cycle. This is not something a CRE manager can control or influence but there is a clear requirement to track changing market values in order to attempt to make appropriate decisions at differing times in the cycle. To some degree this is related to the role of the CRE manager as an investor. Many organizations continue to own a significant amount of their occupational portfolio despite the view that the capital can be more effectively invested in the core business (Woudenberg, 2000). There are often non-financial considerations that are taken into account when a 'lease or buy' decision has to be made. For instance, owning a building ensures that a competitor cannot be active at a specific location or that future presence in the city or region is secured. Other reasons to own might be the availability of space and locations offered in the market at the time or the wish to develop a building that stresses the corporate identity. In all these cases, the result is that a building has a value in the real estate market, and that value should be tracked.

Tracking the real estate cycle is also important for leasing decisions. For instance, during the 1990s, a larger number of UK occupiers found themselves in over-rented property by signing long-term commitments at the top of a cycle (Baum and Crosby,

1995). It is difficult to predict downturns but there are often early warning signs that should make decision-makers more cautious. By testing decisions against a number of different scenarios, some of the risks associated with the cycle will exposed.

The final aspect of market change relates to new product development. It is increasingly recognized that creating a workplace is about more than merely acquiring space. Space needs technological infrastructure, fittings and fixtures, as well as a bundle of services, from cleaning to catering, to make it a fully functional workplace. Traditionally, providers of real estate (investors and developers) have left many of these elements for the tenant to source and provide, however, in the last decade there has been considerable innovation in this area. The serviced office model is a good example of the trend of bundling space and services (Gibson and Lizieri, 2000). These offices operate on a fully functioning basis allowing both quick entry and short-term commitment that is not possible with other forms of real estate solution. They are clearly targeted at organizations or activities that require this type of flexibility; new market entry, special short-term project space and new business development are typical functions that are housed in this type of property (Gibson and Lizieri, 2000).

There are other forms of property solution, which blend space and service, which are emerging, including the total outsourcing of corporate portfolios. The implications for a CRE manager are significant. A role which was solely concerned with finding sites and buildings appropriate for their organizations, and evaluating lease or buy decisions, now includes consideration and evaluation of a much wider range of options. Tracking the development of these new products becomes the only way that CRE managers can ensure that a full range of accommodation options is considered and that the most appropriate solution is ultimately provided.

This section has demonstrated that the context within which CRE managers operate is not only complex and multifaceted, but that it is also ever changing. These managers need to identify the change drivers, to find ways to predict the impact of these changes and to manage their portfolio with change in mind. The following section outlines two tools that can assist with this process.

12.3 Prepare for future changes

There are several tools that can help CRE managers cope with future changes and minimize the risks involved. Corporate real estate asset management implies the management of a stock of buildings at the asset and portfolio level, taking into account both the short- and long-term consequences of change. However, many organizations just focus on object or individual building decisions, or on translating current user needs into finding the right building for a reasonable price. Corporate real estate managers are often pre-occupied with making a deal – getting the best price in the market at a specific point in time. Although the performance of the portfolio is often assessed, the long-term consequences of portfolio strategies on future performance are not estimated.

For individual decisions the operating statement is forecast, but often only in a linear way. Trends are projected based on assumptions about how the organization will grow and function such as: the organization will continue to grow at a specified rate, the functions which the organization undertakes will remain largely unchanged, or the space allocated per member of staff will continue to fall with the introduction of more ICT. Any linear extrapolation is likely to be flawed unless these underlying assumptions are tested. Many organizations do not develop 'what if' scenarios and are, therefore, not paying attention to future risks. In today's fast changing environment, organizations need to focus more on ways to deal with future risks and uncertainties rather than finding solutions that are only valuable for today's problem. This is the essence of developing robust solutions.

Two ways a CRE manager can cope with uncertainty are by scenario planning, to develop an overall portfolio strategy, and by creating flexibility within the portfolio itself. Each of these is considered below.

12.3.1 Risk management through scenario planning

Making a future plan for your real estate portfolio means dealing with a lot of uncertainties. One way to deal with these uncertainties is to use scenario planning to analyse the consequences of certain decisions that follow from the portfolio strategy.

The portfolio strategy should be aligned with the mission and business strategy of the company. As the strategy differs for each organization, the optimal portfolio strategy will differ as well. The portfolio has to be managed in such a way that it supports the bottom line of the organization. It is therefore important to unravel the (often hidden) strategy of the company. Most companies do not have a clear vision or overall strategy. Having discussions or workshops with the CEO, CFO and Board members can make the strategy explicit. A helpful tool is a so-called stakeholders' analysis, in which the needs of the various stakeholders inside and outside the company are assessed. When this analysis is undertaken in co-operation with top management, it is also possible to determine whether there is a difference between the official strategy, written down in a business plan or the annual report, and the strategy in use (Dewulf and Van der Schaaf, 1998).

The impact of different portfolio decisions can be estimated for each stakeholder that the CRE manager has to deal with including the business managers, the employees and the shareholders. The problem is that the impact cannot be estimated in a linear way. There is not one future, but different plausible futures. Organizations are confronted with many uncertainties that could have a major impact on their results. The first step in a scenario approach is to determine the driving forces behind the company's operations. These driving forces can be categorized as:

● predictable forces
● low predictable or uncertain, but highly influential developments – these developments are called the 'critical uncertainties'.

A helpful tool in determining the 'critical uncertainties' is mapping the driving forces in a co-ordinate system, consisting of the axes of 'predictability' and 'impact on the issues addressed'. Figure 12.2 shows the results of a historical analysis and brainstorming sessions organized for the Dutch Government Buildings Agency (GBA). The developments situated in the lower right corner of the figure were the critical uncertainties the GBA had to deal with (Dewulf and Van der Schaaf, 1998).

Most organizations still put a lot of effort into gathering data and assessing future developments that have only a limited impact on their performance. For the highly predictable developments we can make accurate forecasts. The unpredictable and high impact variables have to be put in scenarios. It is difficult to predict at this point, for

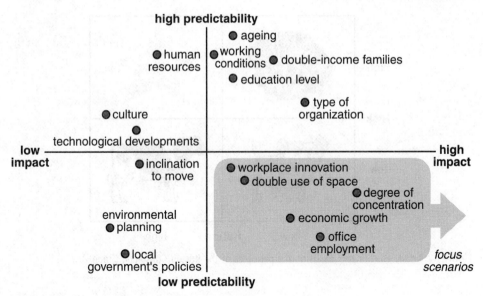

Figure 12.2 An example of uncertainties an organization is confronted with.

instance, whether workplaces will be used by two or more people, but it will certainly have a strong impact on the demand for office space (Depuy and Van der Schaaf, 2000).

After the critical uncertainties are identified plausible scenarios can be developed. For discussion purposes, it is often sufficient to generate only two or three extreme scenarios. Once the scenarios are developed, the impact of the strategies on the performance for each stakeholder can be assessed. Through scenario planning the portfolio managers can gain some insight into the future consequences of certain decisions. Scenario planning is, however, not a tool to predict the future, but a tool for dealing with future uncertainties.

Using the scenario methodology it is possible to bring to light the consequences of certain decisions. This provides the CRE manager with some understanding of the risks and returns associated the different buildings in the portfolio. The owned assets in the portfolio mostly determine the financial risks, although this is not always true; this is discussed further below.

To give a judgement on the degree of long-term risks caused by changes that take place within the real estate portfolio, scenarios can be used to analyse potential developments of the assets within the portfolio. The consequences of different strategies on the performance of the portfolio can be estimated and characterized between low/high risk and low/high (financial or business) returns.

To visualize the impact of a certain strategy on the risk/return ratio of the several buildings in a portfolio, the principles of the Boston Consulting Group growth-share matrix may be used as shown in Figure 12.3.

A building that shows a negative or low return in all scenarios is characterized as a *dog*. These buildings are especially those in peripheral locations with a resulting low exit value. Buildings marked as *problem children* or *question marks* are more or less cost recovered

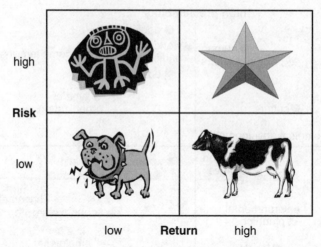

Figure 12.3 Risk/return matrix based on the principles of the Boston Consulting Group.

– one wrong decision with these buildings could result in a negative return. The total result is largely dependent on future developments. The *cash cows* of the real estate portfolio are buildings with a stable income that one expects to be in the portfolio for a long period. To be able to guarantee the continuity of the portfolio, the building should be of a certain size: a small building will just have a small impact on the whole portfolio. The *stars* of the portfolio have a high return in all scenarios but will also have a high risk. Each decision here should be made very carefully.

The results of the risk analysis are a starting-point from which the CRE manager can make his/her decisions. By mapping the buildings in the Boston Consulting Group matrix, the impact of a certain strategy on the different buildings can be visualized. The performance of '*problem children*' has great influence on the overall result of the whole portfolio. By carrying out risk analysis buildings that carry a high risk can be identified. When vacancy occurs in these buildings it will have a negative influence on the overall result, hence looking critically at these building can help to minimize the loss.

Risk management should be undertaken in a proactive manner by identifying the possible risks that have a negative impact on the performance of the real estate portfolio and by taking steps to minimize those risks. The risks can be minimized by trying to adapt the buildings in the real estate portfolio to the standards of the real estate market; this will make it easier to dispose of buildings when necessary. Often, however, a CRE portfolio is characterized by the number of specific real estate units that are plants or production sites. The Dutch GBA, for instance, owns many small buildings in the harbour area of Rotterdam that are occupied by the tax department. These buildings are determined as 'dogs'. They have almost no market value. For the GBA it is very important to have some knowledge of the policy of the board of the tax department so that future demand can be estimated and any decisions regarding the signing of long-term lease contracts with the tax department can be properly informed.

Scenario planning is a helpful tool for generating and evaluating portfolio strategies. It can help to identify and manage the many risks that the CRE manager is confronted with. Many CRE managers are, however, frozen in inactivity when they have to make important portfolio decisions in uncertain times. One thing we know about the future is that it will be uncertain. CRE managers will therefore have to learn how to deal with uncertainties.

12.3.2 Creating a flexible portfolio

As well as understanding the impact of possible change on an organization's portfolio through scenario planning, another strategy adopted by an increasing number of CRE managers is to identify the parts of the portfolio that are likely to be subjected to the greatest amount of change and build flexibility into those assets. Flexibility has become the mantra of most CRE managers, but the real skill is in identifying what type of flexibility is required in each part of the portfolio.

Flexibility has to be examined at both the building and portfolio level. Most literature focuses on the building level and on the physical flexibility of buildings in particular. However, even at the building level, flexibility concerns a much broader range of perspectives. Not only do CRE managers need to consider what the building is physically capable of (e.g., different layouts, ceiling heights, location of columns), they also need to consider both the functional and financial flexibility (Gibson, 2001). Functional flexibility concerns the range of ways a building can be used. For instance, this flexibility could be related to whether the building might support new ways of working, e.g., shared offices, a clean desk policy or hot-desking. Alternatively it could relate to the range of business processes the building could support, e.g., team working, customer service and consultancy. Financial flexibility focuses on the ability of an occupier to exit, that is, to walk away from the building in a minimal time and at a minimal cost. In today's environment, organizations often have limited ability to plan and predict their future space requirements for at least some of their activities and they are therefore in search of 'disposable space'.

But do CRE managers really need all three types of flexibility in all of their buildings? It is recognized in the literature that flexibility has a price: a building with a layout that can be arranged in a wide variety of configurations is likely to cost more than one that is designed to support a single layout. An example of this is the cost of cabling and a cable management system which is highly flexible, yet this in only one of the range of services needed. Similarly, having a building that supports a wide range of functions often leads to different layouts in different areas that change frequently as the organization evolves. Probably the most costly aspect, however, although not yet clearly analysed, is the financial flexibility. What is the real cost of gaining a short contract for space? This is particularly relevant in the UK, where the average length of lease for business space is still well over ten years (Crosby and Lizieri, 1998). If the risk of voids is transferred to the landlord, then the tenant is likely to need to pay a premium.

Financial flexibility is also related to the ability to (quickly) reduce costs or maximize returns when there is a sudden reduction in the space required, or there is a need to raise capital. In order to create flexibility in a real estate portfolio, some organizations choose to lease instead of own buildings. Leasing is, however, not necessarily the most flexible choice as a lease may have several limiting conditions and expose the organization to other kinds of risks. For instance, if the real estate portfolio of a corporation consists of a significant number of leased buildings, it would be important to spread the review and renewal dates of these lease contracts – if all the contracts are to be reviewed within a narrow time frame then the organization is highly exposed to the conditions in the real estate market at that point in time. This may or may not be favourable. As outlined earlier, an organization could find a significant part of its portfolio over-rented. This was the case for a number of organizations in the UK that had gone through fast growth during the late 1980s that then found their rents above market level for much of the early 1990s, Baum and Crosby, 1995).

This is a risk that an organization can minimize by managing their buildings within a portfolio context. The organization might have to pay more for break clauses or shorter leases, but it would be able to reduce its level of property market risk. While this is a difficult trade off, it is one that CRE managers need to make.

If it is agreed that flexibility is likely to come with an increase in costs, although this is still yet to be conclusively proven, then CRE managers must understand what type of flexibility is required, and at what time. This is when portfolio, rather than building level, analysis is most important.

Drawing on the human resource literature, Gibson and Lizieri (1999) developed the concept of a core-periphery business space model. The idea is that different types of flexibility are more important in some parts of the portfolio than others. The model is laid out in Figure 12.4.

The link between human resources and CRE is parallel. At the bottom, the core portfolio is that which adds the greatest value to the organization. These properties are felt to be required for the long-term and, just like core workers, they are expected to be functionally flexible, i.e., to have an ability to be adaptable for a range of uses. An example of a core building might be a landmark headquarters that houses a wide range of functions over its life span or a key research and development unit which is strategically located. For these types of buildings financial flexibility is not important, as it is highly unlikely that the organization would want to vacate them.

Two periphery levels then support this core portfolio. The first level is that portion of the portfolio that would need to fluctuate with the business activity. If times are good, the organization needs more space, and if times are bad then they need the ability to end the contract. There is some focus on physical flexibility in this layer, so that a property can

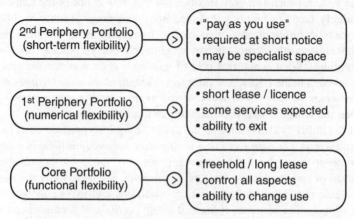

Figure 12.4 Core-periphery portfolios (Gibson and Lizieri, 1999).

support a range of activities throughout a business cycle; however, the need to exit when the economy or market enters a significant downturn is fundamental. By having the ability to exit, an organization can try to avoid having a large number of surplus buildings that need to be managed and paid for. Many organizations have been caught holding surplus assets for a considerable period and this is a significant drain on resources.

The final level is that part of the portfolio required to support high risk and ephemeral activities, such as new market entry, the testing of new products and the setting up of 'special' project groups. These are the requirements that by their very nature will be short-term. For instance, a new product may or may not be successful; if it is successful, additional long-term space will be required, and if not, the space will no longer be required. Similarly, the special project, by its very nature, is bound to come to an end, and therefore only requires space for a defined period. In all of these cases the ability to exit is paramount, therefore financial flexibility is essential, as is the need for speed of entry. Serviced offices provide an effective solution for this layer in the portfolio.

There is evidence that organizations are taking this type of approach. When the Department of Social Security in the UK was reviewing its portfolio prior to outsourcing it to Trillium they undertook a similar analysis, dividing their portfolio into core, flexible and surplus elements. This was done in order to negotiate the amount of flexibility they required within the contract (Evans, 2000). Similarly, a study of 48 large occupiers in the UK found that their preferred split of types of ownership within their portfolio included a mixture of freehold, leasehold and other short term arrangements as shown in Figure 12.5 (Gibson, 2000). In the previously described study for the Dutch GBA, a distinction was made between headquarters and back offices, as well as between specific and non-specific buildings, non-specific buildings being more easily leased than specific buildings.

In the examples described there is an indication that CRE managers are attempting to build flexibility into their portfolios only in cases where it is needed.

There is yet another way that organizations are attempting to gain flexibility in their portfolios; this is driven, in part, by the changes in the way work, particularly office work, is carried out. Staff no longer work from a single location, as the impact of ICT has allowed them greater freedom of movement both inside and outside the workplace. Responding to this change in working practice, organizations have introduced interventions in order to increase the use of space, such as lengthening the operating hours, sharing workspaces and introducing a wider range of settings within the office. These office innovations are often used in the accommodation process as a way of dealing with the constantly changing business environment (Vos and Dewulf, 1999). However, organizations need to understand the applicability of these concepts to their own

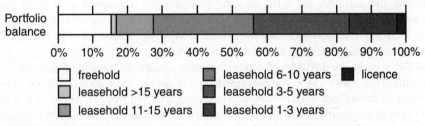

Figure 12.5 Average preferred portfolio split between types of contract (Gibson, 2000).

organizations – they cannot simply copy a concept that has been successful in another organization as each organization is unique and has its own set of activities, context, culture and history.

Nevertheless, these new working practices and alternative workplace solutions are helping organizations gain greater flexibility, particularly within their office portfolios. In some instances these approaches have been able to cope with the expansion and contraction of the workforce without having any significant impact on the total amount of office space required. This is flexibility gained by innovative use of the facilities rather than in the way the space is acquired and held. These approaches have again extended the range of options open to the CRE manager when optimizing their organization's portfolio.

Overall, from a portfolio perspective the objective is to select the appropriate type of flexibility for the range of situations within an organization. Buildings in the portfolio cannot only be assessed in their own right; they are part of a wider portfolio perspective. An ability to understand which buildings can move between functions, business units and applications and which are only suited to a specified use is fundamental to this approach. Additionally, it is necessary to have knowledge of the way in which each of the business units uses its space and facilities as well as their future plans. Therefore, a portfolio approach requires good quality data on both the real estate and the business if the risks and rewards are to be balanced.

12.4 Conclusion

In the last few decades CRE management has been transformed into a professional discipline. It is taught at universities, service providers have developed special CRE advisory groups, and research funds are dedicated to CRE. Despite this growing professionalization, little attention is paid to portfolio management within a corporate setting. Investors see their portfolio as an asset that will produce high returns on investments, however, for a CRE manager the real estate portfolio is a means of support for the core business of the organization. The portfolio strategy has, therefore, to be aligned with the needs of all the stakeholders within the company.

In today's highly competitive world, organizations are confronted with many uncertainties. The demand for space is highly unpredictable and the market developments are equally hard to predict. Modern managers have to find ways to cope with these uncertainties: on the one hand, by developing scenarios, a CRE manager can gain insight into the long-term consequences and risk of certain portfolio decisions, while, on the other hand, the portfolio has to be as flexible as possible.

Despite the fact that the term scenario planning has become a new management buzzword in today's business, few CRE managers are using scenarios to plan their real estate strategy. The examples described earlier show the important role that scenario planning can play in defining a robust portfolio strategy.

Similarly, given the difficulty of making accurate predictions, creating flexibility within the portfolio is another way to combat uncertainty. However, some aspects of a business are more certain than others and it is important to acquire the right type and degree of flexibility in the parts of the portfolio where it is most needed. Additional flexibility often comes at a cost and the CRE manager needs to understand the risk and return trade off.

The point has been made that CRE managers and researchers need to pay attention to portfolio management within a corporate setting, and that more research is needed in this area to supply evidence that will demonstrate the benefits of these approaches. Nevertheless, it is clear that there are real management benefits in treating the CRE assets as a portfolio that, just like a portfolio of investments, can be managed to provide a good return with a manageable level of risk.

References and bibliography

Baum, A. and Crosby, N. (1995) Over-rented properties: bond or equities? A case study of market value, investment worth and actual price. *Journal of Property Valuation and Investment*, **13** (1), 31–40.

Crosby, N. and Lizieri, C. (1998) Changing lease structures – an analysis of IPD data. In: *Right Space, Right Price (Paper 5)* (London: Royal Institution of Chartered Surveyors).

Crosby, N., Gibson, V. and Oughton, M. (2001) *Lease Structures, Terms and Lengths: Does the UK Lease meet Current Business Requirement?* A Report on the Attitudes of Occupiers in the UK for the Royal Institution of Chartered Surveyors (London: RICS).

Depuy, L. and Dewulf, G. (2000) *Sturing op de Portefeuille* (*Directing the Portfolio*). Internal Report of the Dutch Government Buildings Agency (Delft: GBA).

Depuy, L. and Van der Schaaf, P. (2000) Managing a corporation's real estate portfolio. In: Dewulf, G., Krumm, P. and de Jonge, H. (eds) *Successful Corporate Real Estate Strategies* (Nieuwegein: Arko).

Dewulf, G. and Van der Schaaf, P. (1998) Portfolio management in the midst of uncertainties: how scenario planning can be useful. *Journal of Corporate Real Estate*, **1** (1), 19–28.

Evans, M. (2000) The PRIME contract: the Department of Social Security and Trillium. *Journal of Corporate Real Estate*, **2** (3), 208–20.

Gibson, V. (2000) *Evaluating Office Space Needs and Choices*. Report for MWB Business Exchange (University of Reading).

Gibson, V. (2001) In search of flexibility in real estate portfolios. *Journal of Corporate Real Estate*, **3** (1), 38–55.

Gibson, V. and Lizieri, C. (1999) New Business practices and the corporate real estate portfolio: how responsive is the UK property market? *Journal of Property Research*, **16** (3) September, 201–18.

Gibson, V. and Lizieri, C. (2000) Space or service? The role of executive suites in corporate real estate portfolios and office markets. *The Real Estate Finance Journal*, **15** (4) Spring, 21–9.

Keogh, G. and D'Arcy, E. (1994) Market maturity and property market behaviour: a European comparison of mature and emergent markets. *Journal of Property Research*, **11** (3), 215–35.

Lizieri, C., Crosby, N., Gibson, V., Murdoch, S. and Ward, C. (1997) *Right Space, Right Price? A Study of the Impact of Changing Business Patterns on the Property Market* (London: RICS Books).

Vos, P. and Dewulf, G. (1999) *Searching for Data: A Method to Evaluate the Effects of Working in an Innovative Office* (Delft: Delft University Press).

Woudenberg, E. (2000) The impact of corporate real estate on corporate economics. In: Dewulf, G., Krumm, P. and de Jonge, H. (eds) *Successful Corporate Real Estate Strategies* (Nieuwegein: Arko).

13

Project management

Kaye Remington*

Editorial comment

Project management is a profession in its own right. It deals with the delivery of a plan within established project constraints. It is a generic discipline in the sense that it can apply to projects (or events) in almost any field. Nevertheless it has particular significance to the built environment and fits within the general purview of facility management. It is an important area of competency, not only because it applies to acquisition of new facilities, but also because many facility managers operate as quasi project managers or vice versa.

The project management body of knowledge is defined as comprising the following categories:

- scope
- time
- cost
- quality
- human resources
- communications
- integration
- risk
- procurement.

Each area has a particular role to play in the successful delivery of projects, yet it is the linkages between them that are most influential. Major new developments are typically outsourced to external consultants who manage virtually the entire process, but smaller projects or events can be handled by in-house staff where the necessary expertise is available.

* University of Technology, Sydney, Australia

Project management adds value by ensuring that new projects are delivered within stated time constraints at an agreed cost and to the level of quality expected. The field is considered as primarily strategic in its own context, but is more tactical when viewed as a subset of facility management. What is strategic is the decision to embark on the procurement of a new initiative, a decision that clearly lies with the organization hierarchy and is based on advice received. Value can be enhanced by ensuring that appropriate procurement strategies are adopted, that the team is properly briefed, and that adequate opportunity for input and review occur.

On another level, projects are occurring routinely as part of core business activity. These projects must similarly address time/cost/quality issues. They increasingly involve multidisciplined teamwork, short-term spatial requirements, specific equipment and other infrastructure for the organization. By being aware of these projects, their timing and their expected impact on resources, the facility manager can contribute to the productivity of the project team. This is particularly so in office environments where the layout is designed around a project-based culture with shared resources and adaptable technology.

Opportunities for value enhancement exist everywhere, and the smooth delivery of projects and events is no exception. Nevertheless, project management requires specific skills and qualities, particularly an ability to lead and to harness the expertise of others towards achieving a common objective.

13.1 Introduction

Changes in workplace practice can mean radical changes in the way we conceptualize our workspaces. When changes like these are implemented in organizations the application of project management methods can be of great assistance in the success of the project.

13.2 The human factor

Consider a situation where an executive decision is made to convert the work environment to a more flexible one, allowing employees to work from home, coming into the office at staggered times or only for essential meetings. At first glance this seems to be a decision that should please most. It will reduce the need for office space, help lower impacts on peak hour transport and assist people to accommodate family responsibilities, like dropping the children off to school before coming in to the office. Implementation of any change of this kind without a full stakeholder-needs analysis is, however, very dangerous (Cleland, 1998). Take another example, a simple office refurbishment: what are the human factors that will affect the project and how can they best be accommodated, and what are the consequences of not paying enough attention to the needs of the people affected?

Most projects, like the two examples above, require people to change the way they go about doing things. As human beings we are extraordinarily resistant to change. Changes like these can be achieved smoothly and effectively, but they cannot be achieved without good project planning. Not taking the human factors into account at the beginning of the project, and throughout the management of the project, will result in ineffective communications management and poor management of the expectations of the stakeholders. Stakeholders include anyone who could affect, or be affected by, the project. Failure to identify key stakeholders at the beginning of a project and failure to manage the relationships between them and the project is one of the major causes of project failure. The correlation between what stakeholders are led to expect from the outcomes of the project and what they finally receive is crucial. Ed Hoffman, director of the Project Management Centre of Excellence at NASA has suggested that '... project management is all about managing expectations.' (Remington, pers. com., undated).

Communications management and management of stakeholder expectations are inextricably linked in project management. Without good communications management, and that means using a variety of communications means, especially verbal communications, one cannot know or manage the expectations of the stakeholders (Dinsmore, 1990).

13.3 Stakeholder 'mind mapping'

One of the simplest and most effective tools to apply early in projects is the stakeholder 'mind map'. For these exercises it is good to get together as many people who might be affected by the project as possible. This can be the beginning of engaging the stakeholders in the process and it can set the tone for future collaboration and partnering. This kind of approach is preferable to the 'top-down' approach, where people are simply informed about what will happen to them, an approach which tends to engender opposition from the start. All kinds of representatives can be involved in assisting the project manager to identify key stakeholders and key relationships.

The stakeholder 'mind map' shown in Figure 13.1 is that of a simple office refurbishment. Although the initial diagram looks a mess, it can be 'unscrambled' to provide a valuable tool for identifying and managing the stakeholder relationships. When created collaboratively, at the beginning of the project, such an exercise can assist the project manager in establishing the key stakeholders as a cohesive team, working to ensure that everybody's interests are heard and, within reason, accommodated, and to

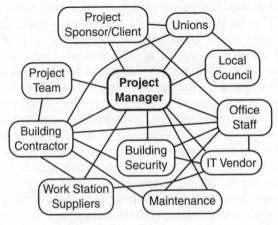

Figure 13.1 Stakeholder 'mind map' for an office refurbishment project.

define the links between the various stakeholders. Often in refurbishment projects it is easy to miss stakeholders who can be the source of problems during the implementation of the project. For instance, the neighbouring office tenants were omitted from the process in the example above, and they proved to be very troublesome because they were severely inconvenienced by the noise and dust associated with the construction. To avoid an adversarial relationship with other tenants in the building or in adjacent buildings, it is better that they be consulted and their needs ascertained during the planning stage, before tenders are let and before the construction begins. Equally, pre-construction meetings, involving all the relevant subcontractors and major suppliers, can solve many issues before they become problems. Often the providers of key service such as security, maintenance or waste management, are forgotten and their vital knowledge overlooked.

13.4 Responding to stakeholders' needs

This is a problematic area. It is essential to uncover what stakeholders *need* from the project. This may be different from what they actually *want*. The project manager has the onerous task of working with the stakeholders to determine what they really need. Often that means not only interviewing them, but also negotiating with them as a group, or individually, so that conflicting needs can be resolved. An example of this is a space reallocation project at a city university (Remington, K., author's personal experience on a project in Sydney, 2001). For universities in most countries, space is one of the scarcest resources, but at a university located in the midst of a prime real estate area, space is a major expense item, and must be allocated and re-allocated with great care. When space becomes available it prompts a rush of applications from various groups within the university, all of whom appear, at face value, to have equal needs. It is the difficult task of the person managing the project to analyse the various applications, make a sound judgement and, in some cases, prevent a major dispute between the applicants. The task requires separation of needs from wants and an important part of the process is the separation of overt requirements, or stated needs, and covert, or unstated, needs. The latter are often related to agendas of which the project manager may be unaware. This kind of analysis takes time and it also requires high-level negotiation skills on the part of the person managing the process (Pinto, 1996).

Often the dilemma for the project manager in this kind of situation is how much it is wise to reveal. Obviously, if the change might result in job losses, the project manager must be very careful about what is revealed, but experience has shown that if all the stakeholders are engaged at an early stage in a negotiated process the key issues that might become problems at a later stage in the project are more likely to be uncovered and collaboratively solved. There is also a danger, during a stakeholder needs analysis, that the stakeholders will be allowed to escape with the project. In this kind of situation the project manager loses all control of the process and the stakeholders take over, with the most powerful voices winning the day. The art of project management in such situations is in preserving a balance between stakeholder participation, which is essential, and project control. This is usually done by setting time limits for responses and applying open, logical processes that allow everyone an equal voice. Time limits for discussion must be reasonable, however, and reflect the complexity of the issues concerned. Often it helps to have a skilled facilitator/negotiator to manage these meetings. Facilitation and negotiation require particular skills that should not be undertaken without extensive training and experience.

13.5 Project objectives and project success

The project objectives should be derived logically from the agreed needs or requirements of the key stakeholders. They should be stated in measurable terms. It is vital that we can measure whether project objectives have been achieved, and how well they have been achieved, so that we can demonstrate whether or not a project has been a success. In some situations measurement might have to be in qualitative terms, for example, a staff satisfaction survey will provide a measurement of whether or not a move from one premises to another has been effectively managed. If the project manager negotiates the criteria for project success with the stakeholders, during the stakeholder needs analysis, and obtains agreement, he/she will be working to objectives that are achievable and relevant to the needs of the key stakeholders (Clarke, 1999).

Objectives are often stated in terms of time, cost and quality. The famous project triangle (Figure 13.2) expresses the relationship between the three objectives and the scope of the project.

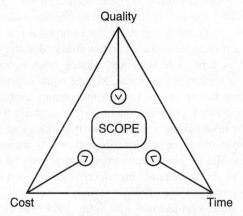

Figure 13.2 Project triangle showing inter-dependency of project objectives in terms of time, cost and quality.

The triangle is sometimes a useful device when the project manager is negotiating with clients, sponsors or key stakeholders, e.g. a project might involve a move from one location to another, with the lease on the existing building terminating at a set date and the new lease commencing at a defined date. In a case like this the key factor is time. To achieve the time objective, which is immovable, the key stakeholders might need to agree to a compromise on the cost or quality objectives. Alternatively the client or sponsor might have to reduce the scope, or content, of the project, or develop a project plan that allows for delivery of the project in a number of stages, as finance becomes available.

13.6 The project plan

Objectives must be clearly defined and agreed before the project can be planned. In some cases, however, objectives are not entirely clear and, due to time constraints perhaps, the project manager must proceed with the implementation of the project or at least provide some kind of plan to senior management. In this case it is more effective to conceptualize the project in stages. It might be possible to plan the first stage with some degree of accuracy but, in the case of a feasibility study, the following stages will be dependent upon the outcome of the first stage. Figure 13.3 suggests a pathway through a project in which the objectives are unclear.

If a project manager needs to provide a project plan for a project with unclear initial objectives it is best to trace a path through a series of scenarios based on the most likely outcomes of each project phase. This will provide a range of time schedules and budgets based on the outcomes adopted. For example, the project manager might take an educated guess that a feasibility study might produce two likely outcomes, B and C, and two

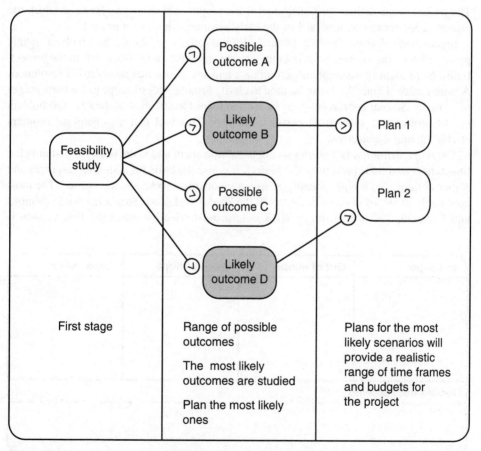

Figure 13.3 Suggested process for 'mapping' a pathway through a project that has a range of possible outcomes, each dependent upon the preceding decisions.

possible but less likely outcomes, A and D. Rather than waste pre-planning time on the less likely outcomes, it is better to concentrate on those that are more likely.

In a project with unclear objectives the aim is to increase the level of certainty so that budgets can be allocated. Project plans can be drawn up for the most likely scenarios; this provides the client with a realistic range of options so that realistic budgets can be set. Rather than planning worst and best case scenarios, which tend to provide a very broad range of outcomes, it is more effective to plan the most likely pathways.

Projects which have unrealistic budgets assigned to them in the initiation phase are doomed to failure. Project plans that involve a carefully developed scope and schedule are more likely also to have realistic budgets from the beginning. The other basis for elimination of unlikely project scenarios is risk. When presenting the options to senior management it is most important to state the assumptions made with each choice and the associated constraints.

13.6.1 Planning project scope

The project scope defines the boundaries of the project. There are two levels at which the project scope should be analysed in the early planning stages of a project.

Broad scope defines the 'big picture'. Key stakeholders should be involved in this process, which determines what is in the project and what is definitely not in the project. It also helps identify assumptions, constraints and any issues that need further resolution. A simple table (Figure 13.4) can be used to clarify broad issues of scope in the early stages of a project. Similar instruments may be used to focus the attention of the key stakeholders on the project and away from personal agendas. This tool can also help stakeholders clarify the major objectives.

The *scope definition* is a much more detailed document and is the basis for planning the time schedule and the budget for the project. It is also the basis for assigning resources and responsibilities and it helps identify areas where quality checks are appropriate. The most commonly used tool in scope definition is the Work Breakdown Structure (WBS) (Simons and Lucarelli, 1998). Creating a WBS is also most effective when the first version is

In scope	Out of scope	Assumptions	Constraints
Issues not resolved			

Figure 13.4 'Big Picture' scope matrix.

created as part of a co-operative process. This is another opportunity to engage the key stakeholders and benefit from their input and expert knowledge. 'Sticky notes' can be used for this process: a 'bottom-up' technique is useful as stakeholders are asked to identify and write down on separate 'sticky notes' all the tasks they would need to achieve in order to complete their part of the project. The 'sticky notes' are placed on a wall or white board and arranged and re-arranged into a workable sequence, then gathered up and converted into a scope definition document for sign-off or amendment by key stakeholders.

It is not strictly necessary to break down subcontracted activities, however, it is useful to do so if the project manager is not fully confident that the subcontractor has the capacity to deliver the subcontract to the required functionality and on schedule. The process facilitates management of difficult subcontracts. Also, during the tender creation and evaluation stages, a well-constructed WBS makes the writing of the documents and evaluation of the tenders much easier (Turner, 1999; Cleland, 2000).

Once the tasks have been identified the project manager can provide the logic that underpins the organization of the WBS. The WBS can be based on the major deliverables, i.e. the products of the project, as in the example in Figure 13.5.

Alternatively a WBS can be organized according to the project phases (Figure 13.6).

The two WBSs shown have been devised from different perspectives and positions in the management of the activities that make up the project.

A question often asked is: How far should one go in breaking down the tasks or activities? Obviously it is possible to go on forever, which would be a waste of time and totally unnecessary. The answer depends upon the planning stage and whether or not the

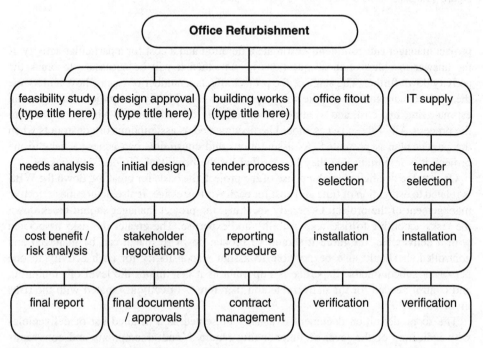

Figure 13.5 Simple WBS defined in terms of major deliverables.

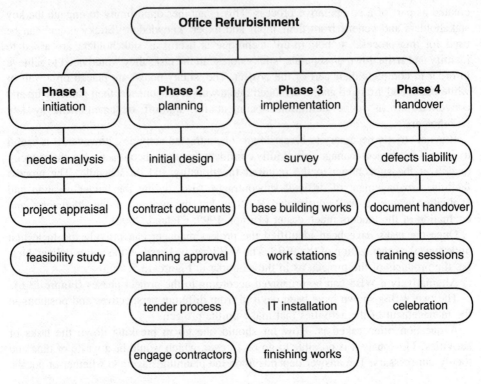

Figure 13.6 Simple WBS defined in terms of project phases.

project manager can confidently estimate a duration and a cost for a particular activity. If the answer to either of those questions is 'no', then it will be necessary to break the activity down further. The stage of the project affects a number of issues: how much time the project manager can devote to the plan, how confidently the project manager can estimate time and cost, and what degree of certainty the stakeholders wish to obtain from the process. All of these issues should be documented as assumptions or constraints when the project plan is presented. As assumptions and constraints can generate risks to the project, it is important that they be clarified for the benefit of any decision-makers.

Once in the detailed planning phase of the project, the basis for breaking down the WBS is related to estimation of time and cost for each work package. It also determines effective management of the project. Generally speaking, the project manager should break down the tasks according to risk, responsibility and expertise. The greater the risks associated with a deliverable, the more it should be broken down so that it can be managed and controlled. It should also be possible to assign responsibility for each activity to one person or subcontractor. Expertise is important as it determines the level of monitoring and control needed for the deliverable and therefore is directly associated with the risks involved.

The scope definition document is usually presented as a detailed list of deliverables with indicators of the level of functionality required, and assumptions and constraints noted. It is important, however, to indicate in the final scope definition document what is not included in the project scope. This will avoid any confusion at a later date as any issues can be negotiated before signing.

13.6.2 Planning the schedule

Three kinds of modelling technique are commonly used for planning and managing time on a project. These are:

- milestone plans
- precedence networks
- bar or Gantt charts.

Whereas milestone plans and bar or Gantt charts are most commonly used for communication of time, a precedence network is the most effective planning tool.

Milestone plans are excellent for communicating a sense of urgency to stakeholders. Simple to construct and easy to remember, they indicate the key dates to be met within a program, which often correspond with the second level of the WBS. Often a milestone plan is all that is needed for a simple project or one that has a short time frame but generally they have role as a communication medium rather than a planning tool. An example is shown in Figure 13.7.

The precedence network (Figure 13.8) is a much more sophisticated tool for planning time as it is based on the WBS and takes into account the relationships or dependencies between tasks or activities (Gido and Clements, 1998). In the initial planning stages of a project it is particularly valuable to involve key stakeholders or those with particular expertise in planning the time. This process, which can be achieved usually in a couple of hours with some user-friendly 'sticky notes', achieves a number of aims: contribution of knowledge by people who have had previous experience with particular tasks is invaluable, the process helps obtain commitment from those involved, and problems associated with sequencing of tasks are often solved in this context.

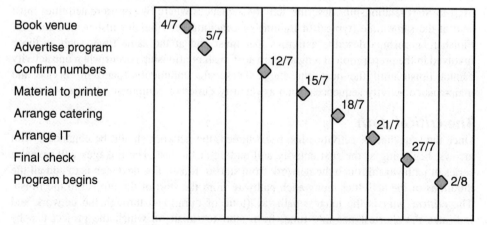

Figure 13.7 Milestone plan for a training session to be held during project handover.

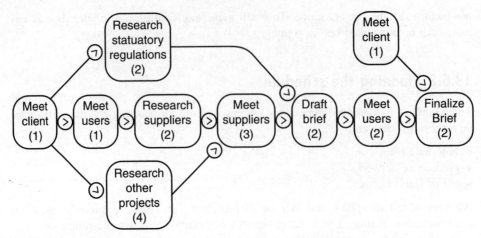

Figure 13.8 Simple precedence network for development of a project brief.

The aim is to re-arrange the tasks defined in the WBS into a sequence that takes into account those tasks that must precede others. Opinions vary regarding how best to proceed. Generally it is preferable to arrange the tasks first, without regard to time constraints, as it is better to arrive at a realistic understanding of the total time needed to complete the project first. Once the tasks have been arranged in sequence, durations must be assigned to them. The duration is the time taken from the start to finish of a particular task or activity. Then, if the time frame determined using a realistic assessment exceeds the overall time objective for the project one can either negotiate for more time, based on some real figures rather than guess work, or 'crash' the schedule by organizing for certain tasks to be done in parallel with others. The latter will usually imply a cost increase as extra resources will be required but either way a reasoned decision can be made.

Types of dependency relationships

The most usual relationship between activities is Finish-to-Start, i.e., one activity must completely finish before the next one can start. Other relationships can exist however, e.g., Start-to-Start relationships describe activity sequences where two or more activities must start at the same time (typical of training or coaching activities for instance). Finish-to-Finish relationships describe activities that must end at the same time, such as those involved in the preparation of a meal. Start-to-Finish relationships occur when one activity cannot finish until the next one starts. These are commonly found in service and maintenance activity sequences, such as security duties or equipment change over.

The critical path

Once the sequence of activities has been agreed, the network should be connected with arrows, beginning at the first activity and ending at the last. The linkages will reveal a series of pathways through the network from start to finish. The next step is to add all the durations of the activities along each pathway from the start of the project to the finish. The *critical path* is the longest pathway (total of durations) through the network and indicates the shortest possible time, from start to finish, in which the project can be completed (Figure 13.9). If the project manager needs to 'crash' the schedule in order to

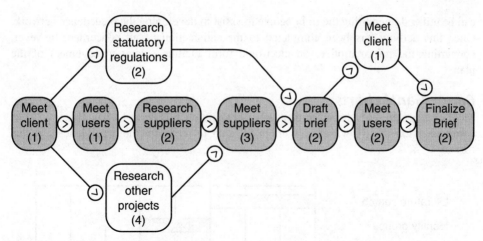

Figure 13.9 Clearly the longest pathway through the network is the shaded one. This is the critical path and indicates the shortest possible time to complete this phase of the project.

reduce the planned time for completion the activities on the 'critical path' should be investigated first to see if they can be shortened in duration, by, for example, adding extra resources. Alternatively the project manager might look for activities that do not have to be sequential but could be completed in parallel.

Contingency

Where to add a little extra time, just in case, is often hotly debated in project management circles. A useful rule of thumb is this: in general, do not add contingency to individual tasks unless a particular task is critical and has particularly difficult resource implications. For instance, a task that only certain people or contractors can perform, that is on the critical path and has several dependencies, might be a possible contender. If time constraints allow, it is always wise to include a small amount of contingency, to be managed by the project manager, at the end of the project.

It practice it is common to have more than one schedule: one to show the client or sponsor, and another working schedule, that reflects the true position at any time. The reason for this is management of expectations – if the client knows that the project is ahead of schedule he/she might make arrangements to let space, for instance, ahead of the time originally scheduled. The project manager must be very sure that the project is ahead of time before the client's expectations are raised in that respect.

Scheduling software – a word of warning!

Several relatively inexpensive software packages are available that are advertised as 'Project Management' software. These can be very useful once the initial planning has been done, but substituting them for the participative processes suggested above is not recommended as the great potential value of the participative process will be lost. Many errors are made simply because one person has created the WBS, by themselves and without consultation, using a software scheduling package. It is extremely difficult to visualize all the complex relationships between activities while restricting oneself to the narrow view of a computer screen. The scheduler also misses the wealth of experience that can be utilized by inviting the right people to assist in developing the precedence network. Once this activity has been completed to the satisfaction of all concerned, however, converting the information to an electronic form facilitates later management of the plan.

Bar or Gantt charts

Most software presents the schedule as a bar or Gantt chart (Figure 13.10). This is an ideal communication too. However, it is more difficult to represent the dependencies in this format.

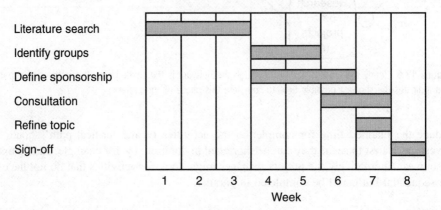

Figure 13.10 Bar or Gantt chart for a part of a sponsorship project.

The master schedule and subplans

In most cases the project is best divided into a series of subplans. These represent relatively discrete subprojects, which together make up the entire project. It is important to develop these plans in such a way that they interrelate correctly. The subplans can then be rolled up to create the master schedule. Cascading schedules in this fashion makes it easier for the project manager to visualize the entirety of the project whilst having the facility to 'drill down' into the detail of the subplans at any time.

One common source of difficulty with respect to the project schedule is failure to relate parallel projects or subprojects via a master schedule. As most large projects are divided into a series of subprojects, which might be separate contracts, dependencies will exist between the subprojects, or indeed between tasks in different subprojects. It is essential, therefore, when planning the schedule to consider all possible subprojects and break them down into sufficient detail to reveal any dependencies that might exist between tasks in different subprojects. Scheduling errors often occur as a result of the omission of the precedence network planning when the project is conceptualized using bar or Gantt charts alone. For instance, a project for the refurbishment of an office space involves several subprojects and corresponding subcontracts; these might include an IT subproject, a furnishings and fit-out subproject (that might involve several subcontracts), a subproject to achieve the relocation of staff from one place to another, a security subproject, a staff communications subproject and a data transfer subproject, so that the business runs smoothly during the transition. The smooth operation of such a project will depend on sequencing all the associated activities in such a way that all dependencies are revealed and accommodated in the master schedule. It is not enough to see these subprojects as separate streams in a bar or Gantt chart, e.g., security will have to be consulted before the fit-out teams have commenced and will need to come in again before the fit-out is completed, again on delivery of the computer equipment and removal of files and equipment, and once again before the staff arrive to occupy their stations so that PIN numbers and staff IDs are allocated. The tasks in the security subproject are thus interwoven into the fit-out subproject, the IT subproject and the staff communications and relocation subprojects.

Resource planning

The resources needed for the project should be planned carefully and a Resource Schedule drawn up so that they are available in time. Resources can take the form of equipment, raw materials, supplies, space and human effort. If resources arrive too early there are storage, security and insurance implications and therefore additional costs. If resources do not arrive on time they can have a major effect on the schedule and often on the final cost of the project. Even when the contract comprises a series of subcontracts, nominally the responsibility of others, the project manager is responsible for ensuring the co-ordination of the subcontracts so that resources are available according to the master schedule. For instance, in the office refurbishment project discussed above, late arrival of the carpet could hold up the installation of the work stations, the installation of the computers and hence the availability of the space as a whole. In the interim the staff must be housed elsewhere, entailing extra cost to the project, lowering of staff morale, causing inefficiencies in the business and possibly loss of business as planned communications transfers are disrupted.

Other activities must be accounted for in terms of human resource implications. These activities are most likely to be associated with the management of the project. They will include such things as time to attend meetings. This will incur a cost to the project not only in terms of project management fees, but also as people associated with the refurbishment and the removal of staff and equipment will need to be able to plan their time so that they can be available when required and their full participation ensured.

13.6.3 Planning the budget

Often the client, sponsor or senior management sets an initial budget, and commonly this will be determined before the preparation of the project plan. It will have been determined by other priorities, such as available funds, or corporate and investment priorities. Often the project will be just one of many competing for scarce funding. The importance of planning a realistic budget for the project as early as possible cannot be over-emphasized – most projects that fail because of over expenditure do so because there is a mismatch between allocated and planned budgets that is not properly resolved at the beginning of the project. It is at this stage that the project manager has the opportunity to negotiate changes to the budget or the scope that will significantly increase the chances of the project meeting agreed objectives. The results of any negotiations that might involve changes to scope, budget or quality at this stage must be communicated to all stakeholders and signed-off by all key stakeholders.

There is a definite sequence that must be followed when planning a project: the first step is planning scope, then time, and finally resources and cost. The time schedule cannot be planned until the scope is defined as the activities cannot be sequenced until they are identified; equally the time schedule will have an impact on the resources and costs for the project, particularly those involving allocation of human resource time or 'effort'.

Duration versus effort

Some activities are subcontracts, in which case they are assigned costs through tender processes and the allocation of work hours is the responsibility of the subcontractors. Most projects also require time for individuals, including the project manager and the project team, to work on aspects of the project or particular activities. This is commonly known as *effort*. One significant cost which falls into this category, and which is often omitted in the cost planning stage, is the cost of the project management activities. These include the time needed to conduct meetings with stakeholders and contractors, to conduct negotiation sessions or value management sessions, time taken to plan the project and manage the schedule and budget, as well as time for co-ordination, monitoring and control of the project.

13.6.4 Planning for quality

Quality in projects, as in almost any other sphere of activity, often involves subjective measures based upon shared meanings and understanding between the various stakeholders. Achieving the desired quality in a project involves establishing and maintaining the quality of the project processes and the quality of the products of the project. Ideally these two aspects of project quality are inseparable. In planning for quality, appropriate measurements must be established that are relevant to the needs of the key stakeholders. In the construction industry, quality standards define minimum requirements for safety, durability and waterproofing, but they may not address other issues of quality that might be important to key stakeholders. These might include issues associated with the image of the business, the ambience of the work environment for staff or efficiency in terms of energy consumption. These quality standards can only be determined by consultation with the key stakeholders. Some quality requirements are relatively easy to measure: it is quite straightforward, for instance, to measure whether or not there is sufficient daylight in an office space or whether the temperature is within a comfortable range. Other quality requirements are particularly difficult to measure, such as those associated with aesthetics. To establish quality standards for project delivery with regard to these more subjective measures it is necessary to allow architects and designers time to spend with the clients. Design schemes should be illustrated with examples and models so that the clients are confident that the project will satisfy their needs with respect to these less objective but equally important aspects of project quality.

13.6.5 Planning communications

Maintaining good communications on a project such as a building refurbishment, an office re-location or workplace re-design is essential to the effective management of the project and realization of the project objectives. Communication failure in projects that have multiple key stakeholders is the most frequent cause of poor project performance and the results can be anything from cost overruns to dissatisfied clients and key stakeholders (Dinsmore, 1990). Planning for good communications at the start of a project is essential. The stakeholder 'mind map' provides an excellent basis as it helps the project manager to visualize not only who the stakeholders are but also the networks that connect them to each other. It is important, first, to establish the communication protocols for the project and this can be achieved as part of a stakeholder needs analysis. The communications protocols should specify who will be communicating with whom, when they should receive information, their responsibilities and what is expected of them, the format of documents and templates to be used when reporting or requesting information, the media to be used, the software platforms, and so on. The communications plan sets out all the information needed to manage the communications for the project.

Informal versus formal communications

Much is achieved in negotiation and reaching agreement using informal communications such as telephone conversations and informal meetings. While such informal communication is like the glue that cements relationships between the various parties, there is an old project management axiom:

What has not been written has not been said!

It is important to confirm all agreements, understandings and the like in writing. It is not only important to maintain project records, particularly in case of any legal proceedings that might ensue, but confirmation in writing provides the other party with an opportunity to confirm or deny what was understood as a result of the informal communication. Communication, as a process, is fraught with potential difficulties and every precaution must be taken so that meanings are clarified before they become misunderstandings.

13.6.6 Planning procurement needs

Project procurement involves the procedures necessary to obtain all the goods and services necessary to deliver the project. It involves preparation of tender documents, evaluation and selection of tenders and suppliers, and preparation of contracts for the supply of the goods and services nominated. Using the WBS as a basis the procurement plan sets out all the goods and services associated with each of the activities or tasks, indicates when they are required, necessary lead time (if appropriate) and the names of the suppliers or subcontractors. It also indicates the type of procurement method (tender and selection process, choice of contract, etc.) and who is responsible for managing each process. In summary, the procurement plan provides the basis for the project manager to manage the procurement needs for the project. In some cases, where there is a series of subcontracts, the project manager does not organize the procurement for those subcontracts personally, but the project manager needs to keep a check on procurement by subcontractors in order to ensure the overall co-ordination of the various subcontracts that make up the project. Successful contract management during the implementation of the project depends upon satisfactory procurement planning. Contracts cannot be effectively managed if the conditions of contract have not been written to reflect the special needs of the project (Cleland, 1994; Pinto 1998).

13.7 Project organization

In some organizational environments, particularly where organizational governance is not by projects, but where the organization is arranged according to functions, approval processes can be unclear or complicated. This is less often the case where projects are managed by organizations that are themselves organized along project lines. It is useful, early in the project-planning phase, to construct some kind of project organizational chart that reflects the reporting and approval relationships between the various parties. Figure 13.11 shows a typical project organizational chart with the project manager at the centre.

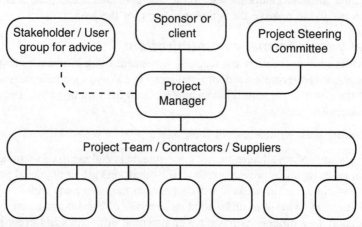

Figure 13.11 Project organization chart indicating relationships between the project team, subcontractors and the key entities responsible for approval or advice to the project.

When approval processes are unclear it is vital that the project manager clarifies which parties need to approve the various stages of the project before the project gets underway. In some cases the project manager will need to negotiate the roles of the various approval committees and gain agreement, particularly where the pre-defined levels of clarity of responsibility are low amongst the various parties. For example, the role of a stakeholder-user group might be simply to give advice, however, the boundaries of their responsibilities must be established and communicated at the outset, otherwise such a group might take over roles which are not appropriate and which other groups within the organization are fulfilling. If boundaries of responsibility and levels of approval are not established before the project gets underway the project manager can find that he/she is trying to manage a minefield (Pinto, 1996; Turner, 1999).

13.8 Project risk

There are many other aspects to project planning which have not been mentioned. One important aspect is risk planning; it is good practice to start thinking about potential risks from the beginning. Potential risks can be associated with particular stakeholders, they arise from assumptions and constraints associated with scope definition and the schedule. Many risks, if triggered, have cost or time implications or both. Early identification of potential risks will contribute significantly to management of risks if or when they are triggered later in the project (Wideman, 1998).

13.9 Implementing and controlling

Project execution is made simpler with a well-developed project plan, however, the plan is not cast in concrete and will be subject to change Once the project plan has been finalized it should be formalized as the baseline against which the project is managed. While the project plan is not a static document, the final agreed plan, produced at the end of the planning phase, must be preserved so that progress on the project can be monitored against an agreed baseline.

Changes can be imposed on the project with respect to scope, time and cost, and/or risks can be triggered, or planning assumptions may have been wrong. An agreed change management procedure must be in place to cope with changes to the project. Most changes to scope will result in changes to the planned budget and/or the planned schedule. The change management procedure must include a process for evaluation of the impact of the potential change, communication of the proposed change and negotiation with key stakeholders. Once the impacts of the change have been assessed and agreed, a process of authorization is essential before the change is implemented. Establishing who, amongst the key stakeholders, has the responsibility to 'sign-off' must be agreed before the implementation of the project begins. Often, due to tight time frames, key stakeholders are left out of the change management process. For instance, an office refurbishment project recently completed was subject to a major change during construction – because of cost over runs in other areas of the project, less expensive workstations were substituted for those originally chosen. The changes were agreed between the project manager and the client but neither the office staff nor the IT vendor was consulted. The new workstations, once installed, did not meet the IT requirements and, in some cases, did not cater for the working conditions of staff who needed special desk areas and storage facilities. The users judged the project a failure.

13.10 Managing non-contractual relationships

Aspects of the project will be managed according to the conditions of various contracts with suppliers and subcontractors. As mentioned previously the ability to manage effectively depends upon the care with which the contracts are written and the tenders evaluated and let. Managing non-contractual relationships depends upon careful identification of stakeholders and effective communication and reporting throughout the project life cycle. Stakeholders, like neighbours, can become potential risks to the project if they are not managed appropriately – that means taking the time to talk with the stakeholders in question, and keeping them informed about progress and how it will affect them. Any informal communication must be backed up in writing, even if only in the form of a simple note for confirmation of what has been said or agreed.

13.11 Project reporting

A reporting schedule should have been created as part of the communications plan. Project reporting requirements will change depending upon the phase of the project but during implementation, reporting drives the project. Different stakeholders will need different kinds of information at different times. Taking the building refurbishment project as an example, the client or sponsor will require regular updates on the schedule and the budget, and contractors will require regularly updated and accurate information regarding any variations to the contract documents. Staff will require information about how the project will affect their work processes, and notification of any changes to previously agreed schedules and lead times so that they can arrange their own work environments, contacts, and so on.

13.12 Project closure

Closure involves administrative handover and commissioning of the project. It should be planned as part of the scope definition but more often than not is left to last. It includes activities such as the transfer of 'as-built' drawings and specifications to those who will be taking on the management of the facility, and transfers of guarantees and warranties. There might also be a defects liability period during which defects may be identified and rectified by contractors or subcontractors. Commissioning of the project might also require preparation and handover of instruction manuals, and development and delivery of training to users or support services. This is also the time to identify additional projects that might have evolved from work shed during scope definition or because of changes to scope, schedule or budget due to the triggering of risks during the project life cycle (Turner, 1999).

13.13 Post project review

A post project review is designed to capture lessons learned from the project and also to examine whether the benefits expected from the project have been obtained. The timing of the post project review is important. Gathering experience related to the project process itself is best done while the process is underway; while it is rarely done, identifying the lessons learned during the process simply requires collation of existing documentation. A post project review that assesses the quality of the outcomes of the project should take place after a reasonable period of operation so that people can make a proper assessment of the viability of the project outcomes. This could be 3, 6 or 12 months after the project handover (Verma, 1998; Pinto and Slevin, 1998). A business benefits review assesses the benefits that the project outcomes have contributed to the organization. An outcome will have different meanings for different people, and might be seen by some as a benefit and by others as a disadvantage. For instance, a workplace change project, involving people working from home or staff sharing desks and other facilities, might be seen as an advantage to staff with young children or a disadvantage to staff who liked their private offices and now feel displaced. The outcomes of the review will depend very much on how the project has been managed. Often it is necessary to think laterally when managing a project – perhaps those that have lost their private domains in the refurbishment can be offered some other recompense by management, which recognizes their contribution to the organization. This means engaging management in devising creative solutions to aid in the management of potentially difficult stakeholders.

13.14 Conclusion

Successful project management processes that contribute to successful outcomes depend not only on selection of the right tools and techniques but primarily on the establishment and maintenance of excellent communications throughout the entire project life cycle. Particularly where projects involve significant changes to work practices or work environments, the successful management of complex relationships between parties and individuals is fundamental to project success.

References and bibliography

Ayas, K. (1996) Professional project management: a shift towards learning and a knowledge creating structure. *International Journal of Project Management*, **14**, 131–6.

Clarke, A. (1999) A practical use of key success factors to improve the effectiveness of project management. *International Journal of Project Management*, **17**, 139–45.

Cleland, D.I. (1994) *Project management: strategic design and implementation* (New York: McGraw-Hill).

Cleland, D.L. (1998) Stakeholder Management. In: Pinto, J.K. (ed.) *The Project Management Institute: Project Management Handbook* (San Francisco: Jossey-Bass).

Cleland, D.L. (2000) *The Project Manager's Portable Handbook* (New York: McGraw-Hill).

Dinsmore, P.C. (1990) *Human Factors in Project Management* (New York: American Management Association).

Evaristo, R. and van Fenema, P.C. (1999) A typology of project management: emergence and evolution of new forms. *International Journal of Project Management*, **17**, 275–81.

Friend, J., Bryant, D., Cunningham, B. and Luckman, J. (1998) Negotiated project management: learning from experience. *Human Relations*, **5**, 1509(3).

Gido, J. and Clements, J.P. (1998) Network planning and scheduling. In: Pinto, J.K. (ed.) *The Project Management Institute: Project Management Handbook* (San Francisco: Jossey-Bass).

Grundy, T. (2000) Strategic project management and strategic behaviour. *International Journal of Project Management*, **18**, 93–103.

Keegan, A. and Turner, J.R. (2000) The management of innovation in project based firms. Part of the series *Research in Management* (Rotterdam: Erasmus University).

Parkin, J. (1996) Organizational decision making and the project manager. *International Journal of Project Management*, **14**, 257–63.

Pinto, I.K. (1996) *Power and Politics in Project Management*, Sylva, NC: Project Management Institute.

Pinto, J.K. (1998) Power and politics in project management. *International Journal of Project Management*, **16**, 199–200.

Pinto, J.K. (2000) Understanding the role of politics in successful project management. *International Journal of Project Management*, **18**, 85–91.

Pinto, J.K. and Slevin, D.P. (1998) Critical success factors. In: Pinto, J.K. (ed.) *The Project Management Institute: Project Management Handbook* (San Francisco: Jossey-Bass).

Simons, G.R. and Lucarelli, C.M. (1998) Work breakdown structures. In: Pinto, J.K. (ed.) *The Project Management Institute: Project Management Handbook* (San Francisco: Jossey-Bass).

Tullett, A.D. (1996) The thinking style of managers of multiple projects: implications for problem solving when managing change. *International Journal of Project Management*, **14**, 281–7.

Turner, J.R. (1999) *The Handbook of Project-Based Management. Improving the Processes for Achieving Strategic Objectives* (London: McGraw-Hill).

Turner, S.G., Utley, D.R. and Westbrook, J.D. (1998) Project managers and functional managers: a case study of job satisfaction in a matrix organization. *Project Management Journal*, **29**, 11–19.

Verma, V.J. (1998) Critical Success Factors. In: Pinto, J.K. (ed.) *The Project Management Institute: Project Management Handbook* (San Francisco: Jossey-Bass).

Wideman, R.M. (1998) Project risk management. In: Pinto, J.K. (ed.) *The Project Management Institute: Project Management Handbook* (San Francisco: Jossey-Bass).

Yeo, K.T. (1996) Management of change – from TQM to BPR and beyond. *International Journal of Project Management*, **14**, 321–4.

14

Asset management

Bernie Devine*

Editorial comment

Assets are defined as inventory, plant, equipment and other facility contents, and are distinct from real property like land and buildings. Assets are managed through registers that list specification details, acquisition dates, serial numbers, monetary valuation, insurance, warranties, location, maintenance requirements and other information to maintain effective control. Asset management is more of a tactical competency, but includes some strategic aspects in relation to buy or lease decisions, security considerations and workplace layout. Computer assets (hardware, software, networks and data) are commonly managed separately as part of an information technology portfolio, but this is changing as facilities and technology become more integrated.

One of the larger asset classes, at least for commercial buildings, is furniture. The choice of furniture can be a strategic decision for open plan office applications given that flexibility, reconfiguration and plan density are primary considerations. Modular furniture capable of multiple arrangements without the need for structural or technology modifications is preferred. Furniture must be ergonomically designed, aesthetically pleasing and, increasingly, environmentally benign. Privacy, noise reduction, personalization and colour are also important. While purchase cost remains an issue, the contribution that furniture might make to workplace productivity is far more relevant.

Call centres are relatively modern inventions that demand unique furniture solutions. The type of work undertaken there and the stresses of dealing with customers (and customer complaints) continuously through the day require workstations that are comfortable, computer-activated, open plan and non-territorial, yet with noise and visual privacy characteristics. Many call centre operators have invested in shared spaces where employees can 'chill out', relax, maybe do some reading or play computer games, before returning to the work area. The high rate of turnover of staff in these environments has put

* Butler and Devine Management Services, Sydney, Australia

pressure on making the spaces more human, as training costs are the key financial factor. Churn associated with asset reconfiguration can be minimized by all data being housed electronically and made available from any workstation.

Graphics-based software is used frequently to manage asset registers. Database details are linked to graphic images as attributes, and images are viewed along with other building elements in a three-dimensional computer-aided design system. This approach has the advantage of allowing information to be accessed easily, as well as providing the opportunity for the entire facility to be co-ordinated through a single interface. Realistic walk-through representations of floor layouts can be generated to assist the management function and to test configuration strategies. All these innovations help to add value and deliver more productive workplaces.

This chapter ranges from the strategic view to process and performance. As such, it introduces a diverse set of issues for consideration. It is aimed more at provoking innovation and improvement than at providing structured learning. The author has spent more than ten years working on over 100 projects with an asset management focus across the Australasian region. Most of the material presented here is based on observations from those projects.

14.1 Introduction

The day-to-day management of the operations of a facility requires a wide range of knowledge, investment, time and patience. Asset management as a discipline can be viewed as a high level strategic function focused on value enhancement and it can be seen at the micro level in the day-to-day operational decisions of running building services. The fact is that asset management concepts permeate every level and every activity in the operation and management of facilities.

All functions within the operation of a facility should focus on the needs of the customer; we are beyond the time when debate or explanation of this fundamental is required. The management of assets requires an understanding of what the client wants, what the organization is able to provide, and the technical needs of the assets themselves.

This chapter examines a number of aspects of asset management with a focus on value creation. Key concerns are:

● understanding what assets are in the portfolio and managing their performance
● how technology acts as enabler and transformer of the asset management function
● how asset management practices and processes can help optimize value in a property portfolio.

14.2 Asset management

The work of asset management began in the defence forces as a structured approach to managing complex equipment and infrastructure. Much of the current literature is based on some core engineering and systems management concepts originally outlined by the US military (USAF, 1992). The 1992 US Military standard was preceded in 1974 by a higher level standard (USAF, 1974) that can be seen as the genesis of the discipline, with the 1992 document providing the detailed definition that is the basis of most of the more recent work.

The second, equally important, component of asset management is an international standard on quality management, the ISO standard on Configuration Management (ISO, 1995). This provides for the transformation of chaos into order and brings the structure of systems engineering management to focus sharply on the needs of the customer. It is necessary to look at the details of these core publications to properly understand asset management.

In the context of asset management a broad definition of assets is assumed, i.e., physical assets, including all physical items ranging from plant and equipment, through furniture and fittings including information technology (IT) assets, fleet, land and buildings. This definition can be expanded to include space occupied or equipment operated under lease as this space or equipment is effectively an asset that needs to be managed.

14.3 Asset registers and configuration management

From an asset management perspective, understanding what assets an organization has and how they should perform is critical. Asset registers have been within the traditional purview of the accounting department with their main purpose being to record the existence of various assets and their cost. These systems began as physical ledgers or card systems many of which have now been converted to computer databases. The conversion process (from paper based to computer based or from one computer system to another) often highlights the inadequacies of existing data and the challenges that will face those who will manage the data in the new system.

In any case the management of the asset register in a large organization is a challenging task. As mobile technology becomes both more pervasive and more valuable, the challenge of tracking these assets becomes increasingly complex. More important than merely tracking assets is ensuring that the organization's overall investment in assets is delivering the value that was intended.

A key element of setting up an asset register is ensuring that appropriate business processes and data structures are in place. Business processes should focus on configuration management activities.[1] Configuration management is a subdiscipline of systems engineering that is focused on how individual components, subsystems and aggregations of assets are 'configured' so that they work together efficiently and effectively. In other words, a configuration item is a set of assets that have an operational relationship. Configuration management activities are both technical and organizational in nature and include:

- configuration identification – determination of structure, selection and documentation of configuration items (subsystems, components and other aggregates of assets), their functional and physical characteristics and provision of an identification schema that shows what the items are and how they are configured together
- configuration control – activities comprising control of changes to a configuration item (this includes changes to actual configurations as well as amending configuration documentation and would include analysis, design, review, approval, procurement, work, reporting and quality assurance)

- configuration status accounting – formalized recording and reporting of established documents, status of proposed changes and approved changes being implemented
- configuration auditing – examination to confirm that documentation matches the configuration item.

Configuration management should be organized with defined responsibilities and sufficient independence of authority to achieve the required configuration management objectives. The purpose of configuration management is thus to ensure that the correct assets are in use and their management is co-ordinated and appropriate in a system context.

Systems engineering models provide a hierarchical structure for the definition of individual elements of a system and the breakdown of asset structures. Figure 14.1 illustrates the concept of asset structures.

Invariably, organizations develop a business case for investment in new areas of business and in new assets. Often the business-based investments allow for a quantum of infrastructure without any real regard for the way this is provided or how it should perform. Asset-based investment proposals typically focus on the return on investment the new asset will provide – but it can be difficult to find an organization that actively harvests this improvement and monitors its achievement.

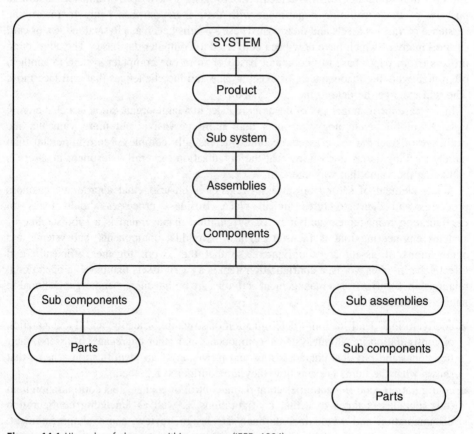

Figure 14.1 Hierarchy of elements within a system (IEEE, 1994).

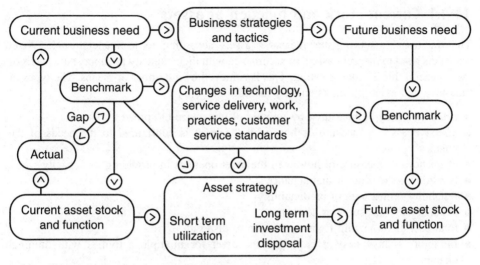

Figure 14.2 Conceptual asset management model.

Effective asset management relies on understanding what the customer needs, both now and in the future, understanding what that means in terms of facilities and services and ensuring that the most cost-effective approach to managing the delivery and operation of those facilities and services is taken. This is shown conceptually in Figure 14.2.

14.4 Current business need for infrastructure

One of the first realizations most people come to in asset management is that an organization's demand for capacity changes in a relatively linear fashion whilst supply can only be provided in a stepped fashion. For example, an organization typically changes its head count one person at a time, excluding mergers and acquisitions. Yet, real estate (say, office space) comes in lumps. Minimizing the mismatch between the step change in supply and the smooth change in demand is one of the keys to success.

For this is to be done well the responsible person requires an intimate knowledge of core business processes. Being able to identify the elements in the total configuration of assets that are the core building blocks and basic constraints to business process capacity is the key. For example, in the hospital environment, each group of operating theatres relies on a recovery area where patients can be monitored while they recover from anaesthetic. The capacity of this area is designed with a specific recovery time and number of cases in mind. To optimize utilization of the expensive capital investment in the theatres and the high operating cost of theatre teams a manager must match surgical case times, the recovery rate for various anaesthetics and length of each shift's activity in order to ensure there is no risk of over-utilization while maximizing cost recovery.

14.4.1 Current stock and function of assets

The traditional asset register measures assets according to a financial scale. Asset managers need to measure assets in accordance with their capacity to support the needs of the business; this is value creation. This is achieved by a number of means. The types of measures that asset managers employ include:

- capacity – may be measured by number of units, size or spread
- effectiveness – a measure of how well the asset is configured to the needs of the business
- efficiency – a measure of how well the asset operates in providing its capacity
- condition – the state of maintenance
- reliability – measures of predictability
- availability – measures of down time
- total cost of ownership from acquisition to disposal
- flexibility – how easy it is to use the asset for multiple activities with minimal change
- adaptability – how easy it is to change the asset
- utilization – the extent to which assets are used
- yield – its ability to generate income.

The trick is not only in individually measuring asset performance but also in understanding an asset's performance as part of a configuration of assets.

14.4.2 Matching current and future requirements with what is available

Based on an understanding of the core business process and the associated infrastructure needs, a good asset manager should be able to define links between business processes and asset requirements. These links can then be measured in terms of actual performance of the assets against performance benchmarks. Benchmarking involves defining a set of metrics to be measured, understanding how they can be measured consistently and identifying a target level to achieve. This may not be the mythical 'world's best practice', but more likely a level of performance that is attainable, sustainable and specific.

The process of defining benchmark levels is driven by constant changes in what the business requires. Business strategies and tactics are constantly changing and this means changes in technology, work practices, processes, customer service standards and people. As a result, what an organization requires as infrastructure is constantly changing. The challenge is to define a management model and plan that responds to constant change in a way that enables delivery of the optimized infrastructure model in a timely fashion.

This means developing a short- and long-term planning process that takes account of the linkages between business objectives and asset requirements. At the same time, the planning model should accommodate the constraints of delivering changes in asset configuration or capacity. Short term is about changing the business process; long term is about changing the asset.

On this basis any planning process must be both conceptual, and thus able to deal with the 'big picture' strategy, and pragmatic, to deal with the finite detail of specific facilities and business units. Asset management requires large amounts of information, brought together in such a way that it provides specific detail and allows the bigger picture to be generated.

One successful planning approach is based on the configuration of basic work modules that match core process requirements combined with synthesis of a global set of measurement tools that support both a broad and specific view. The specific measures are based on those mentioned earlier, calibrated to meet the needs of a specific business. It should be remembered, however, that as the measurement process gets larger so does the expected degree of error.

14.5 IT

Asset registers, as suggested earlier, have traditionally served the needs of accountants. This leaves those responsible for managing assets, including property, to find their own solutions to managing the information they need to support decision-making. This encourages the development of a wide range of solutions across the full spectrum of functional capability. There are some vendors with very popular systems that provide little more than a cataloguing capability for fixed assets using, typically, bar code tags and simple asset descriptors. Other vendors have spent many years developing highly complex systems with broad capability; one worthy of mention is Archibus.[2]

More recently, organizations have recognized the need to provide a more integrated approach to managing information about assets and their operation. This has led to a new approach to the provision of solutions by tier one enterprise resource planning (ERP) vendors such as SAP and Oracle. The benefit that each of these major vendors provides is an integrated solution with a single database of information and multiple applications, all with the same look and feel. The effect of this, as shown in Figure 14.3, is quite stunning. For the first time, all parties in an organization, from property, finance and business units, are working from the same data. This can do more to create strategic alignment than any other initiative in an organization.

There are a range of solutions available from different vendors. These solutions fall broadly into four categories:

- ERP – these systems were mainly developed as financial management systems for manufacturing and financial services organizations. These vendors, such as JD Edwards, Oracle and SAP, have articulated a vision for property and asset management
- single focus solutions – these are financial systems primarily developed for property management from a landlord or agent's perspective. They include products offered by GEAC, Mincom, MRI, Timberline and others
- single function solutions – these are property and asset management specific solutions that do not have financial management capability. Vendors include Ai Software, Assets Works, Datastream, Facility Information Systems (FIS), Indus, Manhattan, MRO – Maximo, Peregrine, Prism, Westmark Harris and Yardi
- specialist solutions – these typically provide computer-aided design (CAD) or other types of specialist functions such as geographic information systems (GIS). There are many vendors in this area with many different solution sets.

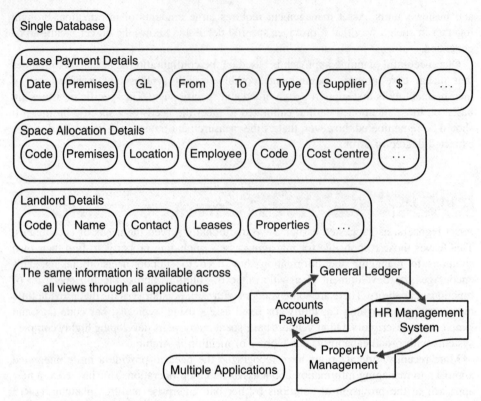

Figure 14.3 Integrated systems concept.

14.5.1 IT strategy

What's in an IT strategy? Specifically, an IT strategic plan should provide an integrated applications portfolio that:

- encompasses the organization's business drivers and future strategic direction
- executes the chief information officer's vision for information technology
- implements technology strategies
- meets investment, cost and risk management guidelines
- comprehends short-term issues and technology constraints.

The strategy should include an applications architecture, technology architecture and organizational architecture. The applications architecture ensures the business processes of the company are:

- advantaged through the appropriate application of technology
- addressed in an order driven by business need and return on investment
- linked together (integrated) as seamlessly as possible to permit an ease of change in response to ongoing business requirements and environmental impacts.

The application architecture will map application solutions to business requirements and drive the direction of and leverage the technological architecture. The technology architecture ensures the necessary IT products and services (infrastructure) are:

- chosen based on an approved investment and integration strategy
- addressed in an order driven by business need and return on investment
- robust enough to support computing requirements
- economically supportable.

It also documents the chosen technology products to form the foundation and support platform for the application architecture.

The rationale for an organization architecture is in ensuring that the necessary management principles, processes, organizational structures, roles and responsibilities are clearly defined and well understood in the context of achieving business objectives. The organization architecture will form the guiding structure for realizing the goals and objectives of the IT strategic plan.

14.6 Industry transformation – corporate real estate (CRE) portal technology

Perhaps the hottest trend in IT is the development of virtual workplaces – software that will enable individuals to personalize their computers according to their role or roles within an organization. These workplaces provide only the information, applications and services that individual employees need to perform their jobs. They also allow organizations to standardize business practices among all their employees, suppliers and business partners, regardless of geographical location. This trend has a major impact on the management of an organization's occupied space (CRE) both in terms of how employees interact with those responsible for its management and what the occupants then expect as service levels.

There appear to be three basic approaches to how this new trend in technology is affecting asset management:

- pure technology solutions – these provide a technology solution with no service offering and assume that the organization has someone who knows how to run the CRE function. Some vendors offer to run the technology by hosting the software but they do not 'run' the CRE service
- technology solutions backed by a service offering – these solutions provide management of CRE service levels as well as the technology. Typically these hosted solutions have come from the US with a few notable exceptions. A number of existing service providers have recognized that their investment in technology and in call centres is out of date and that for major corporate clients a web based help desk and resource centre provides a transparent and more cost-effective service delivery and performance measurement tool
- those technology solutions that are combined with a service offering typically focused on providing service level agreement management or contract management as part of managing the help desk function. A number of the more innovative vendors have diverse offerings such as providing supplier sourcing and procurement as well as contract management services.

An ideal solution in the view of some is a system that would enable a CRE executive to leave individuals and groups to almost self manage their workspace needs. This would include being able to pick a group of people for a new project team from an occupant list (some may not be employees) and having the system help find a location in the portfolio where they could be put together quickly. The system should provide visibility of the space utilization knock-on effects such as the options for maximizing utilization based on the new requirements and how this can be factored into long-term plans for space management. In addition, the system should identify all the assets and infrastructure affected. The system should then automatically draw together what is needed in terms of furniture, what work is needed in adjusting cabling, and so on, and provide a specification of work for a number of contractors selected from the contractor database. Ideally, these contractors could have the specification E-mailed to them and some preliminary estimates obtained for budget approval purposes. The whole process from purchase order to payment would be automated and the occupants notified by e-mail of activities and progress.

A side benefit of this approach is that all the transaction data is captured automatically as a by-product of the process and without extra cost as the data capture process is the process itself.

For this type of solution to work a lot of information, that most CRE managers have 'somewhere', is needed. Most have managed to get it in one place and some have managed to get it all in digital form but very few have been able to integrate it all into a cohesive dataset. One way to do this is via a 'portal' that is Internet friendly and allows all the information to be knitted together. This requires a database that 'understands' space and can interface with CAD drawings as well as integrate with the company's corporate financial system. It also needs to know who the company's employees and contractors are.

The objective is to provide a single point of entry and contact for CRE activities within an organization and enable a self-service solution with links to a supplier base that has market leading business processes and technology. Whilst the transaction-centred activities of procurement, project management and move management clearly have value as on-line categories, the challenge in industry transformation is bringing off-line activities (such as security, cleaning and catering) into an on-line environment in a way that adds value without adding cost.

These initiatives lead to two key areas of focus: CRE value-added services automation and CRE transaction services automation.

14.6.1 CRE value-added services automation

With added value activities, the challenge is to provide enhanced processes and greater richness (depth) of information to a broader group of people. In these categories, we seek to leverage the primary capability of the Internet to provide both richness of content and reach to more users. The key to success is tailoring the solution to the needs of individuals and providing information filtering tools that ensure the right information is provided to each user. Some examples of automation of services to add value would include the following:

Property leasing acquisition and disposal

- research (search tools)
- property review (tools for income generation – to make assets more productive)
- valuation (search tools)
- property search (search tools)
- asset broking (listing tools)
- legal documentation (precedents, contracts, legislation reference and advice)
- sales settlement (electronic settlement tools)
- lease/buy assessment
- approvals process (collaborative space and toolset)
- dispose and leaseback
- funding and financing (modelling tools – feasibility, investment, financial profitability).

Property development

- accommodation requirements analysis and brief preparation (mapping)
- scope of works preparation
- design development, architectural services, interior design services, engineering services (CAD systems)
- site acquisition (demographics and geographic/socioeconomic information tools)
- construction tendering and procurement
- project management and supervision (PM tools)
- development claim review and approval
- funding claim assessment and approval
- building commission – fit-out.

Portfolio management

- workplace management (vacancy and productivity tools)
- space optimization tools (scenario modelling, problem solving)
- management reporting (performance monitoring, benchmarking of users and suppliers activities in relation to contract requirements)
- plan, drawing and image management (CAD systems)
- statutory outgoing review appeal
- risk management
- property insurances
- portfolio planning (buy versus lease)
- property finance restructuring
- depreciation advantage tax maximization
- energy management.

14.6.2 CRE transaction services automation

Transactional activities require streamlining from technology to show a benefit. This should mean reduced cycle time, automation of decisions and approvals based on embedded business rules. Some examples of how technology can be applied to streamline processes and create value would include the following:

Property transaction services

- utilities and energy administration
- statutory charges administration
- lease rental administration
- rent reviews
- lease outgoings
- building outgoings
- facilities management (workflow rules, catalogue management)
- pest control
- property inspections.

Maintenance services

- heating, ventilation and air conditioning preventative and ad hoc maintenance and repair
- vertical travel preventative and ad hoc maintenance and repair
- fire and security systems
- building plant
- general building preventative and ad hoc maintenance and repair (fault logging and purchase order generation), handyman type activities
- cleaning and janitorial services
- security services
- garden and landscaping services
- waste management.

Other sundry services

- telecoms – IT backbone communication cabling and optical fibre loops, PABX and building servers
- corporate signage (new, reassessment, upgrade, preventive maintenance and repair)
- catering
- laundry
- office furniture and fittings
- human resources (parts which could be considered as property related, e.g., janitorial, cleaning, security, trades)
- relocation.

14.6.3 Internet portals

How would these services be delivered? Vendors have recognized that integration is a key issue and some innovative solutions have appeared recently. Portal technology now allows vendors, in conjunction with systems integrators, to deliver highly customized solutions with minimal change to a standard product – mass customization if you will. The key to this is overlaying an Internet portal or 'window' over a set of standard applications. As long as the underlying technology and schema do not change it is possible to manage

Occupant Portal			
Leases	**Space**	**Services**	**Supplies**
Rights	Definition	Contracts	Office supplies
Obligations	Allocation	Demand management	Utilities
Reviews	Assets	Supplier management	Personal supplies
Renewals	Configuration	Preventative/ continuous	
Options	Relocations	Ad hoc	
Deposits	Refurbishment	Insurance	
Guarantees	Resource booking	Personal services	
Insurance	Connectivity		
Contacts	CAD and GIS		
Charges			
Payments			
Financial Management			
CRE Cost management and allocation by unit or process and settlement with suppliers			

Figure 14.4 A CRE portal concept.

upgrades and enhancements without major cost. Figure 14.4 provides a conceptual map of what would be required for a portal to meet the needs of a CRE executive that could be deployed on the desktops of an entire organization. The benefit of such a solution is not only fewer staff in the CRE department, but also that it allows the CRE group to become more strategic and reduces the burden of transactions on the finance department. Above all, there will be reductions in the time it takes for a CRE transaction to be completed – a true benefit to business operations.

14.7 Conclusion

The area of asset management has matured at a rapid pace over the past 10 years and it is expected that, over the next 10 years, the CRE and facility management functions within organizations will be recognized as key strategic differentiators. More organizations will choose to manage the function in house or to form strong strategic relationships with experts to enable an outsourced model that leverages the value of the function and ensures alignment with core business objectives. Keys to success will not be whether or not you outsource, but rather about alignment of processes with strategic objectives, the quality of those processes and how well technology ensures they are in control.

As most parties have already squeezed staffing levels and maintenance activities as tight as possible and technology allows space utilization to be optimized, the next phase in improvement will come through the application of integrated technology to ensure processes are aligned and in control. The future of asset management is therefore closely connected to technology development and innovation.

Endnotes

1 For a detailed and thorough explanation of configuration management see: EIA/IS 632, Electronic Industries Association Interim Standard for Systems Engineering, Washington, DC, December 1994
2 Archibus Inc. www.archibus.com

References and bibliography

IEEE (1994) *Standard for Application and Management of Systems Engineering Process P1220 Trial-Use* (New York: Institute of Electrical and Electronic Engineers).
ISO (1995) *Quality management – Guidelines for configuration management, ISO 10007:1995 (E)* (Geneva: International Organization for Standardization).
USAF (1974) *Engineering Management MIL-STD-499A* (USAF Systems Command).
USAF (1992) *Engineering Management MIL-STD-499B* (draft) (USAF Systems Command).

15

Conflict avoidance and resolution in the construction industry

Linda M. Thomas-Mobley*

Editorial comment

Conflicts readily arise in workplace environments for a whole range of reasons. If productivity is agreed as a primary goal, then it is critical to resolve conflicts quickly. Conflict management is the generic competency for all forms of dispute avoidance and resolution. It is primarily a tactical activity. Conflicts can occur with external consultants and suppliers as well as within organizations, the latter usually being resolved without formal redress to outside expertise. In regard to the facility manager's role, it is not a bad philosophical view to believe that 'the customer is always right', even when this is clearly not the case. However, other dealings need to be based on a proper contractual footing to avoid misunderstanding and possible dispute.

Dispute resolution can be achieved through a range of approaches from litigation, through arbitration, to mediation. The approach selected may be set by organization policy or legal advice, agreed in the terms of contractual engagement, or otherwise chosen by consensus. It is always preferable to try and settle disputes in a co-operative fashion before resorting to more formal proceedings, if for no other reason than to minimize costs. Ensuring that agreements are clear in the first instance is obviously preferable.

Negotiation is a recognized form of dispute resolution used as an alternative to litigation through the courts, and negotiation skills are a valuable attribute for any senior manager. Through negotiation, potential conflict can be diverted into reasonable outcomes before positions become too entrenched. It is not always possible to resolve complex problems in this manner, although it may be worth the attempt if only to reduce the number of issues that remain in dispute.

* Georgia Institute of Technology, Atlanta, USA

Alliances and partnering arrangements have become popular in recent years as a means of sharing expertise for a common goal and reducing disputation. The concept of an alliance or partnership is that co-operation is built on goodwill. Although formal agreements do exist, there is a reliance on mediation or conciliation handled within the alliance framework. This approach applies to facility management, for example, through the agreed sharing of non-core business support services to a group of otherwise disparate organizations, or through the development of new facility infrastructure in conjunction with business partners.

Value can be added by the formation of these alliances and partnerships not only through less formal disputes but also through economies of scale that enable support activities to remain in-house and within the control of the alliance framework. This is usually preferable to outsourcing provided there is sufficient shared expertise available and it can be shown to be cost effective in comparison to the outsourced alternative. Nevertheless, the real essence of any partnership's success probably lies with effective teamwork and co-operation between participants, a desire to make it work, as well as equal sharing of resources and risks. Conflict management is used to keep everything and everyone on track.

15.1 Introduction

Due in part to the labyrinthine nature of the construction industry, conflict is as much a part of a project as bricks and mortar. Seemingly small changes that may cause significant ripple effects throughout the project are made both after careful contemplation and on the spot in haste. Yet, in the majority of the world's construction projects the conflict resolution process runs smoothly enough to prevent the parties from filing formal complaints.

Conflict is a state of disharmony between seemingly incompatible persons, ideas or interests. Excessive disharmony in construction projects is fatal and thus measures that reduce or eliminate conflicts are highly advantageous to all involved in the production of the built environment. The construction process involves several diverse stakeholders with differing agendas, many of which are conflicting. These stakeholders have to interact with each other with relative efficiency for the goal of a successful project to become a reality.

Typically, there are several possible conflict management or conflict resolution methodologies used worldwide. A general overview of these methods is illustrated in Figure 15.1.

This chapter provides an overview of the various methods of conflict avoidance and conflict resolution from the proactive to the reactive, and informal to the more formal. It should be noted that although many of the methods are similar, different countries use them to differing degrees. Also, this chapter is designed to outline and describe conflict management concepts, realizing that terminology differs from area to area.

Six major areas drive the listed methods of conflict management:

- legal system in effect
- contract or agreement language
- willingness of parties to litigate

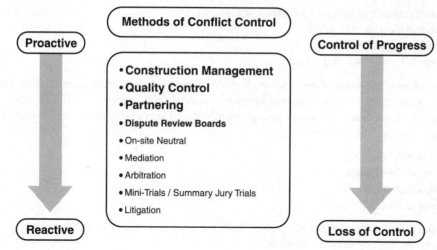

Figure 15.1 Continuum of conflict resolution methods.

- quality management systems
- project delivery system
- conflict resolution techniques available.

These six drivers determine the likelihood of conflict arising in a particular situation. The legal system of the particular country in which the project is constructed, the parties' willingness to litigate and the conflict resolution techniques available – all have a direct effect on a stakeholder's desire to communicate and resolve any potential conflicts or allow a problem to escalate to a point at which that formal intervention is necessary. This desire to communicate is also affected by the language in any agreements that are in place. Finally, the choice of project delivery system and proper attention to producing a quality project may also decrease the likelihood of conflicts arising.

15.2 The genesis of conflict

Conflicts often start as small problems or misunderstandings and in most cases communication among the necessary parties will facilitate a solution. If these disagreements are ignored, or if they cannot be resolved simply, the problem may develop into a full-blown conflict causing time delays and possible expense escalation. Since construction projects are by nature time-sensitive, even in countries where litigation expense is low, early informal resolution is preferred over protracted formal resolution.

There are numerous reasons for conflict: details being overlooked, misinterpretation of a contract clause, a slight error in quantity surveys, unexpected weather changes, delays of all sorts, changes made in the plans and specification, and late payments are a few of them. Studies indicate that there are, however, six leading causes of construction conflicts (Sweet, 1994):

- delays with incorporating changes
- owner driven changes

- unintentionally ambiguous contract documents
- unrealistic risk allocation
- poor communication
- unrealistically low bid (tender) price.

Regardless of the cause of the conflict, the focus should be on its prevention, or at the very least, its resolution. It is useful, for illustration, to divide the continuum of the methods by which conflicts can be resolved into a series of stages. One convenient method divides the continuum into the following stages:

- proactive stage
- project level stage
- informal stage
- formal stage.

As the stage of the conflict progresses from the proactive stage to the formal stage, the cost to the parties tends to increase.

15.2.1 The proactive stage

In the proactive stage it is possible to identify, contain and resolve a conflict without major disruption to the overall project. This is the most efficient level, of the four stages, at which to reach a consensus, and sufficient attention should be given to resolving conflicts and implementing procedures that reduce conflicts at this level. Decisions made with prevention and co-operation as key elements prove to be most successful. This stage encompasses all methods involving pre-planning and encouraging co-operative attitudes among the parties. Vague issues are best addressed at this stage. Clear design intent, and the production of complete and accurate drawings and specifications are essential, and constructability analyses and value engineering may also be attempted. Open communications and the building of a team also heads off possible conflicts in the future.

Management decisions

Management decisions are not usually thought of as a proactive tool for conflict avoidance and resolution, yet proper quality management, contract management and project management are potentially the most cost-effective routes to a smoothly run project. Proper management is an ongoing process that should be implemented at the onset of a project and throughout its duration as a routine company objective.

All parties must read the contract documents thoroughly and become familiar with the terms and conditions; project stakeholders should only enter into contracts with other parties that are deemed solvent and responsible. A decision to enter an agreement should be based on knowledge of the other party's sound reputation, ability to become bonded, and financial solvency, as well as knowing that they are properly insured and have the ability to perform to the quality level expected. If any party fails to meet these basic requirements, the project's potential for conflict that could potentially expose participants to escalated litigation costs increases. A prudent firm would avoid using unsuitable entities by carefully researching potential constructors or joint venture partners and obtaining proper documentation that shows the party's capacity to enter into a specific contract (Ansley *et al.*, 2001).

In addition to exercising savvy management, beginning with the careful selection of the construction team, comprehensive documentation of the project's progress is vital to success. Written notice should be required for changes, time extensions, differing site condition decisions, late payments and termination of agreements. Proper schedule updates are essential as well as cost estimate updates, job site logs and daily reports, meeting minutes and photographs or other methods for updating the changing project. Documentation and notification is an essential part of managing the project successfully and helps keep conflict resolution at the job site or field level. Timely written notices and the documents needed to support them will not only stave off unnecessary and frivolous conflicts during the project, but will also aid in the event that a serious conflict rises above the field level.

Partnering

Manufacturing and distribution industries have used partnering since the 1980s. The American Institute of Architects (AIA), American Consulting Engineers Council and the National Society of Professional Engineers became great proponents of partnering following its success in the manufacturing community (Cushman, 1982). Eventually, the US Army brought partnering into the mainstream as an alternative measure of conflict resolution in construction in 1988.

Partnering is a relationship based on trust (Eilenberg, 1996). As defined by the Construction Industry Institute (Thompson *et al.*, 1996), partnering is a

> long-term commitment between two or more organizations for the purpose of achieving specific business objectives by maximizing the effectiveness of each participant's resources. The relationship is based on team building, trust, dedication to common goals and an understanding of each other's individual expectations and values.

The Associated General Contractors (Carr and Hurtado, 1999) defines partnering as 'the creation of a relationship between the owner and the contractor that promotes mutual and beneficial goals'. Simply stated, partnering is an attitude change involving the significant stakeholders of a specific project. This change requires the parties to treat each other as team members in the project, not as adversaries. Team building is manifested through a series of meetings and workshops run by a facilitator, with all members of the project team assembled together.

Partnering is most beneficial when the parties believe that traditional adversarial relationships may prove to be ineffective. Construction projects with durations of two years or longer are generally preferred as contracts of less than two years are usually too short for the maximum benefits of a partnering relationship to be realized (Eilenberg, 1996).

Once the decision to partner has been made, the total commitment of senior management, as well as that of all stakeholders, such as the constructor, designer, owner, subcontractors and end users, is required. Part of the commitment of an organization to the partnering process is the recognition that financial resources are required in order to achieve the greatest success. Participants who are committed to partnering will need sufficient time to learn about the unique concepts, engage in team-building exercises and attend scheduled workshops. Costs associated with partnering include the costs of conducting the workshops and renting the necessary facilities, and travel-related expenses.

A critical element in the partnering process is the selection of a competent facilitator. The facilitator is a neutral person who helps the participants remain organized and focused from the outset of the process. The facilitator helps develop workshop content and leads the sessions; (s)he is crucial in helping the participants design the charter, identify and resolve potential problems, and in keeping the team energized and focused during the process.

Usually the workshop is held at a neutral location, such as a resort or other meeting facility, in order to enhance the team-building process contribute to a consistent focus on partnering and minimize the potential for participants to be drawn away from the workshop for work-related matters. The length of the workshop depends on such factors as the complexity of the project, the experience of the participants in partnering, the number of partners, and the time needed for team building. Typically this process lasts one or two days.

The focus of the workshop is on creating a document called the partnering charter, which is the focal point of the relationship and the blueprint for success. In the charter, the parties set forth a series of moral commitments, general and specific, such as their mission statement, mutual goals and objectives, and finally their commitment to the partnering process. Signed by all participants and displayed at the job site and offices of the participants, the partnering charter is the codification of the agreement to working relationships based on trust and team building.

In addition to the partnering charter the parties create an issue escalation procedure. This procedure identifies the respective decision-makers for each party, beginning with job site personnel and ending with top management. All parties then agree to deal with issues at the job level first, and dictate procedures and time frames that outline procedures on continuing problem solving, if necessary, at the project level without having the threat of formal litigation looming. The stakeholders also design a framework for conflict resolution that is to be used in the event that an issue cannot be resolved via the issue escalation procedure. This process forces the participants to contemplate potential conflicts and decide prior to actual conflict how these problems will be resolved and who will be responsible.

The end of the retreat does not end the partnering process. The philosophies and procedure adopted at the workshop must be implemented and reviewed on a regular basis in order to achieve the full effect and benefit of the partnering process. Stakeholders must continue to trust each other, communicate efficiently, and adhere to the partnering charter and issue escalation procedures. The parties must also recognize that having an actual conflict arise does not constitute failure, but is merely an opportunity to witness the effectiveness of the partnering process.

A statistical comparison of partnered versus non-partnered projects was conducted at the Kansas City District of the US Army Corps of Engineers. On average, partnering reduced cost growths by 2.65%, reduced modifications by 29%, and virtually eliminated time overruns (averaging 26%) (DeFrieze, 2001).

One drawback to the partnering methodology is that it may be difficult to achieve a sincere commitment from all stakeholders due to stubborn adherence to the traditional adversarial attitudes. Some constructors may see partnering as a waste of time and would rather focus on completion of the project.

15.2.2 Project level stage

The project level stage begins when construction of the project begins. If the proactive stage measures have not been enacted successfully, there are methods that can be adopted by the stakeholders to avoid or resolve conflicts that arise during construction; these are more costly than the proactive measures, yet more advantageous and less costly than litigation.

Dispute review boards (DRBs)

The concept of the DRB has evolved over the last quarter century and is not only effective at resolving conflicts, but in preventing them as well. The first DRB was established in the early 1970s with the construction of the Eisenhower Tunnel at Hoover Dam (Levin, 1998). The first phase of construction on the tunnel was plagued by endless conflicts. The project was, apparently, almost shut down due to the friction at the jobsite. The owners and contractors needed an efficient means of dealing with the conflicts, as they would arise, without waiting until the end of the project (Levin, 1998). As a solution the stakeholders decided that for the excavation of the second leg of the tunnel they would employ three impartial individuals to rule on the conflicts. Both parties agreed to be bound by the board's decisions during the construction phase in order to deliver the project on time. If the parties were not satisfied with the board's recommendations, they were able to appeal the rulings after the completion date. The entire job ran smoothly and none of the board's decisions were appealed (Levin, 1998).

Due to the success of this project some of the world's largest and most complex projects, including the Hong Kong Airport, the Boston Central Artery, the Washington DC Subway and the Los Angeles MTA Subway, have employed DRBs to resolve conflicts (Levin, 1998). The DRB approach has also been used on projects built by the Toronto Transit Commission, World Bank, the United Kingdom Institution of Civil Engineers, the Engineering Advancement Association of Japan, the International Chamber of Commerce and the United Nations Commission of International Trade Law (Levin, 1998).

The function of a DRB is to hear conflicts in an informal arena where communication is encouraged. Recommendations to help with resolution of the conflict are given, and timeliness is considered. Typically the DRB is a panel of three experienced, respected and impartial reviewers. One member is selected by the owner and approved by the contractor, one is selected by the contractor and approved by the owner, and both parties jointly select the third. The board is organized before construction begins and meets at the job site regularly. The DRB members are provided with the contract documents, and are expected to become familiar with the project procedures and participants, and are kept abreast of job progress and developments. The DRB meets with owner and contractor representatives during regular site visits and encourages the resolution of conflicts at the job level.

The DRB process helps the parties head off problems before they escalate into major conflicts. In essence, the DRB creates a type of partnering between contracting parties. Although the decisions of the DRB are non-binding, parties are reluctant to create a bad reputation with the DRB by approaching it with excessive conflicts. This attitude prompts parties to work together to handle any problems that may arise and only to utilize the DRB when absolutely necessary. Finally the cost of DRBs is low compared to potential litigation fees (Morris, 1995).

On-site neutral

An additional project level conflict avoidance and resolution method is the use of the on-site neutral. Although the term on-site neutral is used here, the terms standing neutral or neutral expert are also common. An on-site neutral is actually a mediator hired by the stakeholders to help resolve conflicts before they escalate and cause major delays and disruption. The on-site neutral is typically someone who is familiar with the contract documents and the project, and who will generally attend any project meetings that take place during the construction phase. The on-site neutral is brought in at contract signing.

Care must be taken in selecting an on-site neutral as the position requires experience in identifying potential problems at the embryonic stage as well as expertise in defining the problem, evaluating responsibility for the problem, quantifying the impacts and facilitating a resolution between or among the participants (Steen and MacPherson, 2000).

If a problem does arise, or the unexpected occurs, the on-site neutral is in a position to immediately convene the stakeholders so that they may work out solutions and resolutions. The on-site neutral does not make decisions for the parties or impose an end result, but utilizes mediation techniques to allow the stakeholders to arrive at solutions that satisfy as many of those involved as possible.

15.2.3 Informal stage

When a conflict reaches the informal stage, it usually means all local, jobsite or project level mechanisms for resolving the conflict have failed or been abandoned. This stage includes non-binding and binding methods that tend not to require legal representation by the stakeholders.

Mediation

Mediation has been a common method of conflict resolution for centuries. The concept of mediation stems from the nineteenth century. In 1896 the UK introduced legislation that spoke of the conduct required for industrial relationships. This legislation was entitled the Conciliation Act. In 1913 the US Department of Labor appointed a panel dubbed the Commissioners of Conciliation to dispose of labour and management disputes. Additionally the American Arbitration Association was set up in 1923 as a commercial service for the resolution of conflicts in the private sector leading to the development of the Federal Mediation and Conciliation Service in 1947.

In the UK, the Advisory, Conciliation and Arbitration Service (ACAS) was devised in the late 1940s to solve industrial conflicts. By 1980 mediation as a method for conflict resolution was introduced to the US construction industry via roofing subcontractors. The growth of mediation as a conflict resolution method then developed swiftly in the 1980s and 1990s. In California, claims less than $50 000 were required to be mediated prior to litigation. By 1996 the New York Supreme Court adopted mandatory mediation and in 1997 the mediation clause was included in the standard contract documents of the AIA.

Mediation used to be a public process by which two parties would meet together and solve their problems in one sitting. The parties would sit within arms reach as the mediator expressed his opinion before the parties had a chance to state their entire argument.

Through trial and error, mediation has been modified to its current form. Presently it is part of the growing Alternative Dispute Resolution (ADR) movement. Mediation is defined as 'assisted communications for agreement' (mediate.com, 2001). During the mediation process a neutral, called a mediator, assists the parties in exploring the issues in conflict. The mediator assists in discussion between counsel (if present) and the parties by guiding the parties toward finding their own solutions to the conflict. The mediator does not make the decision, a court reporter is not present, and there are no rules of evidence or other formalities that control the process. A major advantage of the mediation process is that both parties will discover the weaknesses of their own case and of the opposing party's case. Exposing the weaknesses of both sides often drives adversaries to come to a settlement. One drawback to mediation is that if the parties cannot come to an agreement, the matter remains unresolved even though both sides have paid the expenses of a mediation session.

Mediation is completely voluntary; it may be manifested by a contractual agreement or by the parties independent of any prior agreement.

Arbitration

In the fourth, fifth and sixth centuries BC, arbitration was used to settle conflicts among the Greeks. The nature of the conflicts included boundary delimitation, ownership of colonies, ownership of particular territories, assessment of damages suffered through hostile invasion, and recovery of debts (Wehringer, 1966). In the US, the use of arbitration began around 1781 when the Chamber of Commerce of the State of New York set up the first privately administered tribunal of businessmen and became the first administrator of the arbitration process. Eventually a uniform arbitration law was adopted by the National Conference of the Commissioners of Uniform State Laws in 1955, and amended in 1956. Currently, many countries around the world and all states in the US have enacted the procedures that are similar or mirror the American Bar Association's arbitration procedures (Wehringer, 1966).

Arbitration requires all disputing parties to prepare their views, present them, and with a claim of speed and privacy, and a claim for experience and expertness in the person of the arbitrator, achieve a final answer. This method leads to some expense savings while ensuring complete privacy and relative convenience for the parties.

Arbitration specified in AIA standard contracts could be either binding or non-binding; it is the contract drafters who decide which method of arbitration is suited to the particular project. Binding arbitration is one of the oldest methods of ADR. As its name suggests, any decision made by arbitrators under this method is binding on both parties. Non-binding arbitration is less expensive and the issues may be heard again in a substitute ADR forum or formal litigation.

This method of conflict resolution was originally used in deciding construction conflicts to save time and avoid the expense of litigation. Due to the relative formality of the arbitration process, legal representatives prepare for arbitration with the same care as one prepares for actual litigation – this is the reason that the cost for arbitration involving legal representatives approaches the costs for litigation.

Variations on the theme of arbitration exist in other forms of ADR, notably mini-trials and summary jury trials. In the mini-trial a neutral adviser, who is mutually agreed upon, speaks with the senior representatives from the disputing parties and facilitates negotiation. If the senior executives cannot come to a consensus, the neutral observer will pass judgment and render an opinion on how the case would be resolved had it been elevated to litigation. The executives may decide to meet again after the opinion has been stated, and try to formulate a resolution. Additional information about the role of the neutral can be found at the American Arbitration Associations web site. In a summary jury trial, an abbreviated hearing or trial is conducted before a referee or judge and a mock jury. This hearing typically lasts no longer than one day and the mock jury decides the issues of fact. Both of these methods are non-binding.

15.2.4 Formal stage

A complete description of the various types of formal conflict resolution procedures including litigation is beyond the scope of this chapter, but it should be said that by the time a conflict rises to the level of formal litigation the matter is almost completely out of the hands of the original parties. In order to manipulate the various legal systems, a legal representative must fashion the facts in the light most favourable to the particular stakeholder, yet still within the boundaries and constraints of the rules which regulate the process. This manipulation tends to distort the conflict and renders the participants completely dependent upon their legal representatives.

Litigation

In Malaysia, litigation is the preferred process for conflict resolution in the construction industry. The trials of construction disputes are treated similarly to other civil litigation. Generally, matters in which large amounts of monetary damages are claimed, and that contain complex matters, are heard in the high court (Dulaimi, 1998).

In China, litigation is not preferred. The lack of specific construction related laws and the alleged high rates of corruption in the Chinese administration are given as reasons why parties are reluctant to utilize the formal legal courts. Currently, the Chinese government is encouraging the institution of more western legal procedures and promotion of awareness of the Chinese law. Typically, if there are no clauses in the contract that prevent disputes using a formal court proceeding, any party can apply for a court proceeding as their first choice for conflict resolution. The court will, however, try to use mediation before proceeding fully with formal litigation. If the parties reach an agreement during this mediation process any agreements made are binding. Foreign parties to construction contracts enjoy the same rights as Chinese parties if there is a reciprocity agreement between their own country and the Chinese government (Chan, 1997).

The use of litigation in Australia is somewhat favoured, which is surprising in view of the popularity of negotiation as a means for resolving disputes. Litigation in Australia involves adjudication before a judge and is considered binding. The amount of damages in the controversy determines the level of court at which the case will be tried. The legal system in Australia utilizes court-based ADR procedures such as court-annexed arbitration, court-annexed mediation and reference-out-of-court to a referee. This latter method is widely used in New South Wales in complex technical cases. Australian parties can also avail themselves of the use of expert determination or expert appraisal procedures (Watts, 1998).

In Japan there are no special courts or judges for construction disputes in the litigation system. The Japanese court system is three-tiered and adopts an adversarial approach. In

Japanese court proceedings the court has to decide only on the evidence and allegations made by the parties; it has no freedom to find facts not relied on by either party or examine witnesses as in western courts. The judges, without a jury, give court judgment, and an arbitration award is given the same weight and effect as a court judgment (Leung, 1998).

In the US, ADR is encouraged. The court system is overcrowded, and in many instances it can take three to four years before a formal trial can begin. The opposing parties argue their cases in an adversarial procedure and a judge or jury may decide upon the merits presented. As in other court systems with formal rules for court proceedings, attorneys who prepare documents and witnesses usually must represent the parties. Court proceedings are protracted and the courts findings are binding unless the judge commits a technical error that causes the case to be appealed to a higher court (Sweet, 1994).

15.3 Global distinctions

Conflict resolution methods in Australia, Canada, Quebec and the US are somewhat similar, although the laws of those countries differ slightly and the extent to which different dispute resolution methods are used varies slightly. There are two main exceptions to this: the Canadian system tends to use a referee for its large public works projects and in Australia the use of ADR is being driven by the construction industry, as opposed to the US where the changes seem to be driven by the courts.

In Asian countries, the same conflict management and conflict resolution methods used in the US are employed. This can be attributed to the effects of globalization. However, when conflicts occur in an Asian country, they usually encourage less face-to-face confrontation and try to preserve the relationship among the parties. Generally, the Asian systems favour compromise over confrontation.

In Europe, the courts tend to employ similar conflict resolution methods to the rest of the world. However, when conflicts occur, the typical European response is to use various methods of resolution that keep the peace and promote proactive working environments. European methods favour both compromise and, if necessary, confrontation. In Italy, for example, conflicts are resolved almost exclusively through negotiation. Conciliation is the preferred way and is an almost compulsory first step; Italian judges commonly direct the parties to negotiate, and try to arrive at a settlement during the litigation process.

15.4 Conclusion

The construction industry is a wildly exciting, risky and complex business. Cultures the world over have made significant impacts on civilization through their construction projects.

Every culture is identified with significant built works, thus establishing the importance of designers, constructors, artisans, and enlightened owners throughout history. Egyptian, Babylonian and Assyrian cultures developed architecture that distinguished those early civilizations. The Greeks carried architecture and art to levels of perfection never achieved before (Dorsey, 1999).

With such pressure to continue in this great tradition by building higher quality, lower cost buildings in record times, greater exposure to risk develops. Looming large in the assortment of the various risks is exposure to great profit loss due to claims. Controlling the disruption to the project because of these claims lies in management of the inevitable conflict. In ancient times, there was a remarkable integration between the participants, the designers and the builders. Today, this integration seems to be lost. Gone is the concept of the master builder. Project delivery systems like design-build claim to bring this concept back, but the construction industry is still largely adversarial in nature.

Use of proactive preventive measures and ADR procedures is wise and, some argue, necessary for achieving harmony and success. As discussed in this chapter, the earlier the conflict avoidance or resolution procedure is utilized, the better. Knowledge of the various systems accompanied by the freedom to use the various systems is crucial. Careful and close attention must be paid to maintaining the integrity of the team instead of the amount of profit to be made. In a non-adversarial atmosphere, where the team works together toward the common goal of a completed project, sizeable profits can be made while delivering the high quality built product the world has come to expect.

References and bibliography

AAA (1994) American Arbitration Association. www.adr.org

Ansley, R.B., Kelleher, T.J. and Lehman, A.D. (eds) (2001) *Common Sense Construction Law* (New York: John Wiley).

Carr, F. and Hurtado, K. (1999) *Partnering in Construction: A Practical Guide to Project Success* (Chicago, IL: American Bar Association).

Chan, H.W. (1997) Amicable dispute resolution in the People's Republic of China and its implications for foreign-related construction disputes. *Construction Management and Economics*, **15** (6), 1.

Cushman, R.F. (1982) *Construction Litigation* (New York: Publishing Law Institute).

DeFrieze, D.C. (2001) *Partnering: An Overview* (US Army).
www.osc.army.mil/others/gca/partnering/overvw1.htm

Dorsey, R.W. (1999) *Case Studies in Building Design & Construction* (Upper Saddle River, NJ: Prentice-Hall).

Dulaimi, M. (1998) In Malaysia. In: Fenn, P., O'Shea, M. and Davies, E. (eds) *Dispute Resolution and Conflict Management in Construction: An International Review* (London: E. & F.N. Spon).

Eilenberg, I.M. (1996) *Is Partnering Really the Answer?* TC15, ADR Working Commission, Japan (CIB).

Leung, K. and Tjosvold, D. (1998) *Conflict Management in the Asia Pacific: Assumptions and Approaches in Diverse Cultures* (Singapore: John Wiley).

Levin, P. (1998) *Construction Contract Claims and Changes of Dispute Resolution* (Virginia: American Society of Civil Engineers).

mediate.com (2001) What is mediation? www.mediate.com/articles/what.cfm

Morris, M.D. (ed.) (1995) *Construction Dispute Review Board Manual* (New Jersey: CDRB).

Steen, R.H. and MacPherson, R.J. (2000) Resolving construction disputes out of court. *Journal of Property Management*, **65** (5), 58–60.

Sweet, J. (1994) *Legal Aspects of Architecture, Engineering and the Construction Process* (St Paul, MN: West Publishing).

Thompson, Crane and Sanders (1996) *The Partnering Process – Its Benefits, Implementation and Measurement*, CII Publication RR102–11 (Construction Industry Institute).

Watts, V. (1998) In Australia. In: Fenn, P., O'Shea, M. and Davies, E. (eds) *Dispute Resolution and Conflict Management in Construction: An International Review* (London: E. & F.N. Spon).

Wehringer, E.C. (1966) *The Arbitration Process* (Cincinnati, OH: Dogwood Publishing).

16

Quality management

Gerard de Valence*

Editorial comment

Customer satisfaction underlies the basic purpose for facility management. Customers in fact include a wide range of stakeholders, such as employees, consumers, government agencies, consultants, suppliers, shareholders and upper management. The delivery of a quality service is no accident; it requires careful planning, constant monitoring and the introduction of improvement initiatives. Quality management is, therefore, a core competency and integrates all activities by comparing performance against established benchmarks and targets.

The constant drive towards best practice is fuelled by economic gain resulting from increased market share, more sales, higher worker productivity and job satisfaction, better customer relations, community image, and competitive advantage. Higher quality of service not only needs to be achieved but must be seen to be achieved. Regular reporting of performance is useful in this regard, particularly where based on objective independent data. Facility managers need to identify what are the key performance indicators that best reflect the organization's goals and record trends in these indicators over time.

Total quality management (TQM) is an ethos that should pervade all levels of an organization, regardless of its size or line of business. Quality assurance procedures can help to achieve this and ISO certification can ensure that success is communicated effectively to others. There are clear dividends in terms of value where TQM is a routine practice.

Facilities need to exhibit high-quality standards in relation to key performance areas such as occupational health and safety, indoor air quality, fire safety, security, and cleanliness. But quality management has a wider brief – it is now common to undertake a quality assessment of facilities to rank against industry benchmarks and to highlight

* University of Technology Sydney, Australia

areas that either can be improved or are unsatisfactory. Building quality assessment is a methodical audit of premises while a due diligence audit focuses on regulation compliance.

Feedback from facility stakeholders is vital in order to identify areas that are under-performing. Questionnaires and focus groups are used to elicit information, and are effective, provided participants believe that their suggestions will be taken seriously and provided they are given feedback over proposed courses of action. A post-occupancy evaluation (POE) should be undertaken on all new facilities after about one year of operation and presents a snapshot of performance that can be used to fine-tune the original design. There is no reason, however, why POE cannot be implemented routinely to review functionality and customer satisfaction.

No facility is ever perfect following its design and construction, even when it is built exactly to plan. Opportunities exist for further improvements in value throughout the life cycle of the facility, but it is critical that processes exist to enable these potential benefits to be realized.

16.1 Introduction

Buildings are complex artefacts, put together through a process that requires the co-ordination of many different teams working on many different tasks. One of the key determinants of the value of a building on completion is its fitness for use, or functionality. How well a building provides the flow of services that were intended when it was being conceived and designed, the flexibility it allows in its use over time, and the ability to repair, maintain and refurbish the building as it goes through the normal processes of wear and tear, has a big impact on the value that a client and the users of the building get.

When we think about buildings and the built environment, it is clear that the range of stakeholders involved is very large (Carassus, 1999; de Valence and Lauge-Kirstensen, 1998). The users of the building, the employees, workers, visitors and others who go in and out of the building are the ones with the most frequent contact. The building owners rely on the income stream, or the accommodation it provides, for their own purposes. The built environment, and individual buildings in terms of their design quality, is an issue for the community and government. Meeting regulatory requirements is important for government agencies. In the process of delivering the building the consultants, contractors, suppliers and others are also stakeholders.

Given these many characteristics of buildings and their use and the wide range of stakeholders involved, how do we work towards achieving quality outcomes when we hand over a building? The delivery of quality in anything does not happen by accident, it is the outcome of the planning and management processes that have been put into the tasks involved. There are always significant amounts of monitoring and adjustment to circumstance required in a complex process, and seeking opportunities for improvement in the way that work is done and the outcomes of that work is an on-going task in its own right. The overall framework that has often been given to this approach is quality management (QM). In the facility management (FM) industry understanding the issues of QM is seen as one of the core competencies by the industry associations (e.g., IFMA, FMAA) that is required for facility managers, and it links back to important performance measurement issues such as benchmarking, targets and key performance indicators.

Friday and Cotts (1995) see quality FM as grounded in total QM (TQM), which gives them the customer as their starting point. Their view is that 'real' quality within FM begins with the customer, and that FM and TQM have a natural fit. The features of TQM, such as getting it right first time, customer focus and commitment to continuous improvement, are also the keys to delivering quality FM. They also devote a chapter to examples of best practices such as team structure and training, assessment methods, call-out response times, customer feedback, productivity measurement and space use planning. The five pillars of quality that Friday and Cotts (1995, p. 3) identify are:

- meet the service needs of customers because the customer drives the process
- there is no short-term solution
- benchmarking and metrics are essential
- the front line service worker should have the flexibility to make decisions with the customer (empower FM staff)
- quality service has to be marketed inside and outside the company.

The topics associated with QM that this chapter addresses are the ideas behind TQM, one of the most important management ideas of the last two decades, best practice and the importance of performance measurement, the use of building quality assessment (BQA) systems, and feedback on facilities using post-occupancy evaluation (POE).

16.2 What is quality?

When we wonder about the nature of quality, the elements that we would normally expect to see would include some form of excellence (another popular business management word from the 1980s), the existence of a standard, or perhaps the setting of a standard, and meeting the expectations of customers or reflecting their needs. These cover a range of characteristics, and definitional problems have always been an issue in QM. The people who have been instrumental in developing modern ideas about quality have contributed different ideas to the various definitions that relate to what quality is. The common elements to all these definitions are that reliability and quality are closely related, and that the satisfaction of clients or customers with the perceived quality of the products and services they get is crucial.

Probably the most common phrase that you see in the QM literature is that customer expectations are being consistently met by the goods or services that they purchase. Deming (1982) emphasized the nature of a product or service, all those features that reflect the capacity to satisfy expressed or implied requirements or needs of customers and clients. The emphasis that Deming placed on performance was the first step in the quality improvement programmes that the Japanese industry undertook. By contrast, Crosby (1979, 1996) emphasizes conformance to requirements of product. For Crosby, the most important characteristic of quality was that the products were able to meet the requirements that were set for them, and often this involved the specification of performance and production techniques in great detail. Another approach came from Juran and Gryna (1993) who was more interested in the fitness for purpose or appropriateness for use of the product or service. This may be the most relevant approach that we can use for assessing buildings and facilities. Also, Feigenbaum (1991) emphasized that the characteristics of a product or service were a result of the combination of design, manufacturing, marketing, maintenance and after sales service, and that all of these together were needed to meet customer expectations and requirements. Therefore he emphasized the complexity of the relationship between these different aspects of business operations. Finally, Oakland (1995) suggested that the most important things were those that were able to be measured, and that the thing that was going to differentiate a product or service from one merely satisfying customers or clients to something that was perceived as being of excellent quality would be the ability of those products or services to 'delight' those customers and establish the business as one with a reputation for excellence.

From all of this we can define quality as consistently producing what the customer wants while reducing errors. Importantly, however, quality can be seen not so much as an outcome as a process of continually improving the quality of what is being produced. QM, and particularly the ideas that underpin TQM, emphasize the importance of continual improvement and detection of potential problems before they occur. This saves effort on inspection, supervision and remedial work, and avoids customer dissatisfaction. In particular, it has been found that it is in the design phase of product or service development that the quality of the final product is determined.

For example, managing the early phases of a building project is an extremely complex process, as de Valence (1999) and Gray and Hughes (2001) have shown and Ballard *et al.* (2002) discuss. There are many elements involved in getting the design process right and in managing it consistently so as to produce a high-quality outcome when the building is finally handed over. Further, during the process, the co-ordination of supplies and services on the project is required in order to have quality outcomes. The treatment of the suppliers and subcontractors is an important element in this, and a QM approach emphasizes the fact that all these parties are business partners working together to deliver a quality product. The QM approach has to work right down the line so that the contractors, subcontractors and suppliers are working together in order to deliver the product in a co-operative way and meet the requirements of the customers.

QM is a generic title – there are many variations on QM, in fact most organizations that have implemented a QM system have developed a version that suits their particular history, culture and industry. Despite its origins in manufacturing, the QM approach has been extended to all activities, e.g., Albrecht and Zemke (1985) developed a QM system for service industries.

16.3 Top quality management (TQM)

TQM is one of the most significant ideas that emerged from the literature of modern management over the 1970s and 1980s, building on the works of Deming and the success of Toyota (Womack *et al.,* 1995). There were a number of key people involved in the development of these ideas, and they contributed to a new set of ideas and a new way of thinking about how organizations work (see Handy, 1984), and more importantly how organizations compete (Porter, 1980). The elements of quality that are important for businesses, and the way they compete for customers and clients, contracts, resources, market position, funds, continuity and growth include not just the product or service that they produce, but also the service they provide to their clients before and after sales, the prices they charge and the value they give, the reputation of the business, the loyalty of their customers, the reliability of their product or service and any unique selling point that they have that differentiates them from their competitors in the way that they deliver their products. These competitive elements are usually the focus of the QM systems that are put into place by organizations.

QM has developed as a means of making a strategic contribution to achieving business outcomes and improving competitiveness. The lessons from the past show us how and why countries that once dominated industries disappear as new entrants bring new ideas and products to market. Examples include German cameras, British and then Swiss watch-making, and the American television manufacturers. Many managers grasped the idea that without ensuring the quality of their products they would have no markets and no customers left to service. The great impetus that the Japanese gave to the understanding of QM came when they learnt to design and produce high-quality goods, and overcame their reputation for cheap and inferior quality products. They did this so well that in the American markets for electrical goods, photocopiers, and many consumer items, their standards became the benchmark for quality. In this transformation of Japanese industry, two Americans made crucial contributions, Joseph Juran and W. Edwards Deming.

The propositions that quality advocates such as Juran and Deming brought to managers of businesses were not controversial. The ideas that were advocated revolved around common sense propositions such as: if you do not focus on quality your company will lose market share and decline in reputation; good reputations are easy to lose and a reputation when lost is hard to regain; and paying attention to processes and managing them carefully will lead to products and services that have no defects and are high quality.

16.4 Best practice

The company that does a particular process well often comes to be widely recognized for that – the L.L. Bean distribution system is an example (the company makes mail order clothing – see Kane, 1997). In many cases it becomes the basis of the company's image or marketing strategy, as prompt delivery is for UPS, or reliability is for Toyota. The concept of best practice is that there is a best way to manage and undertake a process, and somewhere there is an organization that has found that best way. That organization, for that specific process, is often referred to as the 'best-in-class'.

There are many separate processes involved in getting products or services to the clients or customers. No company does everything well, so finding someone else who is doing something better can identify a 'performance gap'. If a company is serious about improving its own level of performance this helps to set targets and determine how they might be achieved. That process is known as benchmarking.

16.5 Benchmarking

A good description of benchmarking is 'a search for industry best practices that lead to superior performance' (Camp, 1979, p. 3). By regularly measuring and comparing performance against best practice, competitiveness is established and improved. The identification of best practices used by other companies overcomes resistance to change on the grounds that innovation cannot be done or is impractical, because it is a practice that is already being used successfully. The best companies are usually cost-efficient producers of quality products or services with high levels of customer satisfaction.

Benchmarking became an increasingly significant business tool over the 1990s. Many companies in a wide range of industries found it an important method to support moves towards continuous improvement and achievement of best practice. Since its beginnings in the late 1970s at Xerox with Robert Camp (Camp, 1979), benchmarking has become more and more widely accepted and used for enterprise process improvement. The processes and techniques have become formalized and well documented (Leibfried and McNair, 1992).

Benchmarking has been used worldwide to achieve substantial process improvements. By definition, benchmarking is a process of comparison by measuring specific indicators. Leaders in their industry, such as Exxon, IBM, British Telecom and General Motors, all use benchmarking. Leibfried and McNair (1992) describe the use of performance measures as an essential prerequisite for any exercise in benchmarking and, consequently, continuous improvement. Similarly, re-engineering and lean production also need the support of performance measurement.

The process of benchmarking begins with a clear understanding of the organization's mission and objectives, a thorough knowledge of operations, and a commitment to continuous improvement in performance. There are five major benefits from benchmarking. These are (1) that effective goals and objectives based on external conditions are set, (2) customer requirements are defined and met, (3) true measures of productivity are developed, (4) competitiveness is enhanced, and (5) the best ways of doing things are sought, analysed and incorporated into the model (Spendolini, 1992). Although other techniques can achieve these benefits, the innate objectivity of benchmarking sets it apart. Benchmarking is a goal-setting process and legitimizes these goals by basing them on an external orientation. This leads to the development of effective team work as employees at all levels focus on reaching these goals.

The perspective used in benchmarking is stakeholder analysis. Stakeholders are all those with a direct connection to the organization and its activities: suppliers, employees, shareholders and customers, for example. The philosophy behind benchmarking is continuous improvement – rather than seeking a one-off leap in performance, a process of continually lifting standards is used. This makes benchmarking a dynamic process. The technique used in benchmarking is process analysis, where the causal relationships between organizational activities and client requirements are identified. The performance drivers, or causes of activity, are targeted in a mapping of the work process, which, in turn, focuses on those activities that are value-adding or value-creating. Perhaps most importantly, benchmarking allows proactive decision-making. The defining features of benchmarking are its purposive nature, in that it leads to actions and organizational changes that improve competitiveness, and that it is objective and externally focused, because best practice can be found almost anywhere. Also, because benchmarking is based on measurement, it is information intensive.

Performance measurement is similar to benchmarking, in that the measures can be accepted as targets, but is differentiated by the following characteristics:

- performance measures do not necessarily relate to organizational change or the continuous improvement of activities
- the general focus is on absolute levels of performance rather than relative levels

- identification of best practice may not be involved
- these measures are typically less dynamic than benchmarks, and are not revised regularly.

Much of the manufacturing sector's gains in productivity and innovation is based on benchmarking and performance measurement, which ranges from inventory control and process assessment to issues such as relationships with customers and suppliers. Sheldon (1992) discussed the success of a US-based material handling equipment manufacturer in meeting competition and cited performance measurement as being of fundamental importance. He describes it as the way to both drive improvements and to evaluate success. Dougherty and Reicher (1992) described performance measurement as the 'primary vehicle' in the effort to improve resource planning at DuPont.

Objectivity of assessment can provide focus for a process. It can also provide the interface to develop an integrated approach involving all the stakeholders. The clients should expand the objective critical success factors, which define project success, beyond the boundaries of price.

16.6 Building quality assessment (BQA)

Modern buildings have such a high degree of sophistication that a means of objective appraisal and a system for rating the quality of buildings is needed. Such a system should analyse a building's relative strengths and weaknesses, monitor market and technological developments and identify how the asset is being maintained, refurbished and positioned against both benchmarks and competitors. It should also bring together the strategic interests of building owners and operational interests of building users.

Management requires information that is reliable and regularly updated, and so benchmarking of facilities against desired or previous performance serves as an important planning and monitoring tool. The performance of facilities and their contribution to productivity and profitability, in combination with the ability of physical assets to store wealth and create liabilities, make proper measurement and management essential. In a QM approach there is a requirement to evaluate how people respond to design and how they use the facilities they occupy, and the appraisal of buildings and materials performance is a central concept. This type of evaluation is variously known as facility appraisal, facility condition assessment or BQA.

BQA categories listed by Ballesty (1999) include:

- presentation
- space functionality
- access and circulation
- amenities
- business services
- working environment
- health and safety
- structural considerations
- building operations.

The two components of performance relating to facilities are functional and physical efficiency. In economic terms the performance characteristics relating to functional efficiency are most important, because these impact on profitability and contribute to value. Physical performance and the physical efficiency of a building can be measured through:

- deterioration over the life cycle
- building services efficiency
- cleanability and maintainability
- energy efficiency
- environmental impact.

The quality of a building is the degree to which it meets the expectations and requirements of the building users. Users of buildings fall into two main groups: the providers of buildings, including owners, investors, facility and asset managers, and the occupiers of buildings including tenants, visitors, maintenance, cleaning and service personnel. Building providers tend to believe they know what is required and are convinced they meet the true demands of occupants, but occupants frequently think otherwise. A study by Knight Frank Hooker Research (1995) on 'What Workers Want' found the order of importance was air conditioning, lifts, parking, amenities, location, entrance, layout, everything, rental level, windows, and the internal layout. The order of importance as to why tenants move in was: the rental level, image of firm, technically advanced, closeness to client/competitors, staff requirements, more parking, to be in/out of the CBD, and previous management. The BQA assessment criteria should consider all these items.

A good BQA provides a balanced assessment of both a building as a whole, and its component parts, against the requirements of a range of users. It will set standards that are representative of best practice and reflect tenants' and owners' needs. A comprehensive BQA will measure performance in three critical areas. The physical performance relates to the behaviour of the structure, envelope, services and finishes embracing physical properties such as internal environment (e.g., heating, lighting) energy efficiency, cleanliness, maintainability, durability and environmental impact. The functional performance relates to the properties afforded by a building to the benefit (or otherwise) of the occupier, e.g., include space (quantity and quality), layout, ergonomics, image, ambience, amenity, movement/communications, security, flexibility and health and safety. Financial performance is a combination of capital and revenue expenditure, rate of depreciation, investment value, and contribution to productivity and profitability/efficiency. It springs from the physical and functional performance of the building and the way in which it is used.

16.7 Post-occupancy evaluation (POE)

The process of POE provides the basis for facility appraisal. An organization may wish to consult employees and other users about their workplace to improve user relationships with management and the operational performance of the facility. The process of consultation provides a channel for communication and mutual understanding and is essentially good management practice. Flowing from this is the use of POE to evaluate a facility for the purposes of improving the environment to enhance worker productivity, machinery/equipment operation and business performance generally.

Ballesty (1999, p. 204) states that POE should be given consideration during due diligence analyses. He points out that whilst the initial analysis generally relates to the compliance of the property with statutory requirements and its state of repair, ongoing due diligence reporting can be used to ensure that the value of the property is continually reviewed, monitored, maintained and, where possible, improved. Ballesty sees the overall scope of property/facility due diligence auditing as covering four main areas: property market issues, legal aspects, financial accounting and technical reviews. The three main groups involved in POE are the initiators, who are the organization's management personnel, the facilitators, who conduct the on-site analysis and surveys, and the stakeholders, who normally comprise occupants, employees, tenants and visitors who are the end-users that supply the information about the facility and are the key to the effectiveness of the process.

There are commonly three levels of assessment applied to POE. The level chosen is largely dependent on the balance between the forecast benefits to be derived in relation to financial, time and personnel constraints. Each level involves similar approaches in planning the process, conducting the analysis and interpreting the results.

The indicative evaluation is the quickest, cheapest and the most commonly used. It involves walk-through evaluations and structured interviews with key personnel. Inspections reports which summarize the observable building performance indicators are based largely on an assessor's opinions. Information gathering for this type of evaluation can normally be done in a matter of days. As a result, findings are generally limited to identifying the major successes and failures of the facility's overall performance and direct assessment of only the main elements and functions.

The investigative evaluation can take weeks to months, depending on the depth of the investigation. It provides a more in-depth analysis of building performance through the use of interviews, workshop meetings and general end-user survey questionnaires. Inspection reports include a summary rating against objective criteria by comparing the facility to other similar facilities. Physical measurements and photographic records are made and solutions to identified problems are explored in detail.

The diagnostic evaluation is the most detailed form of assessment and is carried out over months and even years. It is a focused, cross-sectional evaluation by a team of specialist consultants, involving sophisticated data-gathering and analysis techniques. Reviewing the facility at a detailed performance level, the team undertakes a comparative analysis of the strengths and weaknesses of the facility. The analysis culminates in strategic planning advice to the commissioning organization to ensure they obtain the maximum possible value from their facility.

The past decade has seen a marked rise in the level of POE studies. Ballesty attributes this to the dramatic growth of FM as a professional discipline and an increasing awareness of the benefits to be conferred by POE. In the long term, he sees POE enabling the establishment of 'facility performance databases and the generation of planning and design criteria for specific facility types' (1999, p. 205).

16.8 Conclusion

There are many books on TQM, covering all aspects of the field in enormous detail. What the quality movement emphasizes is the importance of leadership from senior management if a quality programme is to be effective, measurement and benchmarking, training and employee involvement, and an information system for feedback and assessment (see, for example, Fox, 1991). These are also the key requirements for a quality FM programme.

The synergies possible between TQM and FM that Friday and Cotts (1995), for example, base their work on make a strong case for the use of TQM ideas. Their examples of benchmarks and measures for effectiveness and efficiency of service delivery, responsiveness to customer needs, and the relevance of services provided are important measures for any FM business. However, TQM is not a panacea. There have been a number of books that focus on the problems, rather than the ideas and successes. Brown *et al.* (1994) look at the causes of failure of TQM efforts and how to avoid them. The final chapters in their book look at the importance of an organization's financial, information and planning systems and strategies for creating a learning organization.

Given the range of tasks involved in FM, a QM approach offers a method for developing and maintaining service standards. The opportunity then comes for continuous improvement of the service, which brings the benefits of client loyalty and staff motivation. As the industry develops and matures the use of QM is an opportunity to gain competitive advantage for FM providers.

References and bibliography

Albrecht, K. and Zemke, R. (1985) *Service America!: Doing Business in the New Economy* (New York: Dow Jones-Irwin).

Ballard, G., Koskela, L., Howell, G. and Tommelein, I.D. (2002) Lean construction tools and techniques. In: Best, R. and de Valence, G. (eds) *Building in Value: Design and Construction* (Oxford: Butterworth-Heinemann).

Ballesty, S. (1999) Facility quality and performance. In: Best, R. and de Valence, G. (eds) *Building in Value: Pre-Design Issues* (London: Arnold).

Brown, M.G., Hitchcock, D.E. and Willard, M.L. (1994) *Why TQM Fails and What To Do About It* (Burr Ridge, IL: Irwin Professional).

Camp, R. (1979) *Benchmarking: The Search for Industry Best Practices that Lead to Superior Performance* (Milwaukee, WI: American Society for Quality Control/Quality Press).

Carassus, J. (1999) Construction system: from a flow analysis to a stock approach. In: Ruddock, L. (ed.) *Macroeconomic Issues, Models and Methodologies for the Construction Sector Workshop*, CIB W65–75 Publication 240, Capetown, 17–29.

Crosby, P. (1979) *Quality is Free: The Art of Making Quality Certain* (New York: McGraw-Hill).

Crosby, P. (1996) *Quality is Still Free: Making Quality Certain in Uncertain Times* (New York: McGraw Hill).

de Valence, G. (1999) Project initiation. In: Best, R. and de Valence, G. (eds) *Building in Value: Pre-Design Issues* (London: Arnold).

de Valence, G. and Lauge-Kirstensen, R. (1998) Views on industry structure. *Form/Work*, **1** (2), 77–88.

Deming, W.E. (1982) *Out of the Crisis* (Cambridge, MA: MIT Press).

Dougherty, J.R. and Reicher, R. (1992) Class A MRP II at DuPont: how we measure the results. In: *Conference Proceedings, American Production and Inventory Control Society*, 31–4.

Ellis, Jr, R.D. (1997) Identifying and monitoring key indicators of project success. In: Alarcon, L.F. (ed.) *Lean Construction* (Rotterdam: Balkema).

Feigenbaum, A. (1991) *Total Quality Control* (New York: McGraw-Hill).

Fox, R. (1991) *Making Quality Happen: Six Steps to TQM* (New York: McGraw-Hill).

Friday, S. and Cotts, D.G. (1995) *Quality Facility Management: A Marketing and Customer Service Approach* (New York: John Wiley).

Gray, C. and Hughes, W. (2001) *Building Design Management* (Oxford: Butterworth-Heinemann).

Handy, C. (1984) The Future of Work: a Guide to a Changing Society (Oxford, Blackwell).

Juran, J.M. and Gryna, F.M. (1993) *Quality Planning and Analysis: From Product Development Through Use*, 3rd Edition (New York: McGraw-Hill).

Kane, K. (1997) *L.L. Bean Delivers the Goods*. www.fastcompany.com/online/10/llbean.html

Knight Frank Hooker Research (1995) BOMA leading edge research – tenant demand. *BOMA Magazine*, September.

Leibfried, K.H.J. and McNair, C.J. (1992) *Benchmarking: A Tool for Continuous Improvement* (London: HarperCollins).

Oakland, J. (1995) *Total Quality Management: The Route to Improving Performance*, 2nd Edition (Boston, MA: Butterworth-Heinemann).

Porter, Michael E. (1980) *Competitive Strategy: Techniques for Analyzing Industries and Competitors* (New York, Free Press).

Sheldon, D. (1992) Performance measurement and accountability – a top management perspective. In: *Conference Proceedings, American Production and Inventory Control Society*, 19–21.

Spendolini, M.J. (1992) *The Benchmarking Book* (New York: Amacom).

Womack, J.P., Jones, D.T. and Roos, D. (1995) *The Machine that Changed the World* (New York: Rawson Associates).

PART 3

17

Democracy in design?

Geert Dewulf*
Juriaan van Meel†

Editorial comment

There are many change drivers in the facility management discipline. One important one is the recognition that the physical workplace contributes directly to user satisfaction and productivity. The layout of work areas and the functional relationships between work areas is something that should be resolved in the design stage, but even where it is, activities and processes change, and facilities need to be sufficiently flexible so that they can be adapted without incurring large churn costs. Technology also plays a significant role in freeing up people to be mobile, work remotely and use tools that make them effective.

Workplace ecology is a term coined to describe the *health* of the work environment in terms of making people more effective. It embraces issues of organizational reform, worker productivity, information technology, sustainability, and occupational health and safety. Organizational management is critical in the creation of a productive environment as in many cases it involves breaking down long-held hierarchical cultures that dictate floor space and position, compartmentalization, furnishing status, and territorial demarcation. Contemporary thinking is that workspace should be more task-orientated and interactive in order to support project teams and communication, and less territorial and therefore more flexible, with higher quality shared resources and amenities. Indeed, functional purpose is being blurred by the creation of multifunction spaces such as eating areas that are also meeting spaces, open plan offices with specific task areas, mobile personal storage and communication technology, modular (even smart) furniture, brighter and more pleasant interior designs and the like.

Flexibility is a key strategy. Churn costs have increased in real terms over time due to the expense of moving technology and services when people relocate or when other workplace changes occur. A philosophy of 'move people not technology' leads to

* University of Twente, The Netherlands
† University of Delft, The Netherlands

significant savings and increased value to an organization. Furthermore, advances in personal communication technology have enabled people to become mobile, to work remotely and to use headquarters facilities like 'docking stations' rather than permanent workplaces. Cars, airport transit lounges, hotel lobbies, homes and client offices have all become places where work can be done without losing contact with colleagues and customers. In fact, technology has significantly boosted productivity in many sectors and freed up the need to hold large quantities of expensive floor space.

Workplace ecology also refers to occupant comfort. Indoor air quality, visual privacy, noise reduction, naturally lit and ventilated spaces, external views, and individual control over work area temperature and lighting levels are now taken more seriously. Sometimes this is just a matter of good design, but it is also underpinned by a realization that productivity gains will outweigh refurbishment costs where these features are deficient. Essentially it comes down to a value for money decision, with quite short payback periods and other spin-offs that make this area of facility management highly lucrative.

All workplaces comprise facilities that must be managed. At one end of the spectrum this involves cost monitoring and maintenance, and at the other end it involves strategic alliances and integrated business processes. A 'healthy' workplace is one that performs at its optimum. Workplace ecology is the study of workplace performance, understanding the link between human resources (people) and physical resources (facilities). Productivity is this link. Measurement of productivity is the best means of quantifying improvements in workplace ecology and improving value.

17.1 Introduction

Buildings are primarily objects of use and that should make users experts in what buildings have to do. Office buildings, for example, are built to facilitate knowledge workers and the work they do. In fact, most employees spend more time awake behind their desks than at home. For that reason they probably know more about how offices should look and work than the average architect or consultant. The same holds for urban areas: cities, villages and neighbourhoods are places where people live, recreate and work, and that implies that these areas have to meet the needs of citizens.

Architects, planners and consultants may come and go, but users spend their lives in their creations, so user involvement makes good sense. In many cases, however, the people concerned and affected by the design are never involved. Design and decision-making is centralized, fragmented and something that involves only a small group of experts. More and more consultants and designers, however, are proclaiming a bottom-up or interactive approach wherein users play a crucial role. For example, Franklin Becker (1990), one of the major writers on the subjects of buildings-in-use, promotes the need for 'user participation in design programming'. Likewise, interactive approaches in urban development have taken firm root. As buildings or city plans are artefacts that reflect the interests of the users, direct participation of users in the design process has to be stimulated.

Involving the user is easier said than done. Over the years designers have always struggled with how users of buildings or citizens have to be involved in the design process. The problem is that there is no such thing as *one* user or *one* best way to involve them. In reality the 'user' is not just one person, but an organization comprising different factions. In office design the term 'user' can refer to a diversity of people, varying from the rank-and-file employees to the top management of the organization. All these groups have different ideas, opinions and interests.

User involvement may not always lead to the desired results – when radical new ideas (e.g., flexible office concepts) are introduced, user participation may become an obstacle for change. Traditional ideas may be entrenched and users unwilling to give up 'acquired rights' (Table 17.1).

Table 17.1 Different types of 'users' and their interests

Users	Interests
Top management (CEO, CFO, etc.)	long-term financial interests and corporate objectives
Business heads (management of the various business units)	business objectives, sometimes in conflict with those of other parts of the corporation
End-users (the employees)	the efficiency of their work processes and their individual satisfaction
End-user representatives (unions, workers' council)	the political interests of the employees
Real-state and facility management departments	a balance of the interests of the other parties depending on their position in the organizational chart

Given these problems, the discussion about interactive design is often purely normative and political. User participation is used as a marketing tool by consultants and designers, and fits with current societal belief in new types of governance and modern business. There is, however, no one best method. How to involve users, and to what extent, depends on the specific context, i.e., the level of innovation, the culture and structure of the organization, the size of the project, and the creativeness of participants.

17.2 Approaches to user involvement

There are three different major approaches to user involvement, with attendant advantages and disadvantages depending on the specific characteristics of individual projects. The three approaches are:

- the technical-hierarchical approach
- the interactive approach
- the incremental approach.

17.2.1 The technical-hierarchical approach

The technical-hierarchical approach to design is the most common. The design of office buildings tends to be driven by economic considerations rather than employee desires. This is not really surprising since office buildings are a means of production, just like computers or factories, and as such they have to support and facilitate the core business of the company. The main focus should be the work process rather than 'irrational' employee wishes. During the design process, the designers or consultants analyse work processes and translate these into design. They gather information about the occupants in order to build a basis for informed decision-making by the management (Horgen *et al.*, 1998). In this top-down process, the client/principal and designer communicate with each other without involving the user (Cairns, 1996a).

This technical-rational approach consists of a predetermined and fixed project plan progressing from one decision to another. The rational paradigm starts with the idea that there is only one answer and the task is to find it (van der Heijden, 1996). A typical technical-rational approach consists of the following steps:

- analysis of the mission and strategy of the organization (by talking to the management)
- analysis of the organizational needs by investigation of activities or work processes (survey among employees) and current use efficiency (time-utilization studies)
- design
- implementation
- appraisal and control.

In this process, users are mainly 'informants' who 'deliver' input to the designers and decision-makers. In interviews and surveys, employees have to explain how they work and how much time they spend on certain activities. They may also participate in user groups dealing with specific topics such as interior design and security, but crucial decisions about, for example, the workplace layout are imposed by the company's real estate department or management.

The basic assumption underlying this approach is that the decision-makers (e.g., architects, planners, managers) know by practice and experience what is best for the organization. Not surprisingly many architects seem to be in favour of this approach – after all, they are trained in this work. Interpreting user needs and translating these into a design lies at the heart of their profession and many believe they are good at it. British research (Cairns, 1996a, 1996b) shows that both designers and procurers of buildings are relatively confident of meeting user needs. They consider that user needs are being met regardless of the level of user involvement in the briefing process.

The technical-hierarchical approach is often used in organizations with a hierarchical culture. In such organizations employees are simply seen as the executors of the decisions of the management. The view is that as they are hardly involved in organizational decision-making, so why should they be involved in design decisions? Furthermore, this approach is often used on large projects that have a large number of users. In such cases the reason for excluding users may not be ideological or cultural but merely practical, as involving users when designing an office building for, say, 1000 employees is not easy. Involving many people means accommodating many different ideas and opinions. How can these all be integrated into a single concept? This can be a major problem, especially when trying to implement a radical concept.

Office innovations brings with it an almost revolutionary change in how employees use and experience their working environment (van der Voordt and van Meel, 2000). In many cases office innovation means giving up a personal workspace and getting used to new ways of working. There is often much resistance. Dealing with this resistance requires a lot of energy and social skills from consultants, designers and the managers involved. From their perspective it is easier to predetermine the concept and then sell it to the employees rather than to co-design it with the employees. Intensive user involvement is then regarded as an obstacle as employees often do not (at first sight) appreciate change.

A good example of such a top-down approach is provided by a large American financial company that actually hired marketers to promote a new workplace concept. The concept had been devised by the head office and was to be implemented all over the world. As the company wanted the same concept to be used in all their offices, regardless of location, there was no place for local user participation. In such a case the technical-rational approach may be the best method for achieving the underlying objectives of world-wide uniformity and cost reduction. It is just as clear, however, that a total absence of employee involvement can result in a work environment that does not 'work', i.e., an office in which employees are neither happy nor productive. As the structure of most organizations has become flatter, and employees are the key resource, users have to be involved in the design process one way or another.

17.2.2 The interactive approach

In the interactive approach the designer engages the occupants in the design process. The interactive or collective design is something more than a solution based on contributory influence and 'simple' participation, where representatives are informed but only have the opportunity to react to already formulated solutions, e.g., drawings already finished by an architect (Lindahl, 1996). The basic assumption underlying this approach is that users should be regarded as experts, just like architects and consultants. In particular, Swedish researchers such as Granath *et al.* (1996) stress the importance of 'collective design'. According to them, all actors, including users, have to take part in the design process. The idea is that users can best formulate their demands and wishes for the new workplace, since they are the ones who carry out the organization's core activities.

Using the interactive approach to design does not mean that the whole group of users gathers around the drawing board. Lindahl (1996) writes: 'Of course not everybody who is affected by the design result can participate in the process. That would be an ideal but overwhelming situation. A few actors will represent everybody.' In practice, it is important to make a distinction between the participation of the individual employees – the end-users – and the participation of their 'agents'. In many cases, it is not so much the end-users themselves who have participated in the crucial decisions, but these formal representatives. End-users are often mainly involved at a space planning level (who sits where) while the formal representatives vote in the initial strategic decisions such as location, floor plan and overall design.

Apart from any ethical arguments, involving employees in design is crucial when trying to produce an office concept that is both innovative and regarded by users as 'their' office. There are several good examples of how these can be combined, e.g., the design process for a new research laboratory for Xerox in the US, described by Horgen *et al.* (1998). In that instance, a group of user representatives, designers and experts from MIT went through a series of workshops and design sessions to create a new workspace. Using the ideas and knowledge of the users, the designers were able to create an office that really enhanced the work process of the organization.

In another case, Amstelland Development used the Internet or 'virtual communities' to design a residential quarter, IJ Burg, in Amsterdam. Future residents were prompted to comment and thus alter the design of the quarter. On a larger scale, an interactive approach was used in the redevelopment of Roombeek, the district in the Netherlands town of Enschede that was totally demolished after an explosion in a fireworks warehouse in 2000. After the disaster a clear consensus emerged that the victims should be allowed 'maximum feasible participation' in the design process. Apart from obvious democratic reasons, participation also had to help people in dealing with their loss.

Practical experience (gained directly by the authors) underlines the advantages of the interactive approach. Workshops with end-users generally result in a great number of new ideas about how workplaces should be designed. Users become much more aware of the limitations of a building. When they experience how difficult design can be, their demands about space and privacy automatically go down.

Again, however, it is important to know that combining change and participation can be problematic. One of the crucial questions is whether there is there enough freedom in the decision-making process to allow end-user participation. For example, in a large project for a major Dutch industrial company (which the authors were directly involved in), end-users were interviewed about their work processes and workplace demands. In the interviews, these end-users expressed a clear preference for cellular offices. This was partly based on what people were used to, their work processes and their unwillingness to change just for the sake of cost reduction. Later it was decided that cellular offices were simply not an option in the building because of earlier decisions about the heating, ventilation and air-conditioning (HVAC) services in the building. This led to increased frustrations for the users. In this case it would have been better not to involve the users in such an early phase – the users expected (naturally) that their input would somehow be reflected in the office layout, which, in the end, was not the case. Eventually frustration ran that so high that the department did not move into the new office building. This example makes it clear that an early decision about the extent to which users can actually make design decisions is crucial.

The way users are involved may vary from informing to collaboration. The ladder of participation comprises several stages as shown in Table 17.2.

Table 17.2 Eight rungs on the ladder of citizen participation (Arnstein, 1969)

Citizen control Delegated power Partnership	Citizen power
Placation Consultation Informing	Tokenism
Therapy Manipulation	Non-participation

The level of participation has to depend on the situational characteristics:

- the creativeness of the users – most users think in a very traditional way with the way they were accommodated in the past often used as a reference; the participatory process

in Enschede, for instance, resulted in a very traditional programme as most victims
wanted the district to be the same as it was before the explosion

- the level of experience – several studies (e.g., Cairns, 1996b) have been undertaken
  where communication patterns between clients and professionals were analysed during
  the briefing process; these studies found that the input from the parties varied
  considerably depending on the client's previous experience of construction
- the level of innovation or radical change that the management wants to realize
- the level of tolerance of the user groups towards new ideas – if the user group thinks
  conservatively, or has difficulty in widening its thinking, a more rigid approach by the
  consultant/designer will be needed to force the users to think about new concepts
- the willingness of management to be open for discussion
- the flexibility and the communication skills of the designer – the designer has to be
  open to ideas from the users
- the legislation and culture of a country – in some countries the interactive approach is
  partly driven by legislation.[1]

17.2.3 The incremental approach

While the interactive approach may seem to be the most democratic method it is not
always the best solution. A primary reason has already been mentioned: in many cases
management is not open to new concepts and users are still thinking in traditional ways.
A second reason is that the use of an interactive approach often results in an enormous
variety of opinions about the future of the organization, or the design of the building or
neighbourhood. Participation in design workshops can lead to the introduction of a variety
of ideas – however, this may impose severe pressure on the designers who must then try
to design a building that accommodates the points of view of all the participants.

In the case of Enschede, eight participation sessions were held to provide input to assist
the town planner in preparing a first draft of the redevelopment plan. The organizers of the
sessions provided the participants with cues (in the form of a series of photographs and
accompanying short texts) for reflection, and a list of topics relating to the future district
for discussion. The number of topics and cues was so large that the participants were able
to address almost any topic they might have deemed relevant. The openness of the process
led to an enormous variety of options about the future of the neighbourhood.
Consequently it would not presumably be too difficult to supply advocates for virtually
any concept with a 'useful anthology of statements made by participants to legitimise their
point of view' (Denters and Klok, 2001). Thirdly, an interactive approach may exaggerate
expectations about the real influence of users; in the end it is still the management or, in
the case of an urban area, the city council, who decides what is going to happen.

For these reasons a more incremental, or step-by-step, approach is the best way. The
design process is then seen as a process of negotiation between management, users and
designers. In the incremental approach the official future, as seen by the management, is
taken as the starting point.

Proposing a new way of working often leads to opposition from users or management.
This could halt or seriously delay the process. In cases where radical change is required
the workplace consultant can better start from the traditional concept and confront the
audience, whether this is the user group or the management, with questions related to the

impact on their own interests or results from research or other experiences. Questions such as 'have you thought about . . .?' or using 'what if . . .?' scenarios may stimulate the organization to reconsider their starting point.

Another way to begin is by organizing a presentation, or visiting other buildings or concepts at the beginning of the process in order to stimulate discussion.

The role of the consultant is not to just go along with management but to challenge the official strategy by identifying possible future pitfalls. The consultant's or designer's role is to design an optimal workplace for all stakeholders, but by proposing radical change at the beginning of the process, (s)he may meet strong resistance.

The incremental model is not a linear but an iterative process. Change is slowly introduced but will last for a long time. A clear overview of the elements that should be part of an incremental approach is given by van der Voordt (1999). He also suggests the following ten recommendations for implementing innovative concepts:

- organize a start-up meeting to inform those concerned and to come to agreements about the objectives and approach of the process and the results envisaged
- make sure there are enthusiastic project leaders among both management and users
- set up a project organization whose tasks and powers on behalf of the parties involved are clearly defined and where there is co-ordination between different levels (from top management to daily users) and different disciplines (real estate management, human resources, information technology, finance, etc.)
- make sure there is a balance between policy-oriented guidance from management ('top-down') and the grass roots development of the users' ideas ('bottom-up')
- organize workshops with users to collect data, develop ideas and test design proposals
- plan the number of workshops to achieve a balance between the need for information and discussion, efficient time management and rapid throughput
- alternate workshops with individual interviews and project team meetings during which more detailed discussions and decision-making take place
- involve the architect in the process at the appropriate time, as soon as working processes and trends are clear, and the desired workspace concepts begin to take shape
- ensure that employees/users are informed when the project is completed
- draw up clear agreements about any temporary accommodation and the use and management of the new accommodation; make sure that there is proper training so that users are able to cope adequately with the new accommodations.

17.3 A situational awareness

Of the three approaches discussed there is no one approach that can be said to be best – the choice will depend on the specific situation of the organization.

One might say that employee participation is not necessary in design as long as you build a rather ordinary building that matches the expectations of the future inhabitants – the question is what will happen if organizations start using radically different workplace concepts? In that case the involvement of end-users may become much more crucial for success; on the other hand, it may also hinder real innovation. The underlying idea is that end-users are not likely to give up the space and privacy they have acquired over the years.

The choice of method should depend on:

- the problem that has to be solved – is the new office concept part of a larger organizational change in which new ways of working (e.g., teleworking) are to be introduced into the organization, or in which the office will change radically in terms of privacy, number of square metres per workplace, ownership of the workplace and the like, or is it a 'standard' process in which the current situation is being transferred to a new building with no fundamental changes in work processes or workplace design?
- the culture of the organization – how are decisions generally taken in the organization, do employees/end-users expect to be involved, is it a highly hierarchical or a more 'democratic' organization?
- the size of the organization, the size of the project and the number of users involved – from a practical point of view it is clear that it is quite difficult to involve a large number of an organization's employees in the design process
- the duration of the project and the number of involved users who will eventually occupy the building – there is little point in involving lots of people when many will be gone when the project is finished
- time constraints – in the high-tech sector in particular, organizational changes are frequent and with, for example, rapid unexpected organizational growth there is little time for intensive user involvement. In such cases office space simply has to be available, affordable and ready for move-in; there is little time to think about new concepts and involving users as participation and change requires time for communication, workshops, interviews and feed-back.

17.4 Lessons/'notions' for further research

The issues discussed are, to a certain extent, still open for discussion. To arrive at a comprehensive understanding of the issues, further research is needed and cases have to be studied in more detail. On the other hand, this discussion reveals some new insights into how the design process can be managed and these insights can be used as practical lessons for both architects and real estate managers.

To be able to manage the design process as well as possible the consultant, designer or real estate manager must be aware that the interests of the stakeholders depend on the context (political, legal, cultural and economic) in which they operate. Due to economic pressures, costs seem to be a universal driver in office design, but in organizations with a large tradition of employee participation and decentralized responsibilities many actors will have to be involved in the design process. Additionally one must take into account the fact that the context in which the company is operating may change over time and consequently the key actors and their interests will change as well.

Future research should focus on the use of new technologies to stimulate user involvement. Frank Duffy (2000) sees big opportunities in the unprecedented ways in which users, individually and collectively, can potentially have direct access to the design process. According to Duffy, this intermediation may result in a new, more democratic design, procurement and delivery of the workplace, with participation beyond the wildest dreams of the old believers in user-influenced design. Whether this is true or not is not yet clear. Experiences with electronic means of user involvement are few and advantages have not been proved so far. It is clear, however, that office buildings are made for users. In any project, designers and consultants have to look for the most appropriate way to involve users. What is meant by the term 'appropriate' is still open for debate. The problem is that the discussion so far about interactive approaches or user involvement is a very normative one. More research has to be done to find the best way to build buildings that are productive, pleasant and durable.

Endnote

1 North European organizations are more or less forced to adopt a participative style because of the extensive legislation. In Sweden, for instance, employees have the 'right to negotiate' (*forhandlingsratt*) and in Germany they have the 'right to decide' (*Mitbestimmung*). Likewise, Dutch organizations have a so-called 'workers' council' that must be consulted with regard to every major corporate decision that concerns their work environment. In particular, in government, their influence is clear and, for better or worse, this can indeed hinder change: when famous Dutch architect Hertzberger had to design a new building for the Ministry of Social Affairs he complained about the conservatism of the employee representatives.

They want a lot of room. They want a view. They want rooms . . . the employees' committee has significant rights as far as the working environment is concerned. They can say, 'No we don't want it'. And if so, we (architects) can't do it, whatever it is.

The result was that Hertzberger could not create the open areas he intended to (van Meel, 2000).

References and bibliography

Arnstein, S. (1969) A ladder of citizen participation. *Journal of the American Institutes of Planners*, **8** (3) July, 216–24.

Becker, F. (1990) *The Total Workplace* (New York: Von Nostrand Reinhold).

Blyth, A. and Worthington J. (2000) *Managing the Brief for Better Design* (London: Spon Press).

Cairns, G.M. (1996a) *Client's Input to the Design Process – Communication at Briefing and Feedback Stages*. Unpublished research report funded by RIBA Trust Research Awards.

Cairns, G.M. (1996b) User input to design: confirming the 'user-needs gap' model. *Environments by Design*, **1** (2), 125–40.

Denters, B. and Klok, P.-J. (2001) Rebuilding Roombeek-West: an institutional analysis of interactive governance in the context of a representative democracy. Paper presented at the *EURA Conference on Area-based Initiatives in Urban Policy*, Copenhagen, May.

Duffy, F. (2000) *E-commerce conversations in New York*.
    www.degw.com/cont_dnews_arch_may00.htm ny

Granath, J., Lindahl, G. and Rehal, S. (1996) *From Empowerment to Enablement: An Evaluation of New Dimensions in Participatory Design* (Göteborg: School of Architecture, Chalmers University of Technology).

Horgen, T.H., Joroff, M.L., Porter, W.L. and Schön, D.A. (1998) *Excellence by Design* (New York: John Wiley).

Lindahl, G.A. (1996) Collective design processes as a facilitator for collaboration and learning. Paper presented at the *Fourth Conference on Learning and Research in Working Life*, April 1–4, Steyr, Austria.

van der Heijden, K. (1996) *Scenarios: the Art of Strategic Conversation* (New York: John Wiley & Sons).

van der Voordt, D.J.M. (1999) Spatial implications of new trends in higher education and scientific research. *Boss Magazine*, **10** (December), 40–1.

van der Voordt, D.J.M. and van Meel, J.J. (2000) Lessons from innovations. In: Dewulf, G., Krumm, P. and de Jonge, H. (eds) *Successful Corporate Real Estate Strategies* (Nieuwegein: Arko Publishers).

van Meel, J.J. (2000) *The European Office* (Rotterdam: 010 Publishers).

18

Leading change through effective communication

Tom Kennie*

Editorial comment

Change management stands at the forefront of the facility management (FM) discipline and is a vital driver for organizational change in its own right. It relies on effective communication, an understanding of external constraints, internal goals and infrastructure requirements, as well as an appreciation of project management principles and processes. Change is a necessary part of daily life and should be seen as an opportunity rather than a threat or nuisance.

Value can be added by improving organizational communication concerning change. The major support areas (FM, human resource management and information technology) need to be connected into other management processes so that business effort can be co-ordinated and productive. There is a growing case for linking all support areas either through combined administrative structures or through reporting technology such as enterprise resource planning systems. Organizations need to become smarter and learn through better usage of resources and reduced duplication of effort.

Effective communication covers a broad range of issues, including leadership, motivation, conflict management, networking, and appropriate recognition and reward. In the context of workplace strategies, however, it means an understanding between those that are responsible for FM and the various stakeholders both within and external to the organization. This is an area of potential performance improvement. Contemporary strategies include help desks and call centres for logging user requests, online databases and other computer-based tools, increased public relations, and direct access to boardroom decisions. Most of all, effective communication involves accountability and dissemination of performance levels matched against best practice benchmarks.

Education and staff development are necessary to manage change in the future, and are closely linked with facility performance and value. Modern facilities are so complex that it is critical that their correct operation be well understood. While this expertise can be

* Ranmore Consulting Group/Sheffield Hallam University, UK

outsourced, in-house management teams still predominate, and in these cases the knowledge possessed by FM staff must be preserved and used to best advantage. Often such knowledge is irreplaceable, particularly where facilities require special needs and care (such as historic buildings, nuclear power plants and high security installations).

The facility manager is, therefore, in the unique position of being in touch with organizational decisions and user requirements, and, as a result, can provide a valuable service by co-ordinating much of the change that is required to infrastructure and work practice. Where this is done wisely, lost productivity can be minimized and upper management can be left to focus on core business issues.

18.1 Introduction

'Change' – to paraphrase a well-known advertisement, change is 'possibly the most commonly used word in business today'. Whether you are initiating it, leading it, responding to it, or even resisting or trying to stop it, having the capability to cope with change is a key competence for facility management (FM) practitioners.

This chapter is about the process of leading change. More particularly, it is about the importance of communication during the process of change. It seems so obvious; of course communication is a key requirement for effective change management, surely this is self-evident to everyone? And, of course, it is, but how often have you suffered from the consequences of poor communication? How often have you suffered from the effects of NETMA and NASMAS? These may not be the most memorable acronyms, perhaps but their effects can be long lasting: the 'Nobody Ever Tells Me Anything' (NETMA) and 'Nothing Anybody Says Makes Any Sense' (NASMAS) effects. How can we use effective communication to help limit these effects? What techniques and processes might be worth considering, and is it possible to identify any useful tips and techniques?

This chapter is based around a very simple three-phase approach to change – a diagnosis phase, an implementation phase, and a reflection and learning phase. Before discussing each of these phases it is sensible to ask such questions as:

- what issues should we be considering before we try and implement a change project?
- who might need to be considered?
- what could be some of the primary and secondary effects of the change?

These questions will be addressed later in the chapter.

18.2 Diagnosis phase

In most change projects the pressure to 'get on with it' is generally very high. Confronted with such a demand it is often very easy to rise immediately to the challenge and initiate lots of action. Project plans are drawn up, budgets are agreed, individuals are given responsibility for different parts of the change agenda, and so on, and an apparent sense of progress is created. Applying this model, what might be called the JDI ('Just Do It') approach may succeed – but so often unexpected issues arise, confusion grows and in some instances failure follows.

One of the fundamental principles for anyone when confronted with this dilemma is 'time spent on reconnaissance and planning is never wasted'. The equivalent process in a change project might be that of effective diagnosis, i.e., really understanding the nature of the change, how it might be viewed by different stakeholders and who is going to be affected by the change.

To help illustrate the process of effective diagnosis let us consider a simple example. X is a highly successful firm of lawyers, operating on a global basis. Recognizing the need for more effective team working and knowledge sharing, the firm decides to explore the possible modification of the space in its HQ building from a traditional cellular office layout to one more conducive to the culture they wish to encourage, one involving team working and innovation. The firm also hopes that some savings in space might also be provided, although this is not considered to be a prime driver for the change. Senior management also recognize that the process should not be imposed, should proceed incrementally and should be evaluated on an ongoing basis.

In the first instance, the senior management decides to seek a volunteer team to pilot test the process. To encourage a group to come forward they indicate that the budget for the process will provide for additional information technology (IT) equipment and that the space and furniture design will be selected and implemented in partnership with the interior designers who have been selected to work on the project. Finally, it is also agreed that if the process is not perceived by the volunteer team to have succeeded, then the space will revert to its original design.

So how would the facility manager advise the chief executive on how to progress such a change? One way to start would be by standing back from the issue and considering some of the macro level issues associated with the change, possibly from two different perspectives.

18.2.1 Degree of change

A reasonable starting point involves identifying the degree of change that is anticipated and the extent of the change in terms of the technical and people challenges.

Figure 18.1 provides one framework for considering the first issue. How widespread is the extent of the change? Is it merely a minor change, with relatively low people or

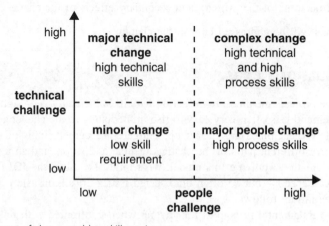

Figure 18.1 Forms of change and key skill requirements.

technical challenges, or is it more significant, demanding a high level of technical and process skills? Clearly, the skills required in both instances will be different and different needs will arise from a communication viewpoint. In the former case the needs might be relatively low and rather informal in nature, whereas in the latter case the requirement will be greater, involving more face-to-face discussions and may require high levels of interpersonal capability.

Figure 18.2 considers the issues from a different perspective. In this instance the diagnosis focuses on the depth and pervasiveness of the change. Does the degree of change merely indicate a degree of adjustment to existing processes (lower left-hand quadrant) or might it indicate a much more radical change to the entire way of operating within a unit or across an entire organization?

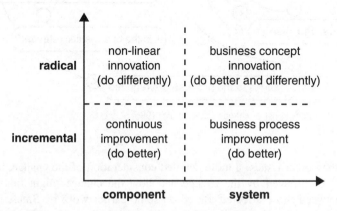

Figure 18.2 Classifying change management initiatives (Hamel, 2000).

Both of the models shown could be applied to the case study. Using the model shown in Figure 18.1 it becomes apparent that whilst the technical demands of the change are not insignificant, they are largely about using existing techniques and technology. On the other hand, the change has the potential to have significant impacts on the relationships between different staff members and requires careful handling.

The second model (Figure 18.2) shows that a change of this type is likely to involve changes in a number of different activities and may ultimately lead to quite different ways of operating. In those senses the change may initially lead to improvements in business processes (lower right-hand quadrant), but ultimately could lead to new and different ways of operating (top right-hand quadrant).

Again the consequences of the diagnosis will begin to identify questions about the approach to change that will need to be considered and the capabilities that the team will need for the project.

18.2.2 Diagnosing relationships and communication issues

A second form of diagnosis that can usefully be considered in the initial phase is illustrated by Figure 18.3. This level of preparation is designed to encourage to give

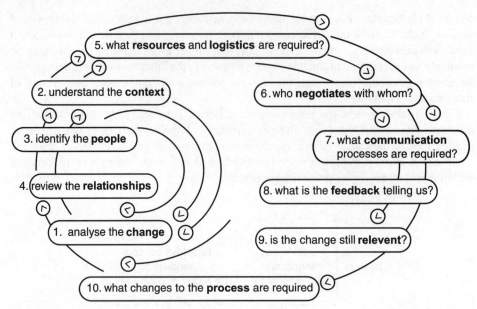

Figure 18.3 An integrated model of change (after Smale, 1996).

thought to (among other issues) a more detailed consideration of the change, identifying those who will be affected by the change and how the change might impact on the relationships between those affected. Based on some earlier work by Smale (1996) the following ten key questions have been developed and might be worthy of consideration at this stage.

Analyse the change

- why change? What are the pressures for change – externally and internally?
- what is the primary issue that requires change?
- what changes are required to tackle the issue at a macro-level? at a micro-level?
- what secondary changes might occur as a consequence of the primary change project?
- what innovations are involved in the proposed changes? (An 'innovation' is a specific practice, method, technology or form of service delivery new to those to whom it is being introduced.)
- who is ultimately accountable for the change?
- who has been given delegated responsibility to project manage part, or all, of the change?
- what should and can stay the same?
- who are the key stakeholders who will ultimately judge whether the change has been successful?
- what does success look like? Can it be defined quantitatively, in SMART (Specific Measurable Resourced Actionable and Time limited) terms, or qualitatively, e.g., in cultural terms?
- what are the most significant risks associated with the change?
- how likely is each and what might be the level of impact for each?

Understand the context – which way is the wind blowing?
- what is going on that is compatible with the change?
- what changes are going in a different direction?
- what other changes can this change be linked to, to gain support?
- are other organizations involved and are they open to the change(s)?
- is the change compatible with the prevailing culture?

Understand the context – what is the impact of change on individuals, working groups and the wider social system?
- who experiences what as 'winning'?
- who experiences what as 'losing'?
- how can you make sure that people have space to mourn their losses?
- what can you do to build commitment to the new solutions and new situations?
- can the change, or part of it, be tried out or 'piloted' before wider application?
- can it be observed in operation elsewhere, and if so, where, and by whom?

Identify the people – reactions to change
- for whom is the status quo a problem?
- who wants change and for what reasons?
- who is likely to resist the change?
- can they be influenced in the short term (and by whom)?
- who may be difficult to persuade in the short term, but who will be likely to become more positive towards the change?
- who may be very difficult to move at all?
- identify all the key players – who has to do what?
- what connections need to be made between people to enhance communication?
- what support might you need to provide to those who are leading the change(s)?
- who will provide support?

Identify the people – who needs to make it happen and. . .
- help it happen by releasing resources?
- support the innovators and change agents?
- change their behaviour in the unit?
- change their behaviour so that their relationship with significant others changes?

Identify the people – who has to let it happen: keep out of the way
- who has to give their consent?
- who could sabotage the change?
- who has to release resources?
- who has to refrain from diverting resources?
- who has to avoid taking counter-action?
- what resources do we have to motivate people to change?
- what other negotiating capital do we have which can be traded?

Review the relationships
- within which relationship does the change imply *a first-order change*? (In first-order changes the task and how it is done change, but the way people relate to each other fundamentally stays the same.)
- who will be affected by a change in relationships?
- within which relationships does the innovation involve *a second-order change*? (i.e., the task changes and so does the nature of the relationship between people)
- who will be affected by a change in relationships?
- how might we need to change our behaviour to change the relationships we have with significant people?
- whose job will be changing?
- who will need counselling and support?

Resources and logistics
- who needs to do what, with, whom, about:
  - physical issues? (e.g., space)
  - personnel? (e.g., are additional staff needed during the change)
  - finance? (e.g., additional financial resources to support the change)
  - quality and standards? (e.g., to ensure both are safeguarded during the change process)
  - administration? (e.g., to ensure additional time is made available from existing resources, or new resources identified)
  - internal/external communication? (e.g., consider when and how to communicate the process of change and ensure 'small wins' are communicated)
  - monitoring? (e.g., what data do we need to track during the change process, how will we access the data, who will communicate progress?)
  - time? (e.g., is the time for the change(s) realistic? if not, who and how will this issue be addressed?)
- are other resources required or available?
- how can missing resources be made available?

Negotiations
- who has to carry out what negotiations with whom to get agreement for changes in working practices and changes in significant relationships?

Communication and engagement – who needs to participate in:
- consciousness-raising events?
- conferences/meetings illustrating the major dimensions of a new approach to those who might be involved in the next phase of implementation?
- forums for information exchange amongst the initiated, to develop familiarity about the change(s)?
- staff development programmes on specific skills?
- team building workshops, for existing teams, new teams or cross-functional teams?

The application of this ten-point pre-change diagnosis can help to identify those who might otherwise not be included in the early stages of the change process and also help anticipate areas of potential conflict as the project progresses.

18.3 Implementation phase

The next phase, after having conducted a thorough diagnosis of the change project and subsequently developed a well thought out plan, is the process of implementation. In this phase a FM professional can play a number of roles. In one role he or she might be acting as project team leader, responsible for leading the implementation of the change. In other instances he or she may be working more as an internal consultant, coaching a senior manager through a change project, he or she may be acting more as a facilitator, working alongside a senior manager who is championing the change. The skills and analytical frameworks required for these roles may be slightly different.

18.3.1 Project leadership

Communication and leadership continue to be of particular importance at the implementation phase of a change project. The following list developed from some initial work by Briner *et al.* (1990) highlights some of the key requirements for a leader of a change project during this phase.

Wiring in stakeholders

In the context of the earlier case study, this requires identification of the relevant stakeholders, which might include the involvement and agreement of other departments and offices that might be affected by the change. Without recognition of this phase, it can so often occur that at some critical phase in the project, resources, which you expected being available, can suddenly be redirected onto other more important projects because of the lack of engagement of critical individuals. In the case study this should include the involvement of the IT department at as early a stage as possible. It is also clearly necessary to inform others within the firm, with whom the team is working, of the timing of the change, and discuss how this relates to the timing of current and future workload.

Negotiating success criteria

A key role for project leaders lies in negotiating a mutually acceptable understanding of success with those who have an interest in the project. Negotiating clear terms of engagement from the start can help minimize any misunderstandings at a later stage.

Building credibility links

Building credibility is a further 'soft' skill required by project leaders. While technically it might be assumed that you have credibility, it may also be important to build credibility in relation to your understanding of 'the people issues'.

Marketing the project

One factor often overlooked when leading a project of this type is the internal marketing required. This should not be confused with self-aggrandisement; instead it is a rather more subtle process of ensuring that the project has a profile within the firm.

Such a profile can be of value because it can help to ensure others are aware of the project and the phase it has reached. At its simplest level it is about internal communication. It is also about ensuring that the project is appreciated and it may also help with the motivation of those engaged in the change if all concerned feel that the project is important and that others value it.

'Helicoptering'

It is important for the project leader in any change project to maintain a sense of perspective across the project. The concept of 'helicoptering' as a metaphor for standing back from the detail may be helpful in considering this issue. If one gets drawn in too much there is a danger of losing that wider sense of context. Equally, standing back too far above the action has equal dangers of not connecting the strategic perspective with the tactical actions. Effective 'helicoptering' demands that often-intuitive capacity to get involved at a detailed level when the project demands an intervention while at other times standing back and trusting the team to self-manage through part of the project.

Constant client communication

Communication is, of course, everyone's duty, but the management of communication is the project leader's responsibility. In that sense it needs to be built into the project plan and the time for internal and external communication needs to be recognized as vitally important to the project. It also needs to be planned, co-ordinated and reviewed to ensure it remains relevant.

Seeking feedback

Finding time during a major change project to seek feedback can be very difficult. Nevertheless, if considered in advance and built into the project, such feedback can be very beneficial.

Continuous review and re-planning

A further challenge facing project leaders is to use regular feedback information to re-plan and modify the original project plan. When the 'client' changes his or her mind on a critical aspect of the project, or a critical team member is ill or leaves the firm, or a major risk factor becomes more significant than originally planned, then how quickly and effectively re-planning occurs can be crucial.

Keeping the whole team informed

Re-planning alone, with no consultation with others and with little effort devoted to communication can, and often does, lead to project inefficiencies. Keeping the team informed of changes is also crucial to developing and maintaining team morale. Nothing causes more frustration and anger during a change project than change without effective communication, i.e., when a change is made to the project plan and it is wrongly assumed that all concerned have been made aware of the change.

There are a variety of ways available for keeping the team informed. First is direct face-to-face communication – this allows the manager to clarify information and gain feedback, and demonstrates that he or she recognizes and values the views of the team. Technology provides fast, effective communication: e-mail is helpful, but it can be overdone, while a project intranet site can provide a dedicated point of reference for all communication regarding the project.

Off-site sessions can also be very useful in keeping all concerned in touch with the project and its progress as well as being useful team building exercises. These sessions involve taking the project team away for some uninterrupted thinking time regarding the project, focusing on each part of the team presenting to the others about progress, problems and resource/risk issues.

Creating a supportive culture

During any significant change project not everything goes right first time. Mistakes occur, deadlines slip, temperatures rise. Creating a culture in the project group that is supportive and avoids the creation of a culture of blame and recrimination is essential.

Providing purpose and direction

Change projects that lack leadership can lose their way for a wide range of reasons: examples include the project leader with a short attention span, who loses enthusiasm for the project; the project leader who wishes to distance him or herself from a project which is not delivering; the project leader who has not built credible links with others and finds him/herself isolated; the project leader who does not realize that they are responsible for leading the group and who would rather walk away from the project – in all of these cases the damage caused by a project leader who lacks purpose and direction can be significant. The project team can quickly sense such a lack of direction. Few projects can survive without someone to enthuse, inspire and act as an effective role model.

Clarifying individual success criteria and reviewing performance

In large complex change projects, success can often be difficult to determine. Of course basic financial targets can be quantified – completing within budget and time is relatively easy, but it is difficult to quantify other forms of success such as having developed a different way of working, or of having established improved working relationships which foster enhanced levels of team working and creativity.

In the rush associated with multiple projects and parallel deadlines it is easy to push on to the next milestone without celebrating the successful completion of the existing change project. It is important to plan for time to celebrate, partly to say 'thank you' to the team members and also to mark the fact that progress is being made.

18.3.2 The internal consultant

A potential role that a senior FM professional might play is that of internal consultant whose function might be to work alongside senior managers during a major change initiative. In a sense the FM professional will be sitting between the vision and strategy being proposed by senior management and the reality of their implementation in the organization. When the project is progressing according to plan the role may be relatively passive and primarily concerned with providing feedback on progress.

A more active role might develop when the project is not delivering the benefits originally intended. Corporate level initiatives can fail to deliver projected benefits and some changes 'derail' despite clear and unequivocal direction from senior management. Darragh and Campbell (2001) have identified three main reasons why corporate level initiatives became 'stuck' and three key questions arising from each of these reasons that corporate level managers (chief executives and senior management) could usefully consider when a change agenda does not appear to be delivering the expected results.

First, the managers in the business units do not see 'the issue' that the change initiative was trying to address in the same way that the corporate managers do, and hence do not give their support to the initiative. The first basic question to be asked is: Does the business team see the same issue that the corporate centre sees? Specifically:

- does the business team agree the facts?
- does the business team agree that the facts point to an issue that needs to be addressed?
- does the business team agree that it is responsible for addressing the issue (i.e., does it feel it owns the issue)?

In the context of the earlier case study this might be expressed as a lack of interest in progressing the project beyond the original pilot phase; the pressure for change is not sufficient. More relevant data might be required and more engagement with the business units is required.

Second, although managers in the business unit might see the issue in a similar way to their corporate level colleagues, they do not give it the same priority; they believe there are more important activities demanding their attention and hence do not take the actions that the corporate managers are expecting. Thus the basic question is: Does the business team give the issue the same priority as the corporate managers? Specifically:

- does the business team judge it has other more important priorities?
- does the business team judge that the benefits to the corporate body are not high enough to change its business priorities?
- are judgements by the business team regarding priorities unduly influenced by incentives?

Again, in the context of the case study this might lead to discussions with senior management regarding the way that they are communicating business priorities to others, and what incentives they are offering to encourage the implementation of the changes that they desire.

Third, the business managers may see the same issue as corporate-level managers and may also have given it priority, but the actions they are taking are not working, and change is not moving forward. The question then is: Are the actions that have been taken producing the desired result? Specifically:

- are there agreed plans and actions to fully address the issue?
- are agreed plans being completely implemented?
- has an unexpected new issue arisen?

With regard to the case study scenario this might lead to a review of the current project plan, clarifying who is progressing each part and reviewing the timescales for completion.

Interestingly, Darragh and Campbell's research indicated that when corporate level managers were asked to suggest reasons as to why a project might not be progressing as well as anticipated, their conclusions were often based around the perception that there was a lack of action plans (the third set of questions) whereas the same question asked of the business unit managers tended to lead to feedback associated with the first and second sets of questions. Clearly anyone acting in the role of internal consultant should look at ways to help senior managers confront such differences.

18.3.3 The internal facilitator

In this third role, which may either be a separate role or form part of roles 1 (project leader) and 2 (internal consultant), the value which a skilled FM professional can offer is to work alongside the business manager who is responsible for the implementation of the change and help facilitate movement during the project.

In the case introduced earlier, the facilitator role might well emerge as the primary key to a successful outcome. Once a pilot group has been identified, who are enthusiastic about the project, the need might arise, in the lead up to the move from the existing space layout to the new more open plan and group working arrangements, for someone to fill the role of facilitator. In this type of situation, often the group (and group leader) recognize the likely changes in working relationships which the new arrangement will create. In addition, on a more practical level, they may also recognize that it will be necessary for the group to agree on a number of 'ground rules' that will apply when the move takes place. The ground rules might range from agreeing to simple booking arrangements for the group space, to respecting each other's space, to negotiating who will sit where, and to more complex issues relating to a change in the relationships which might emerge between the team leader and the team members. A key strategy at this stage might be to act as an 'honest broker' and so help the team to agree on the ground rules and working arrangements, and so avoid any sense of imposition from (in this instance) the partner who was head of the department.

The facilitation role requires a further set of 'process' skills; Schwartz (1994) offers a particularly comprehensive framework for those involved in this activity. In essence the group facilitation process can be defined as a process in which a person who is acceptable to all members of the group, substantially neutral, and has no decision-making authority intervenes to help a group improve the way it identifies and solves problems and makes decisions in order to increase the group's effectiveness. Each of these points is important and each has implications, in that

● the facilitator's client is the entire group
● the facilitator must remain neutral and avoid unduly influencing the decisions that the group may reach
● the facilitator's role is to improve group process (how the group members work together, solve problems, talk to each other, deal with conflict, and so on) as opposed to improving content (what the group is working on).

Often one of the key functions of a group facilitator during the early stages of a workshop is to help group members agree on a set of 'ground rules' which will govern the way in which the group will operate. This might include issues such as:

● test your assumptions
● focus on interests not positions
● be specific – use examples
● explain the reasons behind one's statements, questions and actions
● make statements but invite questions and comments
● discuss 'undiscussable' issues
● do not take 'cheap shots'

- keep the discussion focused
- avoid dominating the debate
- engage everyone in all discussions.

18.3.4 Neuro-linguistic programming (NLP)

Another function of the facilitator might be to encourage clear communication and help individuals to express themselves in a clear and unambiguous manner. One potentially useful framework is the 'meta communication model' developed by Bandler and Grinder (1975) – a summary of this is reported in O'Connor and Seymour (1993). The approach is part of a wider process for understanding interpersonal communication referred to as NLP. The meta model framework is a helpful way of trying to read between the lines of what someone is saying and 'unpacking' what they might be wishing to say. The essence of the process is that everyone, in the process of expressing complex thoughts, first selects only some of the information to express the thought, then simplifies and generalizes. The meta model is a set of questions that is intended to unravel the deletions, distortions and generalizations of language.

Bandler and Grinder created a classification scheme to try and characterize typical ways that people simplify their language and at the same time limit their ability to communicate with others. Whether in a facilitation or consulting role, being aware of these patterns can be of value. Bandler and Grinder identified a number of common language patterns – some of the most common are:

Unspecified noun
Consider the statements:

> *they* don't recognize how difficult this change project is
> or
> *it's* a matter of opinion

These statements can be clarified by asking 'who or what specifically. . .'

Unspecified verb
For example:

> you need to learn more about our approach to . . .

Unspecified verbs can be clarified by asking 'how specifically. . .' – in this instance the important issue is to find out *how* the desired outcome might be achieved.

Comparisons
Sometimes in communication exchanges someone might make comparisons, but fail to specify the basis of the comparison. Any sentence with the words best, better or worse (among others) will fall into this category. For example:

> this change is being handled very badly, much worse than before

Comparisons can be clarified by asking 'compared to what?'.

Judgements

Judgements may be made and stated, but left unchallenged they may lead to misunderstandings. Consider the statement:

   obviously we need to increase our resources in department X

Judgements can be clarified by asking 'who is making the judgement?' (i.e., to whom is it obvious) and 'on what grounds is the judgement obvious?'

Modal operators of possibility

Modal operators are words which set limits by unspoken rules; they are words like 'cannot' or 'must not'. For example:

   I can't change

Modal operators of possibility are clarified by asking questions such as: 'what prevents you?' or 'what would happen if you did?'

Modal operators of necessity

Modal operators of necessity involve a need and are indicated by words such as 'should', 'should not', 'must' and 'must not'. Consider the sentence:

   you should not involve your secretarial team in the early stage of change

In this case it would be worthwhile to make more explicit the consequences of the statement: clarification can be sought by asking questions such as: 'what would happen if you did/did not do this?'

Universal quantifiers

Universal quantifiers are used to generalize about experiences. Words like 'always' and 'never' are examples. Unchallenged they might lead to assumptions being reached which are inaccurate or prejudicial. For example:

   all top-down change is wrong

This statement could be challenged by asking a question such as: 'has there ever been a time when you or someone else has found top-down change to be useful?'

Complex equivalence

A complex equivalence occurs when two statements are linked in such a way that they are taken to mean the same thing. For example:

   you are not looking at me when I am talking and not paying attention

The conclusion reached by the speaker may or may not be accurate; complex equivalence might be challenged by asking 'how does this mean that?'

Presuppositions

A presupposition is often an expression of an assumption about the way the world operates. Sentences containing the words 'since', 'when' and 'if' usually contain a presupposition.

when you have had as much experience as I, you'll understand why this won't work

Presuppositions can be brought into the open by asking: 'what leads you to believe that . . .?'

Cause and effect

Cause and effect is a shorthand way of describing complex relationships. In some instances, however, this can lead to over-simplifications and misunderstandings. It often involves a 'but' in the sentence.

I'd like to delegate this task, but without my involvement it wouldn't get done

Cause and effect can be raised by asking a question such as: 'how exactly does this cause that . . .?'.

Mind reading

'Mind reading' occurs when someone presumes to understand what someone else is feeling, for instance:

I know what makes them tick

Mind reading can be challenged by asking: 'how exactly do you know?'

Summary

In summary the facilitation role is, at one level, very simple, but as one becomes more engaged in its application it becomes a more complex and sophisticated process. It is, however, an increasingly important role for FM professionals.

18.4 Reflection and learning

The third phase in the change process is that of reflection and learning. In this phase the primary function may be to help the team and team leaders to formulate ways of obtaining feedback and to subsequently address the feedback. Typical questions that one might ask during this phase include:

- from whom am I getting positive feedback?
- from whom am I getting negative feedback?
- who is reacting positively or negatively to the change itself?
- who is reacting positively or negatively to the way that the change is being managed?
- what tactics might be adopted to gain movement from those who may be resistant to the change(s)?
- does the change still address the issue it was intended to address?
- what consequences does the change have beyond those intended?
- what are you learning about the relevance of the change(s)?
- do we need to revisit our original objectives?
- what are we learning about the process of leading the change?
- what short term changes are you making to the process?
- what longer terms changes to the process are you proposing?

18.5 Key leadership and communication lessons for change projects

The following list is not intended to be the definitive blueprint for dealing with communication and leadership during change projects. To try and create such a prescription would be foolish – circumstances are very varied, the culture of different organizations varies considerably and the leadership style that is required will of necessity be different. The intention is to offer a list of some of the key issues that might be relevant. Their relative importance and the way in which they are addressed will, however, differ in different contexts and circumstances. Those key issues are:

- senior management commitment is vital – actions speak louder than words
- all staff affected must be involved – consultation must be genuine
- communicate, communicate, communicate
- quality not quantity
- avoid excessive early publicity
- do not assume it has happened – feedback mechanisms need to be embedded in any communication process to provide reassurance that the message has been received and understood
- avoid 'over-organizing' – leave room for input from other participants
- go for some early victories – try and plan some small wins in the early stages
- cuddle the casualties
- anticipate and welcome resistance – be aware of the dangers of the resistance going underground
- respect the past – do not use hindsight to trash what happened in the past
- recognize that change will require additional resources
- think through the steps in the process and clarify clear objectives at each stage
- change agents – consider who will be your champions and support them
- manage expectations – surface client/stakeholder expectations and manage these
- retain a sense of perspective – think about where you will get your support – think about a coach or mentor to assist.

18.6 Conclusion

In this chapter a number of the roles that FM professionals might play in the process of leading or influencing change in their organizations have been explored. As FM becomes an increasingly strategic issue for organizations, so the role of FM professionals as agents and facilitators of change will become even more significant as they manage the critical interplay between the needs of people, processes and place in organizations, and, in so doing, add value to them.

References and bibliography

Bandler, R. and Grinder, J. (1975) *The Structure of Magic I* (Palo Alto, CA: Science and Behaviour Books).
Briner, W., Geddes, M. and Hastings, C. (1990) *Project Leadership* (Aldershot: Gower).

Darragh, J. and Campbell, A. (2001) Why corporate initiatives get stuck. *Long Range Planning*, **34**, 33–52.

Hamel, G. (2000) *Leading the Revolution* (Cambridge, MA: Harvard Business School Press).

O'Connor, J. and Seymour, J. (1993) *Introduction to NLP* (London: Thorsons).

Schwartz, R.M. (1994) *The Skilled Facilitator* (San Francisco: Jossey Bass).

Smale, G. (1996) *Mapping Change & Innovation* (London: HMSO).

19

Productivity improvement

Adrian Leaman*

Editorial comment

While productivity gain is the essence of strategic facility management (FM), measuring such gain and relating it to the resources that were used in its delivery is much more of a challenge. Performance measures are necessary if managers are to be able to compare the performance of their organization with industry best practice, justify budgets, improve service levels and add value. Yet the link between resource input and outcome is not necessarily a direct relationship nor completely manifest. Complex tools are required to identify areas where potential improvements lie, to determine necessary actions and to monitor their success.

It is often said that the bottom line for productivity is financial: return on investment, net profit, share value, and so on. However, performance is multifaceted and other issues, not always objectively measurable, come into play. The triple bottom line approach, which additionally considers social and environmental issues, is an attempt to deal with some of these complexities. The balanced scorecard approach is another popular example.

Facilities contribute to productivity in a number of ways. First, their initial design must be carefully worked out to meet both the present and future needs of the organization; this includes issues such as effective use of space, minimization of travel/ circulation pathways, flexibility of fit-out, choice of HVAC systems, anticipating change, energy efficiency and recycling ability. Secondly, their operation must be monitored and managed so that they live up to design expectations and respond to changing demands; this includes post-occupancy reviews, energy auditing, maintenance strategies, regulatory compliance, new support services and customer satisfaction. Thirdly, and most importantly, facilities must reflect modern work practices by providing support for core business activities – in this case, facilities become strategic assets, not just expenditure centres, and their deployment is more about making people more effective than anything else.

* Building Use Studies, UK

Occupancy costs are a significant and often hidden expense for business. These types of operating costs can dominate combined construction or acquisition, cleaning, energy repair and replacement expenditure by a factor of five over a 10–20 year time horizon. One of the largest categories of occupancy cost is staff salaries. Therefore it is logical to conclude that if facilities can be designed and managed so that they result in a small annual reduction to salaries, then this will make a major contribution to bottom line performance and value. Reducing salary costs may mean employing fewer people as a result of technological efficiencies or re-engineering work processes, but it can also be due to lower levels of absenteeism through provision of a healthy and safe workplace.

The measurement of productivity improvement is a contemporary change driver in FM. It offers the ability to raise the profile of the profession from the traditional operational considerations to the level of strategic decision-making. While pre-design, design and construction are important phases of a facility life cycle, and can add considerable value when done well, they are but a shadow of the value gains that can be found through the pursuit of innovative workplace strategies and productivity improvement.

19.1 Introduction

There are many sources of information on human productivity in buildings. They range from ergonomic studies of, for example, keyboard performance right through to pot-boilers for facilities managers (Oseland and Bartlett, 1999). A useful review is provided by Oseland (1999).

This chapter deals with some of the findings from studies of buildings carried out by Building Use Studies in the UK. It looks at basics first and then unravels some of the implications. We have found that in the UK productivity at work is affected by buildings by about 15% upwards or downwards. To be more exact (see Figure 19.1) our current data

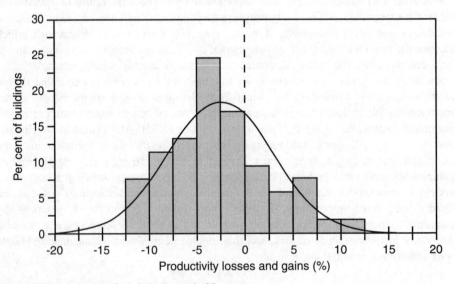

Figure 19.1 Perceived productivity in 50 UK buildings.

suggests 12.5% upwards and 17.5% downwards; in other words, in the 'best' buildings (the top 5% of the buildings we have studied) workplace productivity seems to be lifted by, at best 12.5%, and the 'worst' (the bottom 5%) it is reduced by 17.5%.

We do not use secondary sources or hearsay: we are talking only about buildings that we or our licensees have studied directly. There will be buildings out there where productivity is better and worse than the ones we have found. The normal curve superimposed on the real data in Figure 19.1 gives a rule-of-thumb prediction of where the real upper and lower limits lie.

We have covered about 200 buildings in the UK (115 of which are in our current database) and about 30 in ten other countries. About 80% of these are offices. Statistically, these are paltry samples, but buildings are expensive and (can be) obstinate to study, and the samples are large enough to convince all but the most sceptical.

19.2 Methods

We give a standard questionnaire to occupants with permanent work spaces (this includes part-time and peripatetic staff). The questionnaire has over 40 self-assessment categories covering comfort, health, satisfaction, design quality, user needs and several more; the productivity question is just one of these. In buildings with complex user profiles (such as schools or magistrates' courts) there may be variations on the basic questionnaire for different user groups (teachers, pupils, office staff, judges, prisoners, local users, and so on).

These questions (plus opportunities for comments) fit onto two A4 pages, and can be answered in 5–10 minutes (Building Use Studies, 2001). The questionnaire is kept relatively short because the longer it is, the less chance we have of managers allowing us in to study their buildings, and more questions generate so much information that it becomes costly to manage in a growing dataset.

We use 'need to know' criteria for choice of content, not 'nice to have' and so resist changing the questionnaire too much from one study to the next, because this makes it much harder to manage the benchmarks efficiently. We avoid asking unnecessary questions, as it is vital for good response that all the questions are relevant to the respondents and they see their response as useful.

We use a paper-based questionnaire rather than, say, an Internet-based survey – we could use either but prefer paper because response rates are much higher using 'traditional' hand-out-and-collect methods (we almost always get over 90% response with paper). It is much more difficult to make questionnaires appear short and concise on the Internet because of limitations on the ability to control page layout and font definition, and with hand-out-and-collect there is always a researcher on the spot to deal with queries and, sometimes, to be shown things by respondents.

Why use 'subjective' responses, not 'objective' measures? We do sometimes use physical measurements as checks (e.g., of light levels or temperature). Self-assessments from questionnaires are reliable for what we are doing (which is either feedback or diagnosis, see below). They are much cheaper, and quickly get to the heart of the matter as results can be turned round in days. As one prominent researcher put it: 'People are the most valid measuring instruments: they are just harder to calibrate!' (Raw, pers. com., undated).

Whole questionnaires can be devoted to just one topic (e.g., health or productivity); however, our studies are not normally about finding causes or testing research hypotheses on specific topics. Rather we want to find out where buildings stand and whether they have improved or deteriorated – a kind of quality control writ large. We may also want to diagnose technical or design faults. This approach is 'real-world' research to use Robson's parlance (Robson, 1993).

Building Use Studies licenses the questionnaire. A licence (which is free to supervised postgraduate students, with a fee charged to others) includes the latest version of the questionnaire plus a pre-formatted data file. Licensing helps Building Use Studies to:

- develop a relationship with the researchers carrying out the studies so that they get the benefit of our experience and we, in exchange, get their data files, from which we can extract data and anonymously add it to the benchmark dataset
- keep in touch with the most recent work so that this can be publicized on our website, or if the findings are not to be placed in the public domain, guarantee their confidentiality
- get independent criticism and quality control from researchers on the development of the methodology
- publicize the findings (without necessarily saying which buildings are involved).

Without this type of licensing arrangement we would not know who is using these techniques and the feedback loop would be broken. As the whole point is feedback, the licensing helps 'manage' the loop.

We use the last fifty buildings studied to produce performance benchmarks for UK buildings, but as yet we do not have large enough samples for national benchmark comparisons. For this dataset, however, we are able to:

- examine mean scores for buildings for each of the individual variables in the questionnaire (the benchmarks) against individual scores for the survey building(s)
- compare the changes in benchmarks to see if things are getting better or worse (e.g., are buildings getting more or less comfortable?)
- examine relationships between the variables (e.g., how closely is health and perceived productivity related?)
- examine trends among individuals across all the buildings (e.g., do people sitting next to windows report more favourable perceptions?).

Figure 19.2 gives an example of how scores for individual buildings are benchmarked with the Building Use Studies method. It shows the results for the productivity variable (vertical axis) and the benchmark dataset (horizontal axis) sorted from the lowest (left) to highest (right) and put on a percentile scale (i.e., lowest is at 1% and highest is at 99%). The study building (identified by the diamond shape) has a mean score of minus 4.5% for perceived productivity (vertical scale) which comes at the 36th percentile on the bottom scale. The benchmark is at (or close to) the 50th percentile, which in this case is at minus 2.2%. The study building (a school) thus falls in the bottom 40% of buildings in the current (2001) dataset.

All our work on productivity so far has been reported in journal and conference papers (e.g., Leaman and Bordass, 1998). We have not yet carried out an exhaustive analysis of all the data we hold on productivity as our main funding is for individual building studies.

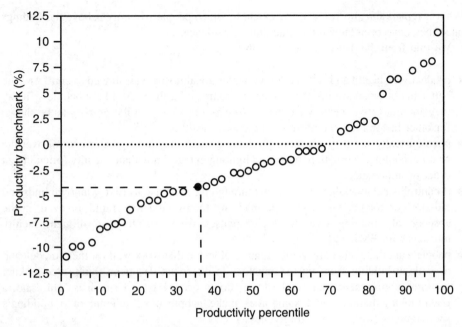

Figure 19.2 Perceived productivity benchmarking.

Although we collect data in a rigorously maintained format, retrospective analysis over (say) 50 buildings is still costly. Also we do not think it is really necessary as most of the conclusions of note can be extracted without detailed analysis.

19.3 Findings

Buildings are hard to study properly. In the jargon they are 'complex dynamic open systems', with hundreds of apparently relevant variables. Not only are they 'multivariate' – which is headache enough – but their contexts change from one building to the next. Circumstances are always different, so not only do you have the complexity of the buildings and occupants to contend with (and the interactions between them), but also site, design, procurement, ownership, history, aesthetics, and so on. It is hard for a researcher to know what to leave out because some of it may be relevant in a particular case and it is difficult to tell in advance. If the data are allowed to dominate, however, the result will be an overwhelming and incomprehensible data mountain. Statistical methods such as multivariate analysis are generally of little help – they will probably make things worse by reorganizing the complexity and repeating it in another form. Usually, we prefer to use stripped-down statistics that tell a simple story clearly, or have good 'question-answering ability' where the data give unambiguous answers to simply-defined questions.

We use a robust method not only to obtain the data relatively efficiently, but also to convince the sceptics. Over the years, we have learnt which of the multitude of variables are most likely to yield useful results in pinpointing whether occupants think buildings work well or not. Many of our findings are 'nearly obvious' – they are common sense.

What is remarkable, however, is the extent to which people will only believe the findings once they have been shown the quantitative evidence.

We find from the dataset as a whole that:

- productivity, health and satisfaction variables are almost always linked to comfort – the better the occupants think the indoor environment is, the more likely people will say they are productive, healthy and happy (see Figure 19.3 – similar graphs can be shown for other factors, e.g., comfort and perceived health)
- people usually say they are more productive when they have greater control over the heating, cooling, ventilation, noise and lighting in their immediate vicinity (often in that order of importance)
- if control is not available to occupants through physical means (like window blinds and radiator controls) it can usually be made up for by proactive, rapid, or even, in the absence of anything else, honest responses from friendly and diligent facility management (FM) staff
- people want things that are usable, manageable and that work well for them on demand or without holding them up too much – despite what designers think, nice-looking working environments tend to be lower down occupants' priority lists. This said, a good-looking building will sometimes make people more tolerant of a building's shortcomings (we call this 'forgiveness')
- simple, naturally ventilated buildings often (but not always) give better results for productivity than air-conditioned buildings, mainly because there is usually more user control, although the 'objective' conditions in them do not necessarily have to be better. The obverse is that 'over-stressed' naturally ventilated buildings (such as those that are

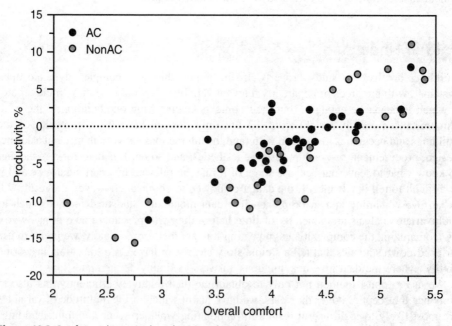

Figure 19.3 Comfort and perceived productivity relationships.

too deep in plan, too densely occupied, or with limited or idiosyncratic user control) can produce dreadful conditions, especially at the height of summer
- the more functions and activities people have to cope with, the less likely they are to say they are productive as well, so open plan often scores worse simply because the number of activities is greater and so is the potential for unmanageable conflicts, although this is not always the case
- noise is a growing problem, especially with random distractions created by activities that are perceived as irrelevant to a particular individual's requirements – this obviously is worse in open plan.

The data points shown in Figure 19.3 are those for 77 UK buildings from the Building Use Studies dataset. The relationship between Overall Comfort and Perceived Productivity is highly significant, with the correlations nearly the same for air-conditioned buildings ($=0.824$) and naturally ventilated or mixed mode ($=0.84$). It is worth noting that the very good and very poor buildings tend to be non-air conditioned, whereas the air-conditioned buildings tend be more clustered in the middle.

These generalizations can also be presented as the aspects of buildings that people prefer; many people will already know most of the answers from their own experiences of buildings. The following list is adapted from the Probe studies (Leaman and Bordass, 2001). High occupant satisfaction is easier to achieve when all or most of the following features are present in the total system (because they help 'virtuous' processes develop or give occupants better control, which ultimately improves their tolerance):

- shallower plan forms and depths of space (usually less than 15 m across the building)
- degrees of cellularization (not necessarily in single-person offices, but at least laid out so that workgroup integrity is preserved)
- thermal mass
- absence of gratuitous glazing
- stable and comfortable thermal conditions
- controlled background ventilation without unwanted air infiltration
- openable windows
- views out
- usable controls and interfaces
- a non-sedentary workforce (e.g., people are not at VDUs all day long)
- predictable occupancy patterns
- well-informed, responsive and diligent management
- places to go at break times inside or away from the building.

Published examples of buildings which meet most of these criteria with high levels of excellence are the Elizabeth Fry Building in Norwich, UK (Standeven *et al.*, 1998) and the Tax Office, Enschede, The Netherlands (The Probe Team, 2001).

The tendency for things to become unmanageable, and thus for occupants' tolerance to decline, can be made worse by some or all of the following:

- deeper plan forms with variable qualities of indoor conditions (e.g., worse towards the middle, better towards the windows)

- senior staff monopolizing the best places, and often leaving them unoccupied when others have to suffer
- areas in use for staff workstations that were not originally intended to be so (e.g., converted storage areas, basements and meeting rooms)
- large open work areas with little variety in them
- larger workgroups (of more than about six people)
- workgroups where people are not sitting within line of sight and earshot of each other, perhaps with people split between different locations
- people sitting too close to sources of noise and random distraction like entrance/exit doors, kitchens, photocopiers and touchdown areas
- people sitting with their backs to colleagues or circulation areas
- too many conflicting activities in one area (especially where people needing to concentrate are mixed in with people needing to communicate frequently)
- higher densities (thresholds differ so there is no rule of thumb)
- longer working hours
- presence of complex technology
- ineffective, absent or bossy facilities management.

It is easier to produce lists of do's and don'ts than it is to make them work in any given situation. For example, in the UK we have found productivity losses of 15% in a building with most of the features in the first (supposedly 'good') list: shallow plan, naturally ventilated, and so on. The reasons why it was so poor were poor thermal performance in summer and occupant densities that were too high for the 'carrying capacity' of the spaces, the high densities having been created by a sudden demand for extra staff spaces produced by a recruitment campaign.

We have also found buildings which, on the surface, seem candidates for poor workplace productivity (with several of the features in the second list) but which turn out to be surprisingly good. Usually the reason is that the occupiers have devoted sufficient resources to FM functions so that problems can be dealt with quickly as they arise and, importantly, the occupants can see the positive results for themselves.

The best results are usually obtained where the indoor environmental conditions are comfortable, stable and predictable and when things go wrong (not just with the ambient conditions) but with other things as well (like office equipment or furniture failures) there is a rapid and effective response system in place. This can be the result of empowered individuals using their initiative and common sense (e.g., with window and blind controls or cleaning things like spilt coffee up for themselves) or a management system which works properly. Rapid response is the key: anything that prevents this happening in practice will reduce perceived productivity.

If, however, a building is basically too hot or too cold, or both, or has some other kind of unmanageable discomfort (like noise from traffic) it is usually pointless trying to apply lipstick on the gorilla. It may, for instance, be better in the long term to sort out a poorly functioning air-handling system than try a new space plan.

19.4 A broader perspective

Looking at buildings from other points of view (apart from perceived productivity, such as comfort, health and energy efficiency) we have found that it is best to avoid:

- unmanageable complexity
- excessive technological and management dependency
- fragile systems (i.e., those that break down easily)
- tightly coupled systems (i.e., those with many interdependencies)
- situations where the occupiers cannot 'own' problems that directly affect them.

This list is a distillation of an approach to building briefing, design and procurement for which more details may be found at www.usablebuildings.co.uk. A shortened version of that approach is shown in Figure 19.4.

PROCESS before PRODUCT
PRODUCT and back to PROCESS
PASSIVE before ACTIVE
SIMPLE before COMPLICATED
BETTER before MORE
PREVENTION before CURE
80 before 20
ROBUST before FRAGILE
SELF MANAGING before MANAGED
EFFICIENT before ELABORATE
TRICKLE before BOOST
INTELLIGIBLE before INTELLIGENT
USABLE before ALIENATING
FORGIVING before DEMANDING
ASSETS before NUISANCES
RESPONSE before PROVISION
OFF before ON
EXPERIENCE before HOPE
THOUGHT before ACTION
HORSES before CARTS

Figure 19.4 Simple guidelines.

Things tend to go wrong when interactions between technology and management break down or do not work properly. Low productivity is one such effect; another is poor energy efficiency. Both are symptoms of chronic building performance problems that tend to endure and sometimes get progressively worse, as is the case with 'sick' buildings. Figure 19.5 divides buildings into four types using technological complexity and management input as the dimensions.

The most effective from an all-round performance perspective are Type A (complex but with high management inputs) and Type B (simpler and with lower or minimal management inputs). The danger zone is Type C – buildings which are relatively complex but do not have enough management resources to service the complexity. The public sector is particularly prone to Type Cs (for further discussion, see Bordass *et al.*, 2002) and they constitute the majority of the office stock in the UK. Type Ds are much less common but are often unrealistically used as exemplars. The key to productivity improvement at

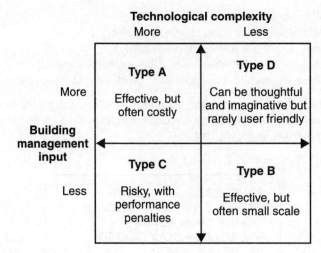

Figure 19.5 Building types.

the strategic level is to ensure that buildings are clearly placed either in the Type A box or in the Type B and to avoid producing more of the sadly ubiquitous Type Cs.

The intensification of many buildings, with more equipment, greater 'space productivity', and longer hours of operation, all requiring high levels of management and support services, ideally leads to the Type A buildings in Figure 19.5. This intensification activity can, however, lead people to think that soon all buildings will be like this.

Intensification is, however, just one part of a larger system. It is inextricably connected to its companion, diversification, as occurs, for instance, when intensified office headquarters support people who also spend some of their time in dispersed locations. Some of these locations may be intensified too (e.g., hotels) but many are simpler. If intensified buildings are just the tip of the iceberg, many uses may fit best in highly robust, adaptable, environmentally benign buildings: ideally Type Bs in our parlance.

If productivity improvement is a serious goal and not just something to which we merely pay lipservice, then it makes sense to try to create a new generation of office buildings which are simpler, smaller in scale and height, more robust in their capacity to deal with change of use, more controllable for their occupants, and more likely to have at least some natural ventilation and be more environmentally benign.

References and bibliography

Bordass, W., Leaman, A. and Cohen, R. (2002) Walking the tightrope: the Probe team's response to *BRI* commentaries. *Building Research and Information*, **30** (1) January, 62–72.

Building Use Studies (2001) www.usablebuildings.co.uk

Leaman, A and Bordass, W. (1998) Probe 15: productivity in buildings: the killer variables. *Building Services*, June, 41–3.

Leaman, A. and Bordass. W. (2001) Assessing building performance in use: the Probe Occupant Surveys and their implications. *Building Research and Information*, **29** (2) March/April, 129–43.

Oseland, N. (1999) *Environmental Factors Affecting Office Workers' Performance: A Review of Evidence*. CIBSE Technical Memorandum TM24 (CIBSE).

Oseland, N. and Bartlett, P. (1999) The bottom-line benefits of workplace productivity evaluation. *FMJ*, April.

Robson, C. (1993) *Real-world Research* (Oxford: Blackwell).

Standeven, N., Cohen, R., Bordass, W. and Leaman, A. (1998) Probe 14: Elizabeth Fry Building. *Building Services*, April, 37–41.

The Probe Team (2001) Probe 22: Enschede Tax Office. *Building Services*, October, 31–35.

20

Integrated building models

Robin Drogemuller*

Editorial comment

Rapid advances in technology have led and will continue to lead to value gains for facility operation. The management of information is vital not only as a means of making effective decisions, but also for collecting live data and using it to its best advantage. Information is power. Unfortunately the sheer volume of information makes it difficult to distinguish those aspects that are important from those that are routine. Computer software enables managers to sort, analyse and interpret data in order to understand what is happening and to underpin strategic decisions.

Computer-aided design (CAD) technology has revolutionized the way facilities are designed and managed. In particular, three-dimensional CAD models form an up-to-date database of building structure and layout, contents, equipment, personnel, maintenance and energy usage capable of describing all attributes of facility performance. It is a common practice these days to require the design team to deliver an electronic model of their work as the basis for ongoing management. Non-computerized approaches to data recording are cumbersome and unproductive.

Information management is firmly rooted in the need to communicate data quickly and reliably. The Internet has opened up new opportunities for dissemination to a range of stakeholders irrespective of their physical location. Web-based tools are being used to generate reports, co-ordinate (even view) remote facilities centrally, have meetings via videoconference and publish regular summaries of performance for consideration by senior management.

At another level, intelligent facilities can automatically monitor and react to changing environmental circumstances without constant human intervention. Building management systems that control air-conditioning and lighting systems, warn against fire, unauthorized access and equipment failure, minimize energy usage, and adjust sun-protection devices

* CSIRO, Australia

are already commonplace in new facilities. The cost of installation is offset against productivity gains throughout a facility's life cycle, and includes, but is not limited to, routine performance measurement.

Even more significant is the management of information technology (IT) resources required by facility occupants as part of daily business processes. The attraction of integrating IT solutions with other aspects of business infrastructure management has led to single point responsibility and greater budget flexibility, which in turn have more closely aligned facilities with core business goals.

The future of information management is impossible to accurately define, but it surely relies on standards and protocols for data exchange and development of systems capable of integrating diverse types of data into a consistent framework. It is likely that graphic-based models will be at the forefront of these initiatives and will underpin other applications that use them as the primary database. Value, in the context of information management, is therefore about effective use of data to provide feedback and inform future actions.

20.1 Introduction

The purpose of this chapter is to describe the development of integrated building models, the concepts underlying their development, and the impact that these models will have on the way the AEC/FM (Architecture, Engineering, Construction/Facilities Management) industry works. These are described from an industry users' perspective; more technically oriented materials are available. Eastman (1999), for example, gives a much wider and more detailed review of the whole area of building product models.

20.2 What is an integrated building model?

For the purposes of this chapter an *integrated building model* defines a representation, that can be stored in a computer, of the information needed to describe a building throughout its life cycle. A *product model* defines the physical elements in a building project. A *process model* defines the processes and possibly the resources required to construct a building. Consequently an integrated building model covers both the product and process model. An *aspect model* defines the physical products and possibly processes required for a building system, such as structural steel or heating, ventilation and air-conditioning (HVAC) systems. Aspect models are normally defined to allow the viewing and manipulation of the objects that are of interest to a particular type of user, such as architect, structural engineer or estimator. See Figure 20.1.

20.3 Why integrated building models?

There are three major reasons for creating integrated building models. The first is to better understand the various systems within buildings and the interactions between them. The second, which is a subset of the first, is to develop better computer software to support the

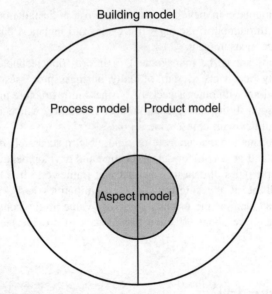

Figure 20.1 Relationships between different types of models.

design, construction, management, refurbishment and/or demolition of buildings, and to enable information exchange between computer applications. The third reason is to provide a method of archiving information about buildings in a way that maximizes the likelihood of being able to access, retrieve and utilize the information at some future time.

If we recognize that the AEC/FM industry is a specialization within the general domain of engineering then it is not surprising that other engineering-based industries have also started developing product models (of which building models are a specialization). When examining the range of models developed across the various engineering industries (e.g., aerospace, manufacturing) it becomes obvious that the AEC/FM industry has characteristics that distinguish it in various ways from other industries. Some of these are:

- buildings are complex
- buildings are large
- buildings are normally unique – the prototype is the finished product
- buildings last a long time
- buildings require continuous maintenance and periodic refurbishment
- the use of buildings changes with time.

None of these factors is unique to the AEC industry; however, it is the particular combination of these factors that makes the development of integrated building models a complex and difficult task.

The second reason mentioned above will have a significant effect on the way the industry works. The seamless exchange of information between computer applications is called interoperability. Software that supports interoperability specifications at an object level for the AEC/FM industry has only become available commercially in the last few years. Already the benefits of interoperability are being enjoyed:

- allowing information to flow between existing application while minimizing the entry of previously entered data (Figure 20.2)
- allowing analysis (e.g., thermal and structural) packages to import three-dimensional geometry directly, saving days of effort converting, for example, DXF files into the internal representation
- making new uses of software feasible, e.g., using ComCheck, an energy code checking program developed at Pacific Northwestern Laboratories.

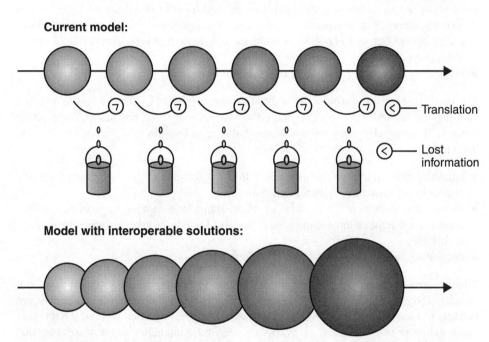

Figure 20.2 Integrated model lifecycle (IAI, 1997).

Not everyone is convinced that fully integrated building models are possible (e.g., Kim and Liebich, 2000); however, adaptations of the shared project model architecture address most, if not all, of the concerns raised by researchers. Presenting user interfaces to more highly structured models in an understandable way is a problem that the software developers will need to address. Such interfaces are already provided to relational databases so the technical implementation should be feasible.

20.4 Interoperability

The definition of interoperability is deceptively simple – the seamless exchange of information throughout the building lifecycle. Underlying this are many commercial, cultural, legal and technical issues. The technical issues, while they are not simple, are the easiest to solve. The significant technical issues that need to be addressed include:

- providing an audit trail that allows the actions of the users to be traced, e.g., who added a wall, who changed the type of construction, who inserted a door and then who shifted the door
- providing well defined aspect models to suit the various players in the AEC/FM industry – some are being defined as explained below
- controlling who can change which pieces of information, e.g., a mechanical engineer should not be able to shift a structural beam to provide a better duct layout
- allowing access to information down to a defined level of detail but no lower, e.g., a quantity surveyor (cost engineer) may allow access to cost plans to an elemental level during design but will not want to share the unit rates that were used to build up the elemental costs
- integrating construction details with the building model.

The legal issues revolve around the ability to track responsibility for changes to a shared building model and also to sign-off at the various stages of the design/construction process. There are also many cultural issues that need to be addressed. The most important include:

- retraining staff to exploit the new technologies – this has implications for power and responsibility structures within organizations
- forcing designers to resolve issues 'up front' rather than leaving them to be resolved during construction – three-dimensional models of buildings give designers little choice in this area
- integrating supply chains with design/ construction software.

The commercial issues are different for the software vendors and the software users. For instance, the major software vendors, such as computer-aided design (CAD) vendors, are unlikely to increase total sales significantly through supporting interoperability with other value adding software. Some CAD vendors already have integrated suites of software that support many (but not all) processes. The major beneficiaries here will be the developers of software that uses the CAD information and the users of this software. Fee structures will have to change to provide some incentive for designers to spend the extra time needed to develop full three-dimensional models of buildings as it is mostly the parties down stream, such as services engineers and cost planners, who gain the benefits.

The development of standard interfaces for the exchange of information assists software developers by allowing 'plug-and-play' compatibility between applications. The move from one-to-one information exchange to shared project models is illustrated in Figure 20.3. This means that software developers only have to write the code for one interface rather than for several, with consequent savings in time and effort. The industry can expect that these savings will be passed on as improved software products.

20.5 Integrated building models and aspect models

One of the assumptions underlying the development of integrated building models is that it is possible to develop a single information model that will meet the needs of all users. It will then be possible to have the full complement of AEC applications sharing

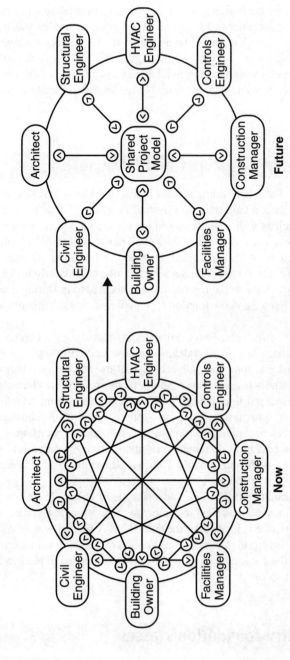

Figure 20.3 Interoperability – moving towards a shared project model (IAI, 1997).

information from the inception of a project, through design, construction and fit-out, to management of the facility, including refurbishment, and eventually to demolition and recycling (Figure 20.2).

The implications for the building life cycle are that information will not be lost at each stage of the procurement process. Ideally, all of the information that was generated during the procurement of a building should be available for the facility manager to use during the operational phase of the building's life (Figure 20.3). If all of the information were adequately maintained it would then be available as a starting point for refurbishments and the revised information would then be available for future operations at the completion of refurbishment.

20.6 Modelling concepts and languages

The key concept underlying building models is that of 'objects'. Objects within a building model normally bear a relationship to physical or abstract objects in the 'real' world. Physical objects such as walls, windows and people are modelled. Abstract concepts also need to be modelled, such as supplier, task and elemental cost. Each object has values (attributes) attached to it that are used to describe its 'state' or characteristics. For example, a wall object will have a thickness and a height, but early in the design process may not have a particular material (construction) allocated to it. During detailed design the external walls will have their construction type assigned as, say, 'aluminium curtain wall' or '230 brick'.

Objects can also have 'behaviours' allocated to them, e.g., a curtain wall will have particular thermal characteristics that determine how heat will flow through it for various internal and external environments. Objects also allow the use of inheritance. A curtain wall is a type of wall, as is a brick or concrete block wall. Some characteristics, such as U-value can be abstracted into a more general 'wall' class from which all of the sub-classes of wall can inherit the ability to store a U-value. A consequence of using inheritance is that the concept of polymorphism can be used. This means that an estimator can specify that the cost of all external walls should be aggregated into the cost for the building shell (along with other costs). This will leave the cost of internal walls to be included under another heading.

Relations are normally modelled as a special type of object. There are numerous types of relations which must be modelled in some way, e.g., a window is placed in an opening in a wall, an office space is located on the fifth floor. A special type of relationship is the composition relationship, e.g., a reinforced concrete beam is composed of reinforcing bars, ligatures and the concrete matrix, a window is composed of the head, sill, mullions, panes of glass, and so on.

20.7 Industry foundation classes

The Industry Foundation Classes (IFCs) owe a heavy legacy to STEP (Standard for the Exchange of Product Model Data – ISO 10303). While the IFCs were started as an internal initiative by Autodesk, their development became an international project in 1996 under

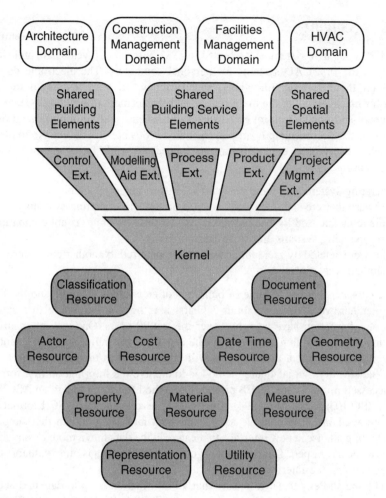

Figure 20.4 IFC release 2.0 model architecture.

the auspices of the International Alliance for Interoperability (IAI). Relevant work from within STEP has been used or adapted by the IAI in order to speed development. The IFC model also uses a layered architecture (Figure 20.4). At the bottom of the diagram is the resources layer – these are generic definitions used throughout the model. Above this is the core layer where the basic concepts are defined. The shared layer is the third layer and defines the concepts that are shared between software supporting two different processes – this is the first layer where 'real' objects are defined.

The top layer defines those objects that are normally only used within one discipline and do not need to be shared with others. For example, cabinets and shelves are defined within the Architecture Domain since information about these objects is not required by other disciplines.

20.8 Archiving

The IBM PC computer system was released in 1981, just over 20 years ago, and was soon followed by the first CAD systems for personal computers. The electronic documents prepared on these systems (drawings, specifications, etc.) are useful to the facility manager for management of the maintenance and replacement of building systems and to the designers and contractors involved in refurbishments and additions/alterations. But how many of the documents prepared in the early 1980s are still accessible in electronic form? In fact, many older documents are no longer accessible for of one or more of the following reasons:

- the operating system and computer hardware is no longer available
- the software that created them does not run on modern operating systems
- the storage media used for backups/archiving (if this was done) is not compatible with modern operating systems and software
- the files were deleted by accident or were not transferred to another computer when the old hardware was replaced.

When we consider that the lifetime of buildings often exceeds 40 years and the life time of computer hardware averages about 3 years, it is not too surprising that significant amounts of information have been lost over the last 20 years. One way of ensuring that information about a building is kept up to date is to make the building model a central part of the facility management process. This will be described in the next section.

A less satisfactory, but adequate method is to archive the information in a format that can be accessed in the future. STEP provides a method for doing this in an ASCII file as defined in ISO 10303:21 (ISO, 1994). Since these files are stored in ASCII format and the schema are available, it is possible to read the data into any program that supports the schema. Failing all else, a new interface that can read the data into a more recent computer program can be developed. This assumes, of course, that the data is more valuable than the cost of developing the interface.

An archiving strategy that will ensure that all information is up-to-date and accessible is to archive the file in ISO 10303:21 format and then apply updates to the files as new versions of the schema are developed. In combination with a strategy for updating the archive storage media this will ensure that base information is available when required. There will still be a need, however, to update information to the current actual state of the building due to refits, alterations, repairs and replacement of building elements, plant and equipment.

20.9 Computer-aided facility management (CAFM)

The management of the operation of facilities is a major cost to large organizations. The preparation of appropriate databases to support facilities management is a significant effort if the databases have to be built from scratch. As discussed above, the aim of both the STEP and IAI activities is to provide as much relevant information as possible through the design and construction processes through to the operational stages of the building. If the facility manager can obtain the 'as built' information for the building in computer interpretable form then this can provide the basis for ongoing management of the facility.

The activities of the facility manager that can be readily supported by building models, in decreasing order of frequency, are:

- room/space allocation, furniture allocation and tracking
- management of services, i.e., telecommunications, HVAC
- interior fit outs and refurbishments
- renewal of services
- renewal/refurbishment of major building components.

The use of IFC compatible software to support some of these activities is described in the next section.

20.10 Information flow through the project lifecycle

This section shows an example of the use of interoperable software based on the IFC release 2.0 specification (IAI, 1999). It is based on the BLIS demonstrations given at numerous IAI meetings and other industry forums. The software described below covers some of the available possibilities. The example diagrams used here chapter were developed and demonstrated by the BLIS software development team co-ordinated by Richard See of Microsoft (BLIS, 2001).

One of the earliest tasks of building designers is to prepare schedules of the spaces in a building in order to derive the initial spatial layouts. This process is supported by Space Layout Editor, which is implemented as Visual Basic for Application (VBA) macros within Microsoft Visio.

The designer sets out a table of spaces (Figure 20.5) which also allows for multiple spaces of the same size and type. These spaces can then be imported into Visio. Figure 20.6 shows the spaces for one floor part of the way through the layout process. A scanned image of a sketch plan can be used as a background. The storey heights can be entered into

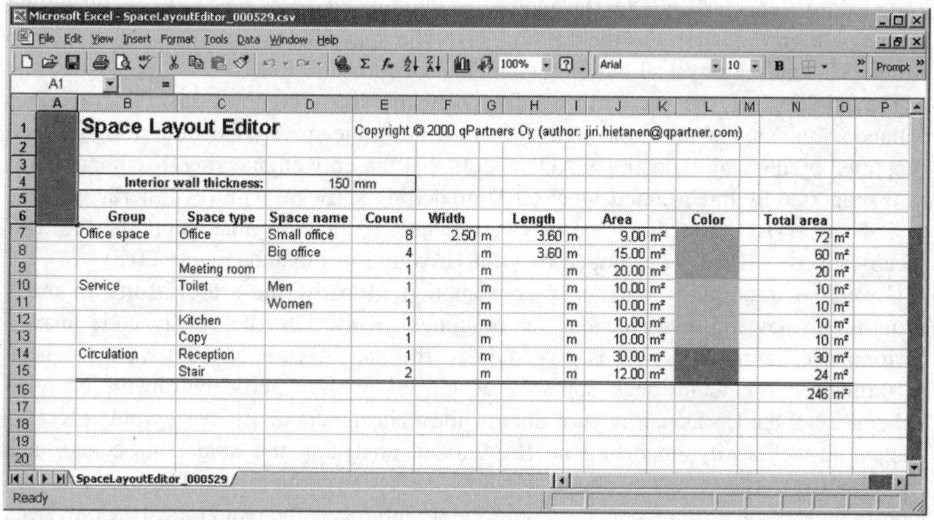

Group	Space type	Space name	Count	Width		Length		Area		Color	Total area	
Office space	Office	Small office	8	2.50	m	3.60	m	9.00	m²		72	m²
		Big office	4		m	3.60	m	15.00	m²		60	m²
	Meeting room		1		m		m	20.00	m²		20	m²
Service	Toilet	Men	1		m		m	10.00	m²		10	m²
		Women	1		m		m	10.00	m²		10	m²
	Kitchen		1		m		m	10.00	m²		10	m²
	Copy		1		m		m	10.00	m²		10	m²
Circulation	Reception		1		m		m	30.00	m²		30	m²
	Stair		2		m		m	12.00	m²		24	m²
											246	m²

Figure 20.5 Space table in Excel.

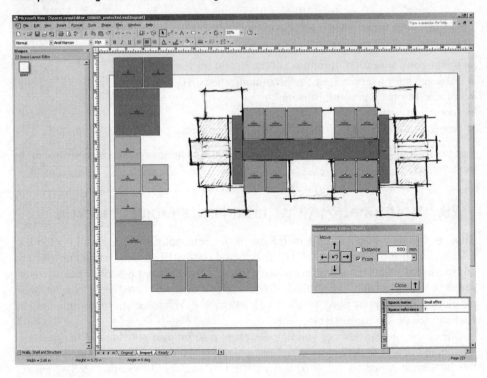

Figure 20.6 Spaces partially placed in Visio – the unplaced spaces are on the left-hand side.

Visio to create multistorey building layouts. Space Layout Editor was developed as a demonstration project and it is unlikely that it will become commercial software.

Once the spaces have been defined for all of the floors of a building they can be imported into other software for the addition of building elements such as floors, walls and openings. Figure 20.7 shows the building after the space layout has been imported into ArchiCAD and the physical building elements added to the file.

The addition of the physical building elements is a complex task that requires collaboration amongst all of the members of the design team. Architectural design is well supported by the major existing CAD vendors, but the only engineering discipline that is well supported by three-dimensional CAD modelling software with IFC interfaces is the design of HVAC systems. One HVAC design application that supports the IFC specifications is Riuska, developed by Olof Granlund in Finland (Figure 20.8).

Given the importance of energy consumption globally, there are efforts in many countries to develop energy codes for buildings. In the US, the energy code takes a performance approach which is amenable to the use of computer software to check performance. The ComCheck software developed at the Pacific Northwest National Laboratory of the US Department of Energy allows designers to achieve a rapid generate–analyse test cycle by importing an IFC model, analysing the model in seconds and reporting the variance, either above or below the code requirements. The user can then modify the design within ComCheck until the design meets the requirements. ComCheck is shown in Figure 20.9.

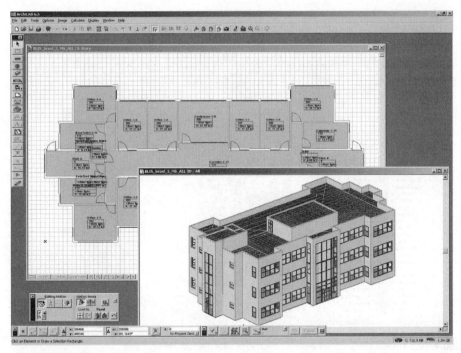

Figure 20.7 The building after addition of physical elements.

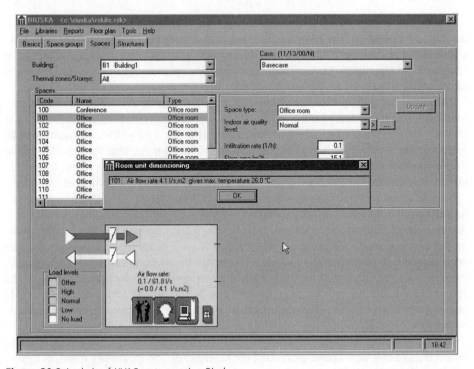

Figure 20.8 Analysis of HVAC systems using Riuska.

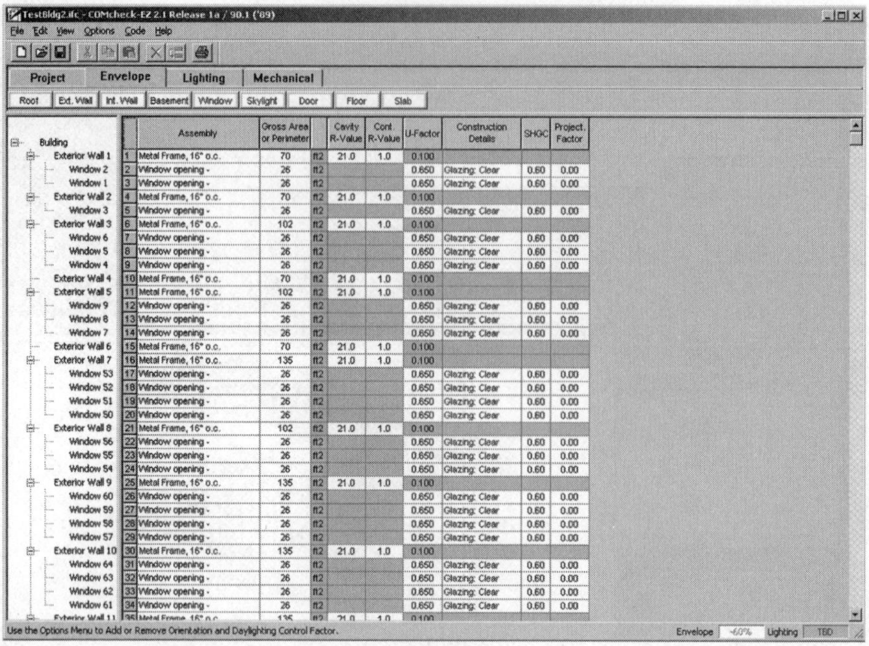

Figure 20.9 Using ComCheck to ensure compliance with the US Energy Code.

The issue of energy use involves more than just the operational energy used on a day-to-day basis in a building. The environmental impacts of embodied energy and the pollutants added to the environment during manufacture and operation are picked up in life-cycle analysis. Olof Granlund's LCA software performs life-cycle analysis on the HVAC system of a building. Analyses are performed against a range of parameters (Figure 20.10).

The taking off of quantities and material characteristics to support HVAC analysis is not much different to the taking off of quantities to support the estimating process. The next release of Timberline's PECAD will read in an IFC file and will start structuring the various costs. PECAD will also export its reports in XML so that the reports can be displayed on a web site without having to provide access to the underlying data. This provides protection for the intellectual property of the quantity surveyor/cost engineer.

The use of three-dimensional models places additional demands on the designers. The models must be accurate and fully specified to be of use to downstream analysis software. A new type of software product has been developed to check a full three-dimensional model for consistency against a specified set of rules. For example, a thermal analysis program needs to have all of the spaces, zones, walls, floors, roofs and windows specified, together with the location of the building and the orientation of the building towards the sun before reliable results can be generated. The Solibri Model Checker (Figure 20.11) allows a user to check a building model against a set of rules; additional rules can be added by the user.

Ideally, while the building is under construction, the building model will be updated to reflect the as-constructed configuration of the building, giving the facility manager a good starting point for the management of the building throughout its operating life.

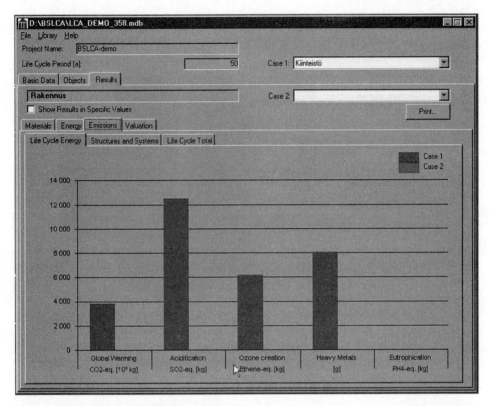

Figure 20.10 Olof Granlund's life-cycle analysis software.

A facility manager does not need all the full capabilities of a CAD system to control the day-to-day operations of the building, however, it would be useful to be able to manipulate and place some objects graphically, such as furniture, fittings and telecommunications devices. A two-dimensional interface to a three-dimensional model would provide the simplicity suited to facilities management operations while still allowing an up-to-date model of the building to be handed on to the designers of the next fit-out or upgrade. This is the target of some of the building-related extensions to Microsoft's Visio product (Figure 20.12). The building elements can be imported through the IFC interface. Furniture and equipment can be allocated to rooms and the details of the person to whom the equipment is assigned, as well as the service history can be accessed and maintained through the object browsing window.

An additional advantage of maintaining the furniture and equipment in this manner is that schedules can be generated directly from the up-to-date information. The visual interface provides a simple means of checking that all of the equipment is allocated appropriately from the spatial perspective. For example, it will be immediately obvious if an office does not contain a desk, while another office contains two desks. Q Partners have developed VBA macros in Excel that generate schedules of the furniture and equipment (Figure 20.13).

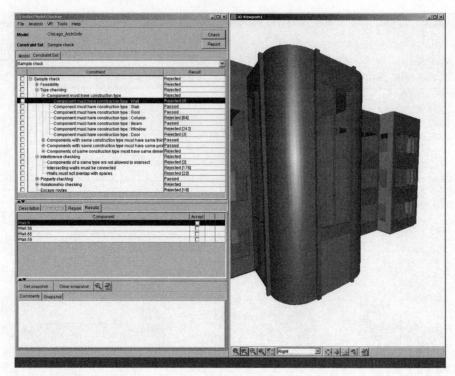

Figure 20.11 Solibri Model Checker during a conformance run.

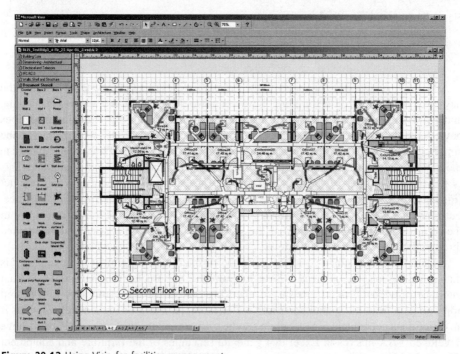

Figure 20.12 Using Visio for facilities management.

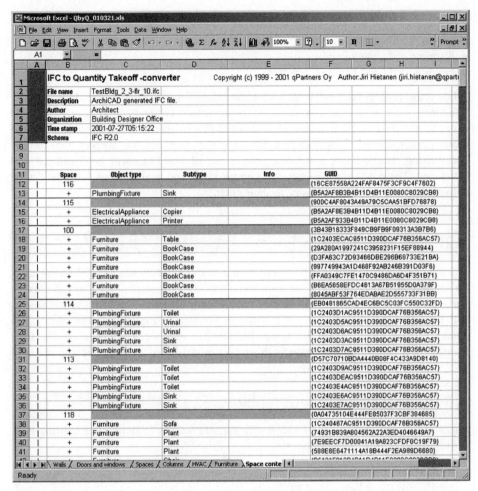

Figure 20.13 Generating a furniture/fitting schedule within Excel with VBA macros.

While the sequence of operations and the simplicity of maintaining the flows of information through the various stages of building design are impressive, it must also be pointed out that there are areas of building design and construction which are not well served by either the IFC models or software to support the models. The disciplines that are not well supported by the IFC models include:

- structural engineering
- electrical engineering
- hydraulic engineering
- mechanical transportation (lifts, escalators)
- site works and landscaping
- linking construction details to three-dimensional models
- some aspects of checking for conformance against codes and standards.

Disciplines that are supported by the IFC models but for which there is little or no IFC compatible software include:

- planning and scheduling
- building code and fire code checking.

While these gaps in the models and software are significant, work is already under way to address these areas, and compatible software should become commercially available over the next few years. It should also be noted that as shared building models become more pervasive, new opportunities for software products to support the design, construction, management and demolition/re-use of buildings will be identified and developed. These will also have an effect on the range and quality of services that can be delivered by the AEC/FM industry to their clients.

20.11 Conclusion

The long held dream of exchanging information seamlessly between AEC/FM software is finally here for some classes of software. How quickly these technologies spread through our industry will depend on how quickly people start using this software in real projects – if there is rapid uptake, then other software vendors will quickly follow suit. If the uptake is slow, then it will be more difficult for software vendors to justify their investment in adapting their products to support model-based exchange of information.

References and bibliography

Augenbroe, G. (1995) *Computer Models for the Building Industry in Europe*, COMBINE-2 Final Report: CEC-JOULE Report, Brussels.

BLIS (2001) *Building Lifecycle Interoperable Software.* www.blis-project.org

Eastman, C.M. (1999) *Building Product Models: Computer Environments Supporting Design and Construction* (Boca Raton, FL: CRC Press).

IAI (1997) *Industry Foundation Classes*, release 1.5. CDROM (International Alliance for Interoperability).

IAI (1999) *Industry Foundation Classes*, release 2.0 (International Alliance for Interoperability).

ISO (1994) *ISO 10303:21 Industrial Automation Systems – Product data representation and exchange, Part 21: Clear text encoding of the exchange structure* (Geneva: ISO/IEC).

Kim, I. and Liebich, T. (2000) An integrated design environment for building data representation and management. Presented at the *International Conference on Construction Information Technology*, Hong Kong, January.

Nell, J. (2001) *STEP on a Page.* www.mel.nist.gov/sc5/soap/

21

Sustainability and environmental assessment

Craig Langston* and Grace Ding†

Editorial comment

Taking a whole-of-life approach to the management of facilities is not only an obvious strategy, but an increasingly necessary one. Built facilities have life cycles in excess of several decades, and in some cases, several centuries. It is no longer plausible to ignore future implications of present decisions and yet there is still plenty of evidence that this practice continues. If market forces do not dictate different actions then surely legislative controls will be introduced to ensure that society's resources are used more wisely.

Sustainable development is often defined as development that ensures that future generations have the same opportunities and amenities as those enjoyed by the present. There is an underpinning philosophy of properly valuing environmental resources, taking a long-term view of decisions and enabling equitable access to benefits. However, sustainable development is an ideal; it reflects a journey of prudent decisions and judgements rather than a goal that can ever be fully realized. Nevertheless, sustainability and environmental assessment is a critical change driver for the facility management (FM) discipline worldwide. It offers enormous potential for delivering value to society in general and to organizations in particular through the efficient use of resources (inputs) without reducing the quality of services (outputs).

While worker productivity can be described as being at the heart of human resource management, sustainability is at the heart of FM. But as the boundaries of these two critical support areas become blurred, new opportunities arise that integrate productivity and sustainability ideals. This is an important outcome – the synergy generated is capable of delivering benefits greater than the sum of the component parts. For example, the sustainable introduction of a naturally ventilated workplace can reduce illnesses from

* Deakin University, Geelong, Australia
† University of Technology Sydney, Australia

airborne diseases, and therefore increase workforce capacity and productivity. The natural ventilation decision and the workplace ecology outcome combine to provide financial reward to the organization in both direct and indirect ways.

The measurement of sustainability is not a straightforward process. These days it is usually undertaken by employing a multicriteria approach that combines a range of performance criteria into a single indicator. There are commercially available evaluation tools that can be used to assist, most notably perhaps the Building Research Establishment's BREEAM system for particular facility types. These kinds of evaluations, some performed during design and others after handover, are likely to be increasingly required. The results can be compared with benchmarks determined from case studies of best practice to further inform and provide guidance.

One of the key tools of the facility manager is life-cost evaluation. Previous reliance on initial costs and short-term benefits has led to suboptimization of assets and a potential burden for future generations that may prove unsustainable. A life-cost approach requires consideration of long-term issues and therefore changes the balance of decisions more in favour of future generations. The technique is invaluable as a means of introducing objectivity into strategic evaluation and should be employed for all significant new initiatives.

21.1 Introduction

The construction industry is one of the largest exploiters of both renewable and non-renewable natural resources, and as a result the impact of building activities on human and environmental health is a major concern of people from all design and construction related disciplines, both locally and internationally. If the construction industry continues to overuse natural resources, a limit on economic growth will eventually be reached and the destruction of the environment will inevitably rebound on the construction industry.

Building activities transform arable land into physical assets such as buildings, roads, dams and other civil engineering projects, and arable land is also lost as a result of quarrying and mining for raw materials used in construction This loss of agricultural land occurs mainly in coastal areas where soil fertility is most suitable for crops. Depletion of forests for the supply of timber products and the provision of energy for building material manufacture also contribute to the loss, while deforestation and the burning of fossil fuels contribute directly to problems of global warming and air pollution.

People are, however, becoming more concerned about minimizing the detrimental effects that the built environment has on the natural world. In view of the environmental impact caused by construction, 'green' building design, designing for recycling and eco-labelling of building materials have captured the attention of building professionals across the world, and building performance has become a major concern of professionals in the building industry, particularly for those involved in FM. Environmental building performance assessment has emerged as one of the main agenda items in the context of sustainable construction. In Northern Europe, people spend almost 90% of their time indoors (Yates and Baldwin, 1994), and consequently building design and construction plays an important role in achieving comfort and wellbeing in people's daily life.

The definition of building performance varies according to the different interests of the parties involved in building development (Cole, 1998). For instance, a building owner may wish his/her building to perform well from a financial point of view, while for the occupants, indoor air quality, health and safety issues, and thermal comfort will be the main performance concerns. An ideal environmental building assessment method should, therefore, address all the concerns of the various parties involved in a development.

Building performance assessment methods are one of the emerging areas of building research and development, and a change driver for FM. Both building professionals and the general public are becoming more concerned about the way that physical assets contribute to environmental concerns, and with finding ways to satisfy often opposing objectives, such as maximizing investment return and functional performance while minimizing energy usage and loss of habitat and other environmental impacts.

21.2 Building environmental assessment methods

Building environmental assessment has been an important concern for building designers, facility managers and occupants for some time. Separate indicators or benchmarks based on single criteria have been well developed to monitor building performance in terms of air quality, indoor comfort, and so forth – however, comprehensive assessment tools are essential to provide a thorough evaluation of building performance against a broader spectrum of environmental criteria.

The release of the Building Research Establishment Environmental Assessment Method (BREEAM) marked the beginning of comprehensive building performance assessment, and it is the first, and most widely used, building environmental assessment tool in the world (Larsson, 1998; Crawley and Aho, 1999). It was developed by the BRE, in collaboration with private developers, in the UK in 1990. It was launched as a credit award system for new office buildings and provides performance labels suitable for marketing purposes. A certificate of the assessment result is awarded to the building owner based on a single rating scheme of fair, good, very good or excellent.

The purpose of the system is to set a list of environmental criteria against which building performances are checked and evaluated. Testing can be carried out during the initial stages of a project, with the results of the investigation providing feedback to the design development stage of building, with changes being made to satisfy pre-design criteria (Johnson, 1993).

Since 1990 the BREEAM system has been constantly updated and extended to include assessment of such buildings as existing offices, supermarkets, new homes and light industrial buildings. The popularity of this system lies in its success in alerting building owners and building professionals to the importance of environmental issues in building construction.

BREEAM is not, however, a full life-cycle analysis method for buildings (Curwell, 1996), and its other weakness, perhaps true for most current building performance assessment tools, is that it is applied only on a voluntary basis. Its successful use depends largely on the recognition and interest accorded it by property developers. Furthermore, buildings can achieve high scores despite scoring poorly in a few key areas, and thus sustainability is viewed as an average of performance rather than the satisfaction of a series of important criteria (Curwell, 1996).

The influence of the BREEAM system has been felt worldwide: Canada, Australia and Hong Kong have each developed their own building environmental assessment methods, based largely on the BREEAM methodology. In Canada, the Building Environmental Performance Assessment Criteria (BEPAC) method was developed and launched by the University of British Columbia in 1993. BEPAC is a more detailed and comprehensive assessment method, which includes a set of criteria spanning global, local and indoor environments (Cole, 1994; Larsson, 1998).

BEPAC evaluates the environmental performance of new and existing buildings and is divided into five main areas:

- ozone layer protection
- environmental impacts of energy use
- indoor environmental quality
- resource conservation
- site and transportation.

It uses a point system (from 0 to 10 points) and places a weighting on the points to reflect significance and priority relative to other criteria within the same topic area. The influence of BEPAC is not as widespread as that of BREEAM and it is mainly used locally.

In the US, Leadership in Energy and Environmental Design (LEED) was developed by the US Green Building Council in 1999 for the US Department of Energy (Energy Efficiency and Renewable Energy, Office of Building Technology, State and Community Program). It is also a voluntary and market-based assessment method that is intended to define and rate 'green' buildings (Crawley and Aho, 1999).

LEED evaluates building performance based on a 'whole building' perspective over a building's life cycle. It has been developed as a self-assessing system designed for rating new and existing commercial, institutional and high-rise residential buildings. It also uses a credit award system at different levels, and points are allocated in accordance with the credit earned for each criterion. It is divided into six credit areas:

- sustainable sites
- water efficiency
- energy and atmosphere
- materials and resources
- indoor environmental quality
- innovation and design process.

Projects are awarded one of four different levels ranging from the lowest level, titled 'certified' (26–32 points), through 'silver' (33–38 points) and 'gold' (39–51 points) to the highest, 'platinum' (52+ points out of a possible 69).

The development of assessment methodology has now moved towards an international collaborative effort aimed at developing a building environmental assessment tool for international purposes. An important aim of the Green Building Challenge (GBC), for example, was the development of a new and common assessment method for evaluating green buildings throughout the world (Crawley and Aho, 1999; Cole, 1999b). An international consortium of 14 countries was involved in developing and testing the method, which covers a wide range of performance issues related to resource consumption, ecological loadings, indoor environmental quality, longevity, adaptability, impact on surroundings, and design and construction process (Crawley and Aho, 1999; Cole, 1999b).

The GBC was also developed as a second-generation green design guideline and as a second-generation tool for building eco-labelling (Larsson, 1998). The overall objectives of the GBC include the development an internationally accepted generic assessment framework, the expansion of the scope of the GBC assessment framework from 'green' buildings to include 'environmental sustainability' and the facilitation of international comparisons of building environmental performance (Cole, 1999a; Kohler, 1999). In order to support the operation of the GBC assessment framework a software tool, the Green Building Tool (GBTool), was developed by Natural Resources Canada on behalf of the GBC group to assess 34 case-study buildings within the GBC process (Cole, 1999a). The objectives of GBTool are to establish international benchmarks for building performance and to offer direction to participating countries in the development of regionally sensitive assessment models (Kohler, 1999).

Other assessment tools also exist. Some are very specialized, concentrating just on energy performance (e.g., the Australian Building Greenhouse Rating Scheme – SEDA, 2002) and apply only to existing buildings, while others are broader and can apply to design development activities as well as to completed buildings. The most broadly based tools have been developed to support full life-cycle analysis of all environmental inputs and outputs, but in most cases these tools are highly complex and time consuming to use in practice. As a result they reside in the hands of a few specialists and are not likely to make a significant difference to the realization of global sustainable development goals.

21.3 The role of building environmental assessment methods in FM

As the problems of natural resource depletion and global environmental degradation have become apparent, building performance has found its way onto the public agenda. Most building evaluation methods focus on narrow criteria such as energy use, indoor comfort and air quality to indicate the overall performance of a building. As environmental issues become more manifest, more comprehensive building assessment methods are required to assess building performance across a broader range of environmental considerations.

These methods reflect the significance of the concept of sustainability in the context of building design and subsequent construction work on site. Designers aim to improve the overall performance of buildings in relation to their effects on both the natural and man-made environments. As such, the primary role of a building environmental assessment method is to provide a comprehensive assessment of the environmental characteristics of a building. It helps to define the direction of environmental progress and to measure the degree of progress. The development of a building environmental assessment method lays down the fundamental direction for the building and FM industries to move towards environmental protection and achieving the sustainability goal.

Assessment methods act as a link, bridging the gap between environmental goals and strategies and building performance during the design and occupancy stages. They comprise a set of environmental criteria that are relevant to buildings, organized and prioritized to reflect desired performance. They address environmental issues at three levels: global, local and indoor. They also include a set of standard guidelines by which individual buildings are assessed and evaluated. The assessments are prepared in such a way as to provide a methodological framework to assess building performance in a broader context of decision making whereby environmental issues take a more significant role in the process.

Environmental building assessment methods serve three broad purposes (Cole, 1998) by:

- providing a common and verifiable set of criteria and targets for building owners and building designers through which they can demonstrate how they are achieving higher environmental standards
- providing direction and guidance for making informed design decisions at all stages during design development leading to the formulation of effective environmental design strategies
- providing a means of structuring environmental information and an objective assessment of a building's impact on the environment, and measuring progress towards sustainability.

Building assessment methods do not just provide a methodological framework for assessing building performance but also collect useful information to form guidelines for remedial work in order to satisfy the pre-design criteria. The collected data can also be used as feedback information for future planning of projects with similar design or offering the same level of service and amenity. The accumulated knowledge and expertise of environmental building design contributes to the greater consideration of environmental issues within the decision making process, thus minimizing the environmental impacts of a building development in the long term.

Building assessment methods also enhance the environmental awareness of building practices, highlighting concern about the design and construction of more environmentally oriented projects. Crawley and Aho (1999) state that environmental assessment methods might provide a means for incorporating more holistic environmental performance requirements into national building regulations, and so reduce the environmental impact of new construction.

21.4 A critique of the building environmental assessment methods

Environmental building assessment methods do contribute significantly to the understanding of the relationship between buildings and the environment (Cole, 1998). However, we are still far from fully understanding the interaction between building construction and the natural world. The tools developed to date have limitations that may restrict their usefulness and effectiveness in assessing the environmental performance of buildings.

21.4.1 Building assessment methods used as design and evaluation tools

Building assessment is of most use during design development – if the design does not meet any of the pre-determined criteria when assessed, then appropriate design changes can be incorporated into the final design solution so that all criteria are satisfied. This enables environmental issues to be integrated with the design process so that environmental damage will be minimized. Even though they were not originally designed to serve as design guidelines, it seems that they are increasingly being used as such. Building design and building assessment guidelines should be kept separate, as design guidelines need to be more detailed and assessment guidelines more operationally focused.

Some environmental building assessment methods may be used to assess existing buildings (e.g., BREEAM 4/93: An Environmental Assessment for Existing Office Designs); however, their usefulness in this situation is in doubt as the remedial work required to enable a completed building to comply with the environmental criteria may be too extensive, too costly and inconvenient. On occasions, remedial work to existing buildings may be very difficult to implement, e.g., the replacement of an existing ventilation system with a more environmentally friendly system or the installation of more windows to allow for natural ventilation.

21.4.2 Project selection

Environmental building assessment tools are less useful for project selection as they are used to evaluate building design against a set of pre-determined environmental criteria. Environmental issues are generally only considered at the design stage of projects where the effects of different development options are taken into consideration.

Environmental matters should be considered as early as possible, as if they are not dealt with before and during the appraisal stage of a project, later alterations to the brief will cost money and cause annoyance (Lowton, 1997). Sustainability should be considered as early as possible in the selection phase in order to minimize environmental damage, maximize the efficiency of natural resource use and reduce remedial cost. Current environmental assessment methods are designed to evaluate building projects at the design stage to provide an indication of the environmental performance of buildings.

21.4.3 Financial aspects

The main issues covered by environmental assessment tools are resource consumption (where resources include energy, land, water and materials), environmental loading, indoor comfort and longevity; financial matters do not form a part of the evaluation framework. This may contradict the ultimate principle of development, however, so financial return must be seen as an important aspect, worthy of consideration. A project may be environmentally sound but very expensive to build; this will affect the primary aim of a development, which is to have an economic return and thus it will become less attractive to developers even though it may be environmentally friendly. Consequently, environmental issues and financial considerations should go hand in hand as part of the evaluation framework in decision-making. This is particularly important when the decision process starts at the feasibility stage, where alternative options for a development are assessed. Environmental and financial aspects should then be considered simultaneously.

Life-cost methodologies do focus on financial implications but seldom link to environmental assessment tools, and therefore environmental considerations are often treated as a separate exercise. While the procedure for calculating life costs is well understood, its place in the sustainable development debate is highly controversial.

21.4.4 Regional variations

Environmental building assessment methods have not generally been developed to allow for national or regional variations. To a certain extent, the development of weighting systems can offer opportunities to revise an assessment scale in accordance with local variations and so reflect the regional importance of various criteria, however, regional, social and cultural variations are complex in nature and the boundaries are difficult to define. BREEAM, for example, has been adopted outside of the UK, where it was originally developed, and adjustments have been made that customize the system to accommodate cultural, environmental, social and economic differences. It is unlikely that any uniform set of environmental criteria can be devised that can be used worldwide without adjustment.

The prime objective of the Green Building Challenge was to overcome the shortcomings of existing environmental assessment tools. GBTool has therefore been developed so that it embraces the areas that have been either ignored or poorly defined in existing environmental building assessment methods. GBTool, however, suffers from other shortcomings: Crawley and Aho (1999, p. 305) point out that 'one of the weaknesses of the GBTool is that individual country teams established scoring weights subjectively when evaluating their buildings'. They go on to say that 'most users found the GBTool difficult to use because of the complexity of the framework'. GBTool is the first international environmental building assessment method and it will be unlikely that it will be used without adjustments that allow for national or regional variations. Curwell *et al.* (1999) believe that the GBTool approach has produced a very large and complex system causing difficulties and frustration for over-stretched assessors.

21.4.5 Complexity

Environmental issues cover a broad area and are difficult to capture in a set of criteria, and consequently environmental building assessment methods tend to be as comprehensive as possible, e.g., BEPAC comprises 30 criteria, while GBTool has 181 (Cole, 1998). Such a comprehensive system requires the assembly and analysis of large quantities of detailed information. To make them more practical, most methods tend to move towards generalization in order to capture most environmental criteria within their evaluation framework; however, this may jeopardize their usefulness in providing a clear direction for decision-making and assessment can become cumbersome. Striking a balance between completeness of coverage and simplicity of use will be a key concern in the further development of environmental building assessment methods.

21.4.6 Evaluation of quantitative and qualitative data

Assessment systems accommodate both quantitative and qualitative performance criteria. Quantitative criteria include annual energy use, water consumption and greenhouse gas emissions, whereas qualitative criteria include the ecological value of a site, local wind effects, and so on. Quantitative criteria can be readily measured, based on the total quantities of inputs and outputs, and points awarded accordingly. For example, in BREEAM, 8 credit points are given for CO_2 emissions between 160 and 140 kg/m^2 per year and more points are awarded if CO_2 emissions are reduced further (BREEAM Offices, 1998). Environmental criteria are, however, largely qualitative and cannot be easily measured and evaluated within the existing environmental assessment framework. They can only be evaluated on a 'feature-specific' basis, where points are awarded for the presence or absence of desirable features (Cole, 1998). This may largely undermine the importance of environmental issues within the decision-making process. The accurate assessment of environmental issues requires a more complex yet operational framework so that they can be handled in a proper manner.

21.4.7 The weighting system

Few, if any, existing assessment methods incorporate a system for weighting criteria. Instead, the overall performance score is obtained by a simple aggregation of all the points awarded to each criterion. All criteria are assumed to be of equal importance and there is no concern for the significance of one criterion relative to another. The main concern here is the absence of an agreed theoretical and non-subjective basis for deriving weighting factors (Cole, 1998) and the derivation of any weighting system is very much dependent on an in-depth understanding of the environmental impact of buildings. The relative importance of performance criteria is an integral part of the decision process as it reflects both intent and satisfaction. Development decisions will differ as a result of public sector and private sector viewpoints. The weighting of environmental criteria needs to be derived on a project-by-project basis, but the absence of any readily used methodological framework has hampered the usefulness and effectiveness of all existing environmental assessment methods in the achievement of sustainability goals.

21.4.8 Measurement scales

The way in which buildings or designs 'score' against the various criteria addressed varies markedly between different assessment methods. Generally the assessment is based on a point award system and the total score obtained for the evaluation reflects the performance of a building in achieving sustainable goals in the industry. However, there is no clear logical or common basis for the way in which the maximum number of points is awarded to each criterion. The use of consistent measurement scales would, however, facilitate comparison of assessment results across countries. Benchmarking baseline performance for assessment is also difficult to assess accurately using existing assessment tools.

21.5 A way forward

The improvement in the performance of buildings with regard to the natural world will certainly encourage greater environmental responsibility within the building industry and place greater worth on the welfare of future generations. There is no doubt that environmental assessment methods can contribute significantly to achieving sustainability goals within the building industry as they provide a methodological framework to measure and monitor environmental performance of buildings and alert building professionals to the importance of sustainable development in the building process.

However, the limitations of existing assessment methods are clearly hindrances to their effectiveness and usefulness. There is a requirement for greater communication, interaction and recognition between members of the design team and those responsible for FM to promote the popularity of building assessment methods. The inflexibility, complexity and lack of consideration of the weighting system are still the major obstacles to the efficacy of existing environmental assessment methods.

Ding and Langston (2002) developed a unique solution to the measurement of sustainability that has the advantage of relative simplicity and the inclusion of life-cost calculation. The model determines a Sustainability Index and can be used not only to compare options for a given problem but also to benchmark projects against each other. The model applies to both new design and refurbishment situations, and can be used to measure facility performance.

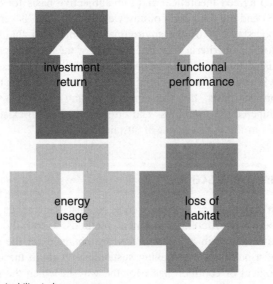

Figure 21.1 The Sustainability Index concept.

The Sustainability Index has four main criteria (see Figure 21.1):

- *Maximize wealth.* Profitability is considered as part of the sustainability equation; the objective is to maximize investment return. Investment return is measured as benefit–cost ratio (BCR) and therefore includes all aspects of maintenance and durability.
- *Maximize utility.* Functional performance, including social benefit, is another clear imperative. Designers, constructors and users all want to maximize utility. Utility can relate to wider community goals and can be measured using a weighted score.
- *Minimize resource usage.* Resources include all inputs over the full life cycle, and can be expressed in terms of energy (embodied and operational). When viewed simplistically, resource usage needs to be minimized as much as possible. Energy usage can be measured as annualized GJ/m^2.
- *Minimize impact.* As loss of habitat encompasses all environmental and heritage issues, the aim is to minimize impact. Assessment scorecards provide a useful method for quantifying impact which can be expressed as a risk probability factor.

These criteria can be combined to assess the performance of new projects and changes to existing facilities using a multicriteria approach. This provides a design tool that can be used to predict the extent to which sustainability ideals are realized, and also as an aid in ongoing FM. Criteria can be individually weighted to reflect particular client motives.

Value for money is defined as the ratio of wealth output to resource input and is investor-centred. The higher the ratio, the more attractive the proposal. These criteria are expected to be inversely proportional to some extent. Quality of life is more community-centred; it can be measured as the ratio of utility to impact, and includes externalities. High ratios are preferred and an inverse relationship is again expected.

When all four criteria are combined, an indexing algorithm is created that can rank projects and facilities on their contribution to sustainability. The algorithm is termed the 'Sustainability Index'. Each criterion is measured in different units, reflecting an appropriately matched methodology. Criteria can be weighted either individually or in groups to give preference to investor-centred or community-centred attitudes. Each criterion is measured and combined to give an index score. The higher the index, the more sustainable the outcome.

The Sustainability Index may be calculated using the following formula:

$$\frac{\text{investment return} \times \text{functional performance}}{\text{energy usage} \times \text{loss of habitat}}$$

While this is fine for comparative ranking, it gives no indication of absolute performance. A benchmarking approach set around a notional base can solve this – if each of the criteria is expressed against a base of 100, different projects/facilities can be compared and evaluated. The Sustainability Index formula can now be modified as follows:

$$\frac{\text{investment return}}{\text{energy usage}} + \frac{\text{functional performance}}{\text{loss of habitat}}$$

A value of 1 is considered to be the notional performance benchmark.

Investment return is calculated as:

$$\frac{\text{tangible discounted benefits} \times 100}{\text{tangible discounted costs}}$$

Benefits and costs are measured over the economic life (minimum 30 years) and benefits are to include all sources of funding. A benefit–cost ratio of 1 is translated into a score of 100, but a value higher than 100 is obviously preferred. This is essentially a life-cost calculation.

A weighted scoring matrix is used to assess functional performance. Any number of non-monetary criteria are scored (1–5), weighted (1–10) and summed. Functional performance is calculated as:

$$\frac{\text{weighted score} \times 100}{\text{maximum score}}$$

The benchmark is set at 50 (i.e., satisfactory).

Energy usage is benchmarked against notional/legislated maximum energy limits using the following formula:

$$\frac{\text{total annualized energy} \times 100}{\text{maximum energy allowed}}$$

A benchmark of 100 is used. Where no maximum limit is known, a reasonable target is used for comparison.

Loss of habitat is assessed across a maximum of five areas, namely design, manufacture, construction, disposal and site context. An assessment scale is used reflecting minimal (20%), moderate (40%), significant (60%), extensive (80%) or unacceptable (100%) risk probability. The result of each relevant area is then averaged to give an overall risk assessment, with a benchmark of 50.

Criterion weighting can be introduced, but if not, then it is assumed that all criteria are of equal weight. It is allowable for value for money criteria to be weighted differently to quality of life criteria according to project motivations. Only three weighting choices are recommended (25:75, 50:50 and 75:25). Therefore all four criteria play a part in all cases.

Following are some examples of how the Sustainability Index works. All examples are hypothetical:

Example 1

This example assumes a public project (such as a hospital or school). It indicates that the project has above average performance in functionality (>50) and environmental impact (<50).

$$\frac{100 \times 0.25}{100} + \frac{80 \times 0.75}{40} = 1.75$$

Example 2

This example assumes an office tower or similar commercial development. It indicates that the project has above average performance in investment return (>100) and functionality (>50), with poorer environmental impact (>50).

$$\frac{500 \times 0.75}{100} + \frac{60 \times 0.25}{60} = 4.00$$

Example 3

This example assumes an eco-development (such as an island tourist resort). It indicates that the project has above average performance in functionality (>50), energy usage (<50) and environmental impact (<50).

$$\frac{100 \times 0.50}{20} + \frac{80 \times 0.50}{20} = 4.50$$

Example 4

This example assumes a new road project (such as a tollway or freeway). It indicates that the project has above average performance in functionality (>50) and poorer performance in environmental impact (>50).

$$\frac{100 \times 0.25}{100} + \frac{100 \times 0.75}{100} = 1.00$$

Example 5

This example assumes a chemical factory or other industrial development. It indicates that the project has above average performance in investment return (>100) and poorer performance in functionality (<50) and environmental impact (>50).

$$\frac{200 \times 0.75}{100} + \frac{20 \times 0.25}{100} = 1.55$$

This new model for measuring sustainability can be used as an absolute accept/reject tool (an index < 1 indicates that the project should not go ahead) that replaces traditional net present value (NPV) models. Its advantage is that it uses a multicriteria approach that measures performance in units best suited to its quantification. It can rank and benchmark projects and facilities in different sectors.

21.6 Conclusion

Ecologically sustainable development is a major concern, and embodies both environmental protection and management. The concept of sustainable development is broad: generally, it concerns attitudes and judgement to help ensure long-term ecological, social and economic growth in society. Applied to project development, it involves the efficient allocation of resources, minimum energy consumption, low embodied energy intensity in building materials, reuse and recycling, and other mechanisms to achieve effective and efficient short- and long-term use of natural resources. Traditional project appraisal does not adequately and readily consider environmental effects in a single tool, and therefore does not assist in the overall assessment of sustainable development.

Buildings have a long life so any improvement in the appraisal techniques used for choosing the best option among alternatives will significantly reduce their future environmental impact and a methodology that embraces various criteria in relation to project development is crucial. The development of a Sustainability Index is a way to address multiple criteria in relation to project decision-making. Use of a Sustainability Index will greatly simplify the measurement of sustainable development, and thereby make a positive contribution to the identification of optimum design solutions and facility operation.

References and bibliography

BREEAM Offices (1998) http://products.bre.co.uk/breeam/default.html
Cole, R.J. (1994) Assessing the environmental performance of office buildings. In: *Proceedings CIB Congress*, Watford, UK.
Cole, R.J. (1998) Emerging trends in building environmental assessment methods. *Building Research & Information*, **26** (1), 3–16.
Cole, R.J. (1999a) Building environmental assessment methods: clarifying intentions. *Building Research & Information*, **27** (4/5), 230–46.
Cole, R.J. (1999b) *Changes to GBC Framework & GBTool*. Report submitted to Buildings Group/ CETC (Natural Resources Canada).
Cooper, I. (1999) Which focus for building assessment methods – environmental performance or sustainability? *Building Research & Information*, **27** (4/5), 321–31.
Crawley, D. and Aho, I. (1999) Building environmental assessment methods: application and development trends. *Building Research & Information*, **27** (4/5), 300–8.
Curwell, S. (1996) Specifying for greener buildings. *The Architect's Journal*, January, 38–40.
Curwell, S., Yates, A., Howard, N., Bordass, B. and Doggart, J. (1999) The Green Building Challenge in the UK. *Building Research & Information*, **27** (4/5), 286–93.
Deelstra, T. (1995) The European Sustainability Index Project. In: Trzyna, T.C. (ed.) *A Sustainable World: Defining and Measuring Sustainable Development* (London: Earthscan).
Ding, G.K.C. and Langston, C. (2002) A methodology for assessing the sustainability of construction projects and facilities. In: *Proceedings of Environmental & Economic Sustainability Cost Engineering Down Under, ICEC Melbourne, 3rd World Congress*, Melbourne, April.
Holms, J. and Hudson, G. (2000) An evaluation of the objectives of the BREEAM scheme for offices: a local case study. In: *Proceedings: The Cutting Edge 2000*, UK.
Johnson, S. (1993) *Greener Buildings: Environmental Impact of Property* (Basingstoke: Macmillan).
Kohler, N. (1999) The relevance of Green Building Challenge: an observer's perspective. *Building Research & Information*, **27** (4/5), 309–20.
Larsson, N. (1998) 'Green Building Challenge '98': international strategic considerations. *Building Research & Information*, **26** (2), 118–21.
Leadership in Energy & Environmental Design (2001) *Rating System Version 2.0* (US Green Building Council).

Lowton, R.M. (1997) *Construction and the Natural Environment* (Oxford: Butterworth-Heinemann).

Moavenzadeh, F. (1994) *Global Construction and the Environment: Strategies and Opportunities* (New York: Wiley).

SEDA (2002) *Australian Building Greenhouse Rating Scheme* (Sustainable Energy Authority). www.abgr.com.au

Spence, R. and Mulligan, H. (1995) Sustainable development and the construction industry. *Habitat International*, **19** (3), 279–92.

Taylor, D.C., Abidin, M.Z. and Nasir, S.M. (1993) Creating a farmer sustainability index: a Malaysian case study. *American Journal of Alternative Agriculture*, **8** (4), 175–84.

Todd, J.A. and Geissler, S. (1999) Regional and cultural issues in environmental performance assessment for buildings. *Building Research & Information*, **27** (4/5), 247–56.

Yates, R. and Baldwin, R. (1994) Assessing the environmental impact of buildings in the UK. In: *Proceedings CIB Congress*, Watford, UK.

22

Resource efficiency

Kirsty Máté*

As the environmental crisis worsens globally, facility management (FM) finds itself in the frontlines of the fight against waste. Built facilities consume nearly half of the world's resources, generate pollution and cause other environmental damage. Greenhouse gas emissions, for example, arising from electricity generation used to manufacture materials and operate buildings, are having damaging effects on atmospheric ozone and global mean temperatures. Other facilities, like paper mills and chemical factories, discharge contaminates into local ecosystems that have detrimental effects on flora and fauna, water quality and air quality.

Energy conservation, a misnomer in itself, is attracting the attention of government agencies and is beginning to become an important issue for facility managers. While the price of energy remains subsidized in many countries, there is little market incentive to find ways to use less. However, as this anomaly is reduced, design and operational decisions will pay more regard to the energy demand of built facilities, and be willing to invest funds to make things more efficient.

Recycling initiatives have infiltrated most developed countries, yet there is significant potential for improvement in this area, especially for commercial buildings. Apart from operational activities, recycling applies to refurbishment and replacement processes that have traditionally delivered large amounts of building waste to landfill sites. Demolition of buildings that have reached the end of their economic life is progressively giving way to innovative reuses to accommodate new functional uses, in many cases delivering highly attractive facilities while preserving historical landmarks. Facilities can, therefore, be recycled through the application of adaptive reuse strategies.

Technology investment can assist in reducing resource use where payback periods are reasonable and the risk of financial loss is low. Facilities are being designed to be more

* EcoBalance, Sydney, Australia

durable and sustainable through the use of smart systems and monitoring tools. Pollution controls, for instance, can result in less pollution being discharged to the environment, or treating it before it is discharged. Today this might sound socially responsible; tomorrow not doing so may be financially unattractive if new fines and abatement incentives are introduced.

Resource efficiency links to sustainable development in a readily objective manner. Savings in resources directly benefit an organization with little disadvantage. The identification and implementation of initiatives to reduce resource usage is a key issue and another important change driver for FM. Every advantage found adds proportional value and may bring additional indirect benefits such as image enhancement, stakeholder satisfaction and higher living standards.

22.1 Introduction

Both government legislation and perceptions of good business management are changing throughout the world, challenging business practices of the past and the present. Issues related to sustainability are amongst these changes and they are the responsibility of all decision-makers within an organization, not only upper management. This shared accountability for the greater principles of sustainability (financial, environmental and social) will provide creative and thorough solutions in this changing and developing social climate.

The issues of solid waste and wasted energy are issues that smart facility managers should be addressing if they are to be of value to their organizations, both now and in the future. Waste created through resource inefficiency is resulting in large financial costs to organizations as well as costs to the environment through pollution, reduced availability of non-renewable resources, increased greenhouse gas emissions and global warming.

This chapter will address the issues associated with resource efficiency in the workplace as they relate to waste and energy at all stages of the life cycle of a tenancy.

22.2 Waste minimization

Waste occurs at all stages of the life of an office tenancy: during relocation, use and refurbishment. Office tenancies have an average life span of around 5–7 years while shopping centre interiors last only around 5 years. These short lives create a variety of waste issues including unwanted furniture, fittings, fixtures and finishes, construction waste on site, as well as manufacturing and production wastes off site. The life of an average multilevel office building is 50–100 years; if the entire contents of these buildings are turned over every 5–7 years the effects of waste and resource management are enormous. US companies, for example, create about five times more waste per dollar of goods produced than Japanese companies and over twice the waste of German companies (Townsend, 1997). Construction waste accounts for up to 40% of the landfill in Australia (Andrews, 1998) and 15–20% in the US (Kincaid, 1995).

Case study

In Portland, Oregon, 47% of all construction and demolition waste was diverted from landfills in 1993. One example of this was the recycling of 76% of the waste from new construction of a 5000 square foot restaurant: 61% of the waste was recyclable or reusable wood, 11% was recyclable cardboard and 4% was recyclable gypsum wallboard. Estimates are that up to 90% of construction and demolition waste is potentially reusable or recyclable, depending upon the type of project and existence of local markets for waste materials (Kincaid *et al.*, 1995).

There are methods that should be adopted by companies and facility managers in order to reduce this unnecessary waste generation and inefficient use of resources. By adopting the waste regime of 'reduce, reuse, recycle', much can be done to achieve this goal.

22.3 Relocation and refurbishment

22.3.1 Adaptive reuse

Buildings are not only long-term assets, but they also represent large investments in both energy systems and materials. Poorly designed systems and/or poorly constructed buildings may need upgrading, or in the worst case, removal or demolition. Inefficient energy systems are one of the main drivers of building renovation.

As the value of raw materials increases and the availability of land decreases, there will be greater value in the reuse of older buildings rather than in their demolition. Adaptive reuse of older structures can result in financial savings to both sellers and purchasers (Gottfried, 1996). Buildings, being valuable assets, should only be demolished if that is unavoidable, with a preference for upgrading systems for better efficiency.

Case study – National Audubon Center, New York

Completely refurbished in 1993, the National Audubon Building is a recycled 100-year-old, eight-storey building. Conservation of the building's shell and floors saved approximately 300 tonnes of steel, 560 tonnes of concrete and 9000 tonnes of masonry – choosing restoration instead of demolition and new construction resulted in savings of around US$8 million.

As well as financial returns there are other hidden environmental benefits. Studies have shown both in Australia and internationally that the embodied energy within the fabric of a building (the energy required to extract and process raw materials into finished products) can be greater in commercial buildings than the life time operational energy requirements (Gottfried, 1996). By reusing or recycling old buildings, the inherent embodied energy of the structures is utilized and the amount of material to be disposed of is reduced.

As the use of existing buildings can lead to obvious environmental benefits such as saved energy, reuse of existing materials (and so a reduction in use of new resources) and in some cases enhanced aesthetic value, it is also recognized that the life of new buildings can be extended into the future through greater flexibility in design. The concept of 'long-life – loose fit' reinforces the worth of a structure as a valuable asset, providing greater financial security for building owners for extended time periods.

22.3.2 Reduce

Reducing waste through clever purchasing and design before it has the opportunity of becoming waste, is the most efficient means of reducing waste to landfill. This may occur through a change in culture, such as introducing notions of hot-desking and increased opportunities for employees to work from home, or by ordering exact quantities and not adding in contingencies for waste.

Facility managers have many opportunities for reducing waste during a major relocation – some of the ways that facility managers can be effective include the following.

Reduce product quantities

Often the quantities of products and resources ordered for an organization are not thought through efficiently. This is mainly due to a lack of planning. Companies have the opportunity when relocating to new premises to address the very working culture of the business. Is this business running efficiently? Should the layout and dynamics of the business remain the same? What is the future of the company and how does it see itself performing in five and ten years? These types of questions should be addressed by managers and directors within businesses before moving, for an efficient and productive business, now and in the future, and can limit the amount of resources actually required. Areas to be addressed include the physical working environment, looking at alternatives such as hot-desking, reducing enclosed office spaces and physical filing requirements, ordering exact quantities of materials/products and reducing products through multifunctioning.

Hot-desking

By performing an in-depth review of how the business operates, opportunities such as hot-desking may become viable options, as do the opportunities for people to work from home. This reduces the amount of furniture and equipment that is required, and not only reduces resource use but also reduces the future possibilities of waste when desks and chairs become obsolete.

Filing

A thorough assessment of filing requirements can also lead to resource savings, as often what is kept and how it is stored can be rationalized. Filing and storage takes up valuable floor space that could be occupied by a productive person rather than unproductive storage systems. Nowadays many companies are also choosing to use digital archiving as their main source of storage; this reduces storage facilities and associated floor space considerably.

Reduce enclosed office spaces

Enclosed office spaces with full height walls, doors, larger furniture and more storage facilities use more materials than open workstations. Most enclosed offices have framed walls, glazing with aluminium framing and timber or similar doors. By reducing the number of enclosed offices, or making them completely obsolete, use of these resources is reduced and the capacity for waste at the end of life is significantly reduced. Replacing offices with workstation configurations also provides the ability to reuse more readily and also reduces waste as they can be used for longer.

Order exact quantities

Often products and materials are ordered with an allowance for waste or without consideration of layouts that conform to manufacturers' production sizes. Addressing these issues can greatly reduce the amount of waste during construction. While facility managers are often not responsible for, and have no direct control over these types of issues, suggestions can be made to construction teams to include such criteria in environment management plans.

Reduce equipment through multifunctioning

Most brands of office equipment currently offer multifunction units where, for example, the same machine functions as scanner, copier, fax and printer. This has a multilayered effect on office space – less space is required for individual machines, initial cost outlay is reduced, the quantity of office equipment needed is reduced, and the use of resources and waste at end of life is reduced. There is one disadvantage: if these machines do break down it can mean all functions of the machine are lost, not just one as would be the case if the functions were separate.

Reduce quantities of materials

Smart thinking can greatly affect the amounts of materials used when relocating and ordering new products, e.g., by ascertaining where full-height workstations are actually required. Areas which require more visual or acoustic privacy may well require full-height walls, while other workstations may not. By reducing wall heights even by 900 mm in an office fit-out with 200 workstations, hundreds of metres of fabrics, metal plates, boards and steel/aluminium framing can be saved. Similarly wall heights can be reduced to meet needs rather than wants or by simply not conforming to norms.

Reducing the range of finishes used on a project can assist with future requirements for disassembly and separation for recycling as well as improving resource efficiency. Using similar finishes and materials throughout a project lessens over-ordering, enables bulk ordering and results in simplified re-ordering and storage procedures.

Reduce floor space

The benefits of reducing floor space within a tenancy are compound – not only does it reduce the floor space required (and so reduce rental costs), it also affects the amount of materials required for fit-out, decreases energy requirements for both lighting and air conditioning, and reduces the demand for the construction of new office buildings.

The floor space assigned to areas that are not used for great lengths of time by people, such as utility rooms, conference/meeting areas and reception areas can often be reduced without loss of amenity. Similarly the size of corridor spaces, staff areas (such as coffee and lunch areas and executive suites) can also be addressed without compromising their environmental quality.

One space may also perform two functions, e.g., a wider corridor space could also house a coffee making facility; a staff area may be configured so that it can easily be turned into an extra meeting room if and when required.

Reduce packaging

When selecting materials and products, consideration should also be given to the packaging used for transport. Requirements for packaging, and how it is produced and disposed of differs throughout the world. Purchasers and specifiers can influence how packaging is created through their power of selection. Company policy can also demand reusable, recyclable and/or biodegradable packaging for all purchases, including furniture, construction materials and equipment. If none of these options is available, policy can demand that suppliers take back their packaging after goods have been delivered and installed.

22.3.3 Reuse

Apart from reusing an entire building to save on resources, the materials and products within the building itself can also be reused to further make resource savings.

Reuse, refurbish and remanufacture products

Before relocating it is advisable to prepare a detailed list of all furniture and other loose fittings and fixtures. An example of a list for a designated department is given in Figure 22.1.

Repairing existing items when relocating, rather than buying new furniture of comparative quality, can save companies large amounts of money, e.g. a prominent law firm in Sydney was able to make 50% savings by refurbishing their 1200 chairs when relocating to new premises (Complete Office Refurbishment, pers. com., 2000). This enabled the company to save money as well as saving valuable resources and reducing waste.

Accounts department

Item	As new	Refurbish	Repair	Irrepair	Relocate	Auction	Charity	Dispose
Workstation 1	✓				✓			
Workstation 2		✓			✓			
Desk chair 1	✓				✓			
Desk chair 2		✓	✓		✓			
Visitor's chair		✓	✓			✓		
Filing cabinet 1		✓			✓			
Filing cabinet 2					✓			
Storage cupboard	✓					✓		
Screen divider				✓				✓
Return 1			✓				✓	
Return 2	✓				✓			

Figure 22.1 Typical list of furniture and fittings for a designated department.

As well as relocating with existing furniture, buying second-hand items instead of new items saves on resources and delays eventual disposal either to landfill or recycling. The US has a large second-hand furniture refurbish, repair and resale industry available to customers. Companies such as Herman Miller will also take back old Herman Miller furniture pieces, repair and refurbish them, and sell them back onto the market, re-branded as 'As New'. Customers are guaranteed of quality furniture, with the Herman Miller branding and reputation, and they save resources through buying second-hand items.

Computers and other electronic office equipment and their components can also be purchased as second-hand or remanufactured items. While obsolescence may be an issue here, some manufacturers are producing equipment such as photocopiers, which once remanufactured, have components which are actually better than they were when they were installed into the machines brand new, due to thorough testing and the repairs made to the various parts.

Reuse materials

One German study, which compared dismantling and reuse or recycling of materials with more conventional demolition and disposal, found that dismantling resulted in a cost saving of 18% even though it took more time (Lawson, 1996).

As well as products, materials can also be reused, repaired/refurbished, sold or given to charities before being disposed of at landfill or destroyed through incineration. Care should be taken when dismantling walling, flooring and ceiling systems so that damage to the materials is minimized, and that they are separated into reusable, repairable, recyclable and disposal storage piles. An inventory similar to that for products should be established so that the quality and quantities of materials that can be reused are clearly understood.

Reuse systems

Most new electrical and lighting systems are now computer based and designed to reduce energy use. It makes good economic as well as ecological sense to make sure that such systems can travel with the company when relocating.

Investments may also have been made through the purchase of energy-efficient lighting hardware such as fittings and tubes. Enquiries should be made at the time of relocation to determine whether these systems can also be taken with the company and established in the new premises.

Electrical wiring systems that facilitate reuse are also available on the market. Wiring can account for a large proportion of waste in the relocation of a fit-out. Ensuring that an electrical system can be reused is therefore a cost saving as well as an environment saving exercise.

Select materials and products which can be reused or remanufactured

Many future waste issues can be avoided with some forward thinking and the purchase of materials and products that are more readily reusable, upgradable, refurbishable/repairable and/or able to be remanufactured.

Carpet tiles, for example, can be very easily removed if laid correctly and relocated to the new office environment. If, however, the tile is not suitable for the new fit-out some companies can take back the carpet tiles for recycling or can upgrade the tiles, or even re-colour them to suit the new fit-out. Throughout Australia and New Zealand around 7000 tonnes of commercial carpet are disposed of every year. More than 90% of that is sent to landfill (Ontera, undated).

Many furniture items can also be selected for ease of reuse and adaptability. Modular systems for office furniture and workstations can usually be modified to suit changing circumstances; however, it is advisable to check how versatile the system is and just how easily these adaptations can be made as there can be vast differences between manufacturers. Factors to be considered include how smoothly a system can be configured from one layout to another, how easily a system can be dismantled and reconstructed, and how readily various items can be upgraded, e.g. textiles changed, colour of framing systems modified, surface finishes altered, adaptability for changing technologies and rewiring capabilities.

'Repairability' is also an issue, particularly in the case of office chairs where adjustable mechanisms can often require repair. Issues of longevity and responsibility are also important; manufacturers' guarantees can often be good indicators of potential longevity. Extended producer responsibilities are becoming increasingly important requirements in government legislation, particularly in Europe and the US. Selecting products where manufacturers are responsible for their product throughout its entire life cycle is beneficial to the purchaser as well as the environment and can also mean that the product is more easily repaired and upgraded.

There are walling systems available that enable purchasers to dismantle and relocate them easily. Finishes can also be altered to complement a new interior. These systems are far superior to standard plasterboard walling systems, which cannot be easily reused unless great care is taken during installation and dismantling. There are cost implications for such systems, but costings should be calculated on a lifetime basis with due consideration given to potential company growth and therefore how often it may relocate and or refurbish during the expected life of the system. The more a company is expected to move, the more cost effective such a system becomes.

Electronic equipment is available which can be upgraded through computer software without the hardware itself being changed. While this concept is still mainly applicable to computers, other electronic equipment will surely follow suit. This may occur in response to legislation, particularly in Europe where manufacturers in the electronics industry are legally responsible for their products during the whole-of-life cycle; manufacturers may be forced to upgrade systems rather than hardware so as to eliminate some of the waste issues that they will otherwise face with obsolete models.

22.3.4 Recycle

The recycling of products and materials that can no longer be used, forms the next tier of the waste hierarchy. Recycling prevents wastage and produces raw materials for other uses, extending the life of the material or the product. Most recycling endeavours also save energy as well as slowing use of virgin resources and reducing waste. Recycling waste is, however, only half of the equation – in order to benefit fully from the recycling process purchasers must close the loop by buying recycled goods.

Specify recycled content products

While we may diligently recycle our waste, whether consumer or industrial waste, the effort we go to is useless unless this recyclate is then manufactured into something else that we buy. Material may be used to make the same product again, e.g., a PET drink bottle recycled into a new PET drink bottle or a completely new product such as a PET bottle being turned into an upholstery fabric. By making a concerted effort to buy products that contain recycled content, purchasers and specifiers are assisting in creating markets for recycled products and so closing the loop.

There are many products and materials that are currently available throughout the world that are manufactured from recyclate. These include recycled plastic lumber to replace timber products; recycled glass slabs to replace granite or marble floor tiles and bench tops; recycled car tyres for floor tiles, landscaping pavers, roof tiles and new car tyres; recycled cardboard and paper made into walling systems and furniture, and mixed with other recyclates to form board products to replace timber and even granite products; plastics such as PET are turned into fabrics for use as upholstery and clothing textiles. There are many other examples of useful products that have been created from existing post-consumer and post-industrial wastes.

While both consumer and industrial wastes would normally go to landfill or incineration, there are distinct differences between the two. Post-consumer wastes are from products that have already had a useful life and so have already completed a full life cycle. Post-industrial wastes on the other hand are wastes that have 'come off the factory floor' and been put back into the same system, i.e., glass offcuts used to make glass products. They can also be waste materials or by-products used to make a new product such as the use of fly ash in the concrete industry. These waste materials, therefore, have not yet completed a full life cycle. Another form of waste is agricultural waste, created through agricultural processes, such as straw waste from wheat, corn and rice fields, rice husks, shellfish shells, nut shells and sugar cane waste (bagasse). These wastes have also been used to make new products such as flooring tiles, board products and mouldable materials.

Separate wastes on site for further recycling

While the purchase of items with recycled content is important, so is the recycling of waste materials. Most construction wastes including steel, aluminium, timber, brick, plasterboard and glass can now be recycled in many countries. These wastes, however, must be separated and disposed of appropriately.

Waste products should be separated on the basis of their recyclability and taken to appropriate disposal depots. As noted earlier, in some countries many of these products can be returned to the manufacturer for appropriate disposal. Facility managers should make themselves aware of the different laws and regulations appropriate to their country and region to ensure best disposal methods.

Specify recyclable products/materials

As well as selecting materials and products with recycled content, products and materials that facilitate disassembly, separation and recycling as well as reuse and or remanufacturing can be selected. Simplicity is the key – products which have not been manufactured from a variety of different materials and are therefore 100% 'pure', facilitate the recycling process. Likewise the complex issue of separating many different materials within one product hinders disassembly, separation and recycling – products containing the least variety of materials should be the ones chosen. The ease of separation of materials is also a key issue: the separation should be simple and quick. Separation of the covering and the core of covered materials, such as plastic-coated wiring, is both difficult and expensive.

A further concern is the ease with which these materials can actually be recycled. Many imported products, for instance, may be labelled as recyclable but if no recycling facility or system exists in the destination country then the advantage is lost. Specifiers obviously need to be know whether the products they choose can indeed be recycled.

Biodegradable products and materials

Biodegradable products and materials that biodegrade safely are also an advantage. Some new plastic products are now available that are made from biodegradable starch materials; while they look and feel like conventional petroleum-based plastics, they are in fact plant based. These products are not yet widely available and are often expensive, but with further development they should become more common and less costly in the future.

Specifiers need to be careful when selecting natural and biodegradable products that may, through their degradation, release toxic chemicals. Seemingly benign products such as textiles may be dyed with chemicals containing damaging components such as chromium; as the fibre of the textile, perhaps wool or cotton, biodegrades the dye chemicals leach into the environment as they are no longer contained by the fabric.

Suppliers

Suppliers should be able to supply certification of an environmental management system (EMS) such as ISO 14001 and/or have in place waste minimization policies and programmes. Demonstration of such actions by a company provides facility managers with greater certainty when selecting suppliers that are trying to achieve sustainability through their work practices and their products.

Suppliers who are also able to take back products and materials for reuse, recycling or remanufacture should also be preferred by the facility manager; these suppliers are showing a commitment to their product through its entire life cycle and this eliminates any problems the facility manager may encounter when disposing of used products themselves.

22.4 Occupation

During the occupation of commercial premises, waste is created through consumables. These include paper, envelopes and packaging, office equipment consumables such as toner cartridges and computer disks, beverage and food containers and incidentals such as sugar sachets, food and other compostable items, cleaning equipment and incidentals such as hand towels in toilets as well as the waste resulting from ongoing maintenance and repair of furniture and equipment.

While it is important to have in place systems to deal with the correct disposal of these items, education on how to use these systems is equally important. There also needs to be company-wide support for such strategies if they are to be successful. Setting up an EMS within commercial premises and gaining 'buy in' from each organizational level, establishing an educational programme and addressing each of the waste issues associated with the main consumable areas outlined above is a major undertaking.

22.4.1 Setting up an EMS

An EMS assists in achieving targets but does not set levels of performance. It will assist in decision-making for a facility manager not only during the occupation of a building but will assist at every stage of the life of the company, in regard to its resources, assets and functions. The international standard for environmental management systems is ISO 14001. The International Organization for Standardization defines an EMS as:

> ... that part of an overall management system which includes organizational structure, planning activities, responsibilities, practices, procedures, processes and resources for developing, implementing, achieving, reviewing, and maintaining the environmental policy. (AS/NZS ISO 14004: 1996)

An effective EMS as stated by SAGE (Strategic Advisory Group for the Environment, ISO, 1996) is:

> ... operating activities (that) meet the needs of present stakeholders (shareholders, employees, customers and communities) without impairing the ability of future generations to meet their needs.

An integrated management approach will address not only the environmental policies but also the product, service and health and safety qualities of the organization and will have supporting procedures, guidelines and criteria in the areas of auditing, customer service, life-cycle analysis, impact assessment, codes of practice, management review and performance evaluation.

Companies set up an EMS for many different reasons, some of them being:

- competitive advantage
- community confidence
- customer request/demand
- marketing tool
- interest in the environmental stakeholders such as regulatory agencies, customers, shareholders, local communities, environmental interest groups
- competitive pressures such as compliance from customers; to get ahead of the competition; to catch up with the competition; industry pressure
- regulatory drivers
- internal benefits of registration with an EMS such as ISO 14001.

The key steps for setting up an EMS include (Wastebusters Ltd, 1997) (Figure 22.2):

- get initial commitment from management
- review current position of company and its environmental impact
- establish an environmental policy for the organization
- identify environmental objectives and a programme to implement the objectives
- implement documenting procedures and training staff
- monitor – audits and reviews
- report to the wider community.

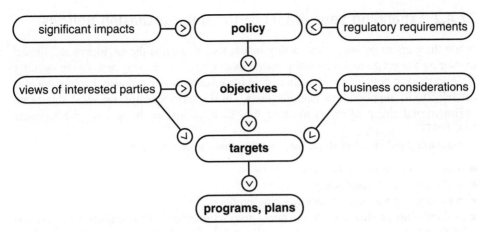

Figure 22.2 EMS planning process (Jackson, 1997).

According to Harding *et al.* (1997), the benefits of a well-designed EMS can include:

- the ability to identify and correct problems internally
- a consistent framework for monitoring environmental performance
- a sound basis for performance reporting
- continuous improvement of environmental performance
- a reduction in resource use, pollution and waste production and therefore greater cost control
- a reduction in corporate environmental liability and risk leading to improved access to capital and lower insurance costs
- a positive market image
- meeting with vendor certification criteria and therefore increasing potential market share
- links with other internal organizational systems.

The success of the development and implementation of BP Australia's EMS has meant that the team has been asked to establish a similar programme at all of BP's main offices internationally, including head office in London. Their EMS is known as the Environmental Improvement Program (EIP) and the main challenges of the initial project were to improve BP's Melbourne Central tenancy's environmental performance by (BP Australia, 1997):

- increasing energy efficiencies
- reducing waste to landfill
- reducing paper usage and dependence.

The implementation of the EIP has had the following results at the Melbourne office (BP Amoco Australia, 2000):

- 24% reduction in energy use
- 59% reduction in paper consumption per head
- 79% of rubbish diverted from landfill on any single day.

22.4.2 Expanding your influence through purchasing policies

While there are many ways that facility managers can change the environmental impact of their own organization, their influence on other organizations may be even stronger. By their purchasing policies the influence of an organization's environmental commitment is extended to its suppliers and manufacturers. 'In the service sector, the most significant environmental effects of an organization may be those that are bought in' (Wastebusters Ltd, 1997).

Purchasing policies can state, for example, that the organization:

● will not buy rainforest timbers
● will only buy recycled paper
● will only buy the highest energy rating products
● will only buy products which are guaranteed returnable to the manufacturer at the end of their life.

Any purchasing policy should, however, be directly related to the areas of greatest environmental impact, which will generally be related to areas of the greatest purchasing.

In many countries eco-labelling systems have been established, which makes the task of purchasing environmentally preferable products much easier. These labels are found mainly in Europe, America and Japan. Organizations such as the Buy Recycled Business Alliance (BRBA, 2002) can also assist with purchasing recycled products and materials.

The Green Office Manual (Wastebusters Ltd, 1997) provides this set of guidelines for assisting in environmental preferable purchasing:

● recognize the importance of purchasing to your environmental performance
● identify areas of purchasing with high environmental impacts
● plan your purchasing strategy to fit with existing structures
● identify the approach that will be most effective for you
● use purchasing tools to help identify environmentally preferable alternatives.

22.4.3 Establishing an education and communication programme

Education and communication are essential for meeting the targets and objectives of an EMS, gaining staff and supplier support, and improving resource efficiencies. Support for such an undertaking must exist across the whole of an organization, from upper management to on-floor teams.

All of an organization's environmental policy – the aims and objectives, the targets, and how they are going to be met – must be clearly understood by all the stakeholders. Staff and supply chain involvement in setting these objectives and strategies is highly recommended as this not only communicates the concepts at an early stage but also establishes ownership of solutions and awareness of environmental issues which may not otherwise be clear to everyone within an organization.

Educating staff and suppliers is an ongoing process. As people change so do the educational requirements and objectives. A system is needed where education about the relevant key environmental objectives, targets and solutions are provided to all stakeholders. For example, in a major banking corporation in Australia, a monthly training routine was put in place to educate the staff on key environmental objectives, including energy use, paper usage, recycling and composting. Each month one of the environmental objectives of the bank was addressed, with educational material provided to team managers/leaders and goals set for that month. At the end of the 12-month programme it was started again, continuing the education of new recruits and providing refresher courses for older staff.

22.4.4 Commercial consumables

Most commercial organizations have very similar waste issues; however, every organization is essentially different and should be assessed on an individual basis. The following is a guide for the more popular consumable items within the average commercial organization.

Paper and cardboard

Even with a more technologically reliant community, which was supposed to create a paperless office, we consume and waste greater amounts of paper than ever due to improvements in photocopy technology and the ease of producing professional looking documents using computers and laser printers (Harding *et al.*, 1997).

Saving paper through the waste hierarchy of 'reduce, reuse, recycle' is highly relevant and plausible. Reducing paper usage through the use of e-mail, for example, is one way; another is reducing margins and font sizes in documents, formatting pages to two columns can also save paper as does double-sided copying and printing. Reusing single-sided copy paper for note pads, faxes, personal document copies, and so on, reduces the use of virgin paper. Using recycled paper, and recycling paper, closes the loop and reduces the need for paper to be made from wood pulp, some of which may come from old growth forests. Enhancing these practices requires education, storage facilities, and provision of recycling and reuse containers.

Packaging and cardboard can be dealt with similarly – cardboard needs to be separated for recycling and purchasing requirements need to make clear what types of packaging are acceptable to the organization. Non-recyclable plastic packaging and polystyrene packaging may be deemed unacceptable, and alternative packaging materials may be demanded or another supplier sought.

Office equipment

While office machines are usually associated with energy consumption they are also great consumers, and great producers of waste, including toner cartridges, printer cartridges, drums, ribbons, filters and computer disks as well as the machines and parts themselves. In 1991 alone, over 1 billion computer disks were sold in the US (Townsend, 1997). Many companies now remanufacture and recycle most of the consumable and formerly disposable items so purchasers should enquire whether a potential equipment supplier offers such services and, if not, look for other companies who do, or who may use another company's consumables for remanufacture and/or recycling. Recycled and or remanufactured items such as recycled computer disks and cartridges can also be bought.

Fuji Xerox Australia's Eco Manufacturing Centre produces over 10 000 remanufactured printer cartridges per month and has the capacity to remanufacture 18 types of Xerox cartridge. The Centre has become a '... world leader in the development of technology and product programs related to the remanufacture of spare parts, print cartridges and office machine components' (NSW Government, 2001).

Beverage and food containers

The use of disposable containers for food and beverages is difficult to avoid. However, many of these items are now manufactured from recyclable materials that should be separated and disposed of in the most appropriate manner. Commercial premises, whether they be offices, shopping centres or government organizations, need to address this growing waste problem by installing recycling systems appropriate to their requirements that will deal with this type of waste.

Some organizations have their own cafeterias where the control of one-use food and beverage items is easier; however, items of this type may still be brought into organizations from outside. While it is impossible to monitor the supply of all these items and ensure that they are recyclable, staff should be encouraged to

- buy take away food in their own reusable containers
- buy coffee in a reusable mug
- buy drinks in recyclable containers.

Any recycling system adopted will depend upon the waste collection services available. Some services will accept mixed recyclables which means a receptacle can be allocated for all recyclable items, while other services will only accept separated recyclables – this means that individual receptacles will need to be placed to accept various materials such as glass, plastic and metal. Obviously these receptacles should be strategically placed so that they get maximum use. They should also be very clearly marked, preferably with words and pictures, or just pictures. These are the most problematic in public spaces where there is not a 'contained audience' to educate. In these cases the clearer and more obvious the graphics, and more appropriate the receptacle, the better the system will work. Cleaners and others who are responsible for emptying such receptacles should be advised of how and where they should be emptied so that, as has happened in the past, the sorted recyclables do not end up in ordinary waste bins at the end of the day.

Rowe and Maw, a firm of solicitors in London with a staff of 360, were using and disposing of (to landfill) approximately 6750 mineral water bottles per year. By installing a water purification system and using sterilized refillable bottles the company not only saved the bottles going to landfill but also saved £1886 per year (Wastebusters Ltd, 1997).

Compostables and other waste materials

If possible, waste that can be turned into compost ('compostables') should also be separated during the waste process. Installing a worm or vermiculture composting system or ordinary composting system is possible in many commercial premises and, depending on the system, need not take up much space. These are particularly useful for premises that have food preparation areas as they can very easily separate compostable waste during preparation as well as disposing of waste food. It is more difficult to contain and control waste in public spaces and open office areas; however, in designated kitchenette or similar areas, where recycle receptacles should also be placed, compostables can have a separate disposal unit. Like other separated waste the receptacles should be clearly marked and their purpose known to everyone concerned.

Hazardous waste such as batteries also needs to be considered. Some countries have appropriate recycling and disposal systems set up to handle the chemicals and toxins contained in batteries and similar products. Finding the most suitable disposal method available requires some investigation and, as is the case with other special waste, appropriate receptacles should be provided for their collection.

Accommodate waste separation in loading docks

As the need for separating wastes, recycling and appropriate disposal increases, so will the allocation of areas for waste collection and disposal in loading dock and similar spaces. When installing a waste management system, the necessary spaces and facilities for carting the waste from the collection point to the recycling/waste depot must be made available. Additional space is required for separation of recyclates in the waste disposal area/loading dock. If sufficient space is not available it may be that mixed recyclates will be accepted by another waste contractor or perhaps the loading dock of an adjacent building can be used for storage purposes. Alternatively recyclates may be picked up more frequently thus reducing the need for larger storage areas.

Cleaning equipment

Cleaning also has its attendant waste issues – this applies not only to the chemicals and detergents used to clean surfaces and equipment, but also to some of the hardware used. Cleaning and catering contractors can be encouraged to select environmentally benign substances; this not only reduces the environmental risks, such as the risk to aquatic life if chemical cleaners find their way into waterways, but also reduces health and safety risks. Chemicals that are of major concern are acids, alkalis, bleaches, phosphates and solvents (Wastebusters Ltd, 1997). Likewise consumables for cleaning equipment should be assessed – are they recyclable, biodegradable, returnable or reusable? Other strategies include using cleaning products which are concentrates, using refillable containers and/or using products with packaging made from recycled material (Wastebusters Ltd, 1997).

Toilet areas also consume a range of products: toilet paper, soap, women's hygiene products and hand drying systems. Environmental impacts can be reduced through the selection of recycled toilet paper, biodegradable soaps which are not individually packaged, and appropriate disposal methods for women's hygiene products. Different hand drying methods have varying implications: electric dryers use a lot of energy, paper hand towels create waste, and reusable cloth towels need to be washed and dried. All of these methods have environmental impacts that vary by country and manufacturer, depending on energy sources, cleaning methods and disposal methods.

22.4.5 Maintenance and repair

Commercial premises require regular and continual maintenance and repair work. Products and materials that require minimal maintenance or repair during their life cycle are obviously the most preferable, however, if and when maintenance and repair work is required, preferable products and materials are those which:

- are easy to repair, e.g., office chairs that have easily removable components such as cloth coverings/upholstery
- do not require the use of products that have a negative environmental impact
- have services available for repair and maintenance within the local area and so do not need to be transported great distances for servicing
- do not cost much the same to repair as it would to buy a new item
- do not require constant repair/maintenance work.

22.5 Energy

Buildings worldwide use 40–50% of the world's energy. In the UK, for example, buildings are responsible for about 50% of annual emissions of carbon dioxide (Wastebusters Ltd, 1997). In the US, buildings are responsible for nearly a third of total energy consumption (RMI, 2001). Commercial buildings in Australia account for 35 million tonnes of carbon dioxide per year; if business continues as usual this output will double by 2010 (SEAV, 2001).

Increasing energy costs, blackouts in the US, the world's most developed nation, and the economic returns associated with improved energy efficiency (Schepers, 2001) are sound reasons why organizations should be looking for energy-efficient buildings and alternative energy solutions.

The energy efficiency of a building is greatly affected by its design and its energy systems. While retrofit solutions may improve the performance of existing buildings it is in the design of new buildings that the most energy-efficient solutions can be implemented. Astute facility managers can influence many of the energy management decisions made within an organization and not only reduce energy-related impacts but save money as well.

22.6 Relocation and refurbishment

It is particularly during relocation and refurbishment exercises that facility managers have the opportunity to put in place new systems that can greatly influence the many aspects of an organization's efficiency of operation. This includes limiting the impacts that the organization, through its operations, has on the environment at local, regional and global levels.

22.6.1 Selecting new existing premises

There are various organizations and systems now available throughout the world to assist in selecting a tenancy within a building or indeed a building itself which is more energy efficient than another. Australia recently launched a new Building Greenhouse Rating

Scheme through the NSW Sustainable Energy Development Authority (SEDA, 2001), which provides buildings and tenancies with an energy star rating with a maximum possible rating of five stars. Owners and tenants can select new premises with energy efficiencies based on the star rating. Other rating schemes around the world include the Building Research Establishment Environmental Assessment Method (BRE, 2002) and the US Green Building Council's LEED Rating system (USGBC, 2002). By using such tools and schemes facility managers will be able to select premises with lower energy consumption and reduced energy bills.

Premises with higher levels of natural lighting will also have lower energy costs. Office premises with a smaller, narrower footprint provide more employees with access to natural light and so reduce the need for artificial lighting. Skylights also increase natural light levels and can therefore be a beneficial addition in a building undergoing a major overhaul.

Research in the US by Pacific Gas & Electric Company (Energy Design Resources, 1999) into the operation of a particular retail chain showed that 'sales in a typical store with skylights averaged 40% higher than in a comparable store without skylighting'. The study indicated that the presence of skylights was the second most important factor, after the number of hours open per week, as a predictor of higher sales across the various stores in the chain.

Premises with access to internal stairs use less energy as a result of reduced use of elevators for travel between floors. Stairs also provide exercise for staff and function as alternative communication spaces.

22.6.2 Designing new energy-efficient premises

The design of new premises naturally provides the greatest opportunity for producing an energy-efficient building. There are many computerized tools now available that can assist the designer/architect to produce an energy-efficient building. The Queensland University of Technology in Australia identified more than 25 computer programs available internationally that are designed to assist with improving the energy efficiency of buildings (QUT, 1999). These tools provide designers with information about how different designs will perform under different climatic conditions.

The main factors in energy-efficient building design are:

● passive solar design – the use of the sun to warm the building, and the reduction of unwanted heat gain by use of sunshading
● the use of materials that enhance the principles of passive solar design by storing and/or distributing heat throughout the internal spaces so reducing heating requirements
● the use of natural ventilation and air movement to assist in distributing fresh air and cooling through the building thus reducing the need for air conditioning
● the use of natural lighting to reduce the reliance on artificial lighting during the day
● the use of energy-efficient mechanical systems where such systems are required.

There are many ways in which the principles of energy efficiency can be incorporated into a building and while a facility manager may not have the opportunity to influence how this is undertaken, he or she should be aware of these principles and encourage management and the design team to incorporate them wherever possible.

22.6.3 Selecting heating, ventilation and air-conditioning (HVAC) and lighting systems

Selection of appropriate engineering systems is an important component of any building project, whether new build or refurbishment. In most commercial and institutional buildings the bulk of energy use is related to space conditioning and electric lighting.

HVAC

In Australia, heating, ventilation and cooling in commercial buildings accounts for 70% of their energy requirements (SEAV, 1999). While many new 'green' building designs are less reliant on HVAC systems, most are still reliant or least partially reliant on these systems to provide acceptable control of their environments. Sophisticated computerized systems can now be purchased that not only increase energy efficiency, but can also be integrated with natural heating, cooling and ventilation systems.

These types of systems are increasingly being used in buildings such as hotels where different customers are demanding both naturally ventilated spaces and fully air-conditioned spaces. Integrated HVAC systems can provide customers with individual control of their spaces without disrupting the efficiency of the artificial system.

In some countries, apart from building energy rating systems, there are similar systems available that enable more informed decision-making in relation to selection of machines and appliances. In the US, for example, the US Congress passed the National Appliance Energy Conservation Act in 1987 which sets minimum efficiency criteria for 12 types of residential appliances including HVAC systems, refrigerators, freezers and other appliances (Pilatowicz, 1995). New standards on clothes washers, clothes dryers, and dishwashers were issued in 1991, affecting units manufactured from 1994. Under the Act an Energy Efficiency Rating (EER) is supplied with each product, which provides an estimate of its yearly operating costs.

Direct digital control (DDC) systems utilize microprocessor/computer technology to provide leading edge control systems for building systems. DDC systems are usually accompanied by building management systems (BMS) and utilize desktop PCs with graphical user interfaces that provide building operators with an interface to the DDC system. These systems are able to integrate mechanical, lighting, security and fire safety systems allowing interaction between the protocols. DDC systems have the flexibility to allow more energy-efficient control systems than those previously available (SEAV, 2000). These include:

- supply air temperature reset (multizone, constant volume terminal reheat and VAV systems)
- terminal regulated air volume (VAV system)
- night time free cooling
- cooling setpoint reset
- optimum start
- condenser water reset
- chilled water temperature reset
- hot water temperature reset.

However '. . . the choice of HVAC system and appropriate zoning of a system has far more impact on energy efficiency than the decision of whether to add a demand controlled ventilation system or high efficiency motors' (SEAV, 2000).

Lighting

As with DDC and BMS systems, computerized lighting control systems can be used to reduce the energy consumption of lighting within commercial premises. In Australian commercial buildings lighting accounts for 15% of total energy consumption (SEAV, 1999) while in the US artificial lighting accounts for 20–30% of all energy use in commercial buildings and approximately one-fifth of all electrical energy use (Gottfried, 1996). Reducing energy use for lighting can therefore have significant results.

Computerized lighting control systems allow additional control of lighting levels according to varying inputs from occupancy sensors. These features include:

- photocell dimming for daylighting controls
- integration with occupancy sensors
- switched control for zoned groups related to lux levels, occupancy levels, and after hours zoning.

Indicative energy savings that can be made by using such systems include (SEAV, 2000):

- luminaire maintenance: 5–15%
- daylight dimming: 0–30%
- occupancy sensor control: 0–10%
- out of hours lighting control: 10–30%.

The selection of energy-efficient lamps and fittings adds to the efficiency gains achieved through the use of computer systems. There are some highly efficient lamps available that, when used with highly reflective fittings and appropriate ceiling treatments, produce major energy savings in themselves. Fluorescent and compact fluorescent lamps are the most energy efficient, followed by (in order of efficiency at similar lighting levels) 35 W white sodium globes, 50 W low-voltage globes and 150 W incandescent globes.

The lighting engineer is responsible for designing appropriate lighting levels throughout a space; if this is correctly done, lower design lighting levels will be provided in areas such as hallways and common areas which do not require high lux levels, while areas such as office spaces, factory floors and other areas where people are doing tasks have higher lighting levels, appropriate to those tasks.

A building with a high efficiency lighting system will have an internal lighting load below 10 W/m^2 in an office environment; with a combination of daylighting controls, lighting power densities of 5–7 W/m^2 are achievable (SEAV, 2000).

Programmes and organizations have been established around the world to assist building designers and owners achieve these goals – one of these organizations is the Energy Star Programme of the US Environmental Protection Agency, which has assisted participants in the programme to achieve annual rates of return of over 30% for lighting retrofits (Energy Star, 2001).

22.6.4 Updating HVAC systems and lighting

One of the first reasons a building may be deemed to be redundant is poor energy design (Edwards, 1998). Updating inefficient building systems is one way to keep a building in working order rather than demolishing it and building a new building. Retrofitting an HVAC system improves efficiency and lowers energy requirements (the energy required by a typical commercial building can be 40–60% of the overall energy consumption of the building) and can also improve occupant comfort and productivity (Gottfried, 1996).

The following actions to improve an HVAC system, suggested by Gottfried (1996), can provide environmental benefits:

- eliminating the use of chloroflurocarbons (CFCs) in the system
- replacing outdated systems or components
- addressing and correcting past problems with ventilation and indoor air quality (IAQ)
- resizing components to current requirements
- installing a new building control system.

When retrofitting a lighting system, it is important to consider all appropriate light levels and quality, architectural and furniture layout and room cavity optics, the replacement and proper disposal of older ballasts containing polychlorinated biphenyls (PCBs) and also, particularly in major renovations, to consider daylighting options (Gottfried, 1996).

22.6.5 Transport

Transport to and from the workplace can be a hidden factor in energy consumption. When relocating premises it is worth looking at locations which are close to good public transport and/or that may encourage staff to ride bicycles to work. Car parking facilities should therefore be reduced to promote the use of these resources.

Encouraging staff to utilize public transport and low impact private transport may be achieved by making these resources more convenient and easier to use, e.g., by providing secure bicycle racks for storage and installing showers so that staff can shower after arriving at work.

A major bank in Sydney recently relocated 1000 staff to a suburban location directly adjacent to a railway station (Máté, pers. research, 1998); to encourage the staff to use the public transport system, even on wet days, an undercover walkway was constructed, linking the station to the front door of the bank's premises.

22.7 Occupant concerns

In most cases organizations do not have the luxury of major retrofits, or the construction of buildings from scratch, that would allow them to achieve highly energy-efficient tenancies. However, there are smaller achievements that can be made that will save energy and reduce costs without large upfront spending.

22.7.1 Lighting

The use of energy-efficient lamps and fittings, as well as setting appropriate lighting levels, is the easiest and most cost-effective way of improving the energy efficiency of a lighting system, without extensive redesign and replacement. Compact fluorescent and fluorescent lamps are the most energy efficient. The simplest improvement involves finding the most energy-efficient lamps that can be used to replace existing lamps without the necessity of replacing the fittings. Some of the more efficient fluorescent lamps currently available also require their own fittings so while these do increase the efficiency of the system, the replacement of all existing fittings may not be cost effective in the short or even the long term. A cost analysis that looks at energy saved and waste caused by replacing existing fittings is necessary before undertaking such a manoeuvre. It is worth noting that approximately 80 million fluorescent tubes are disposed of each year in the UK alone (Wastebusters Ltd, 1997).

Technicolor Ltd, located in Middlesex in the UK, achieved energy savings of 30–50% by installing new high-frequency luminaires and retrofitting existing luminaires with high-frequency ballasts and triphosphor tubes (Wastebusters Ltd, 1997).

Lighting levels within different areas should be checked and compared with the lighting levels recommended in building standards. An exercise of this type was carried out in a faculty building at the University of Technology, Sydney and it was found that 90% of the office areas were overlit and were controlled by an automatic switch system that turned lights on hours before staff arrived to work, and off long after they left to go home (Máté, 1997). If lux levels in certain areas are too high, simple exercises such as taking out some lamps can increase energy efficiency. Areas such as storerooms and bathrooms often have lights that are left on continuously; setting these areas up with separate switch facilities, so that lights can be turned off when unoccupied, or with sensors that respond when someone has entered and automatically turn lights on, can be cost-effective ways of minimizing lighting energy consumption.

Another strategy involves decreasing overall lighting levels in areas such as general office areas and using task lighting where it is most needed. This not only saves energy in the overall space, but puts the individual in control of their own lighting requirements. It is important to recognize that people with poor eyesight may need higher lighting levels than those with good eyesight.

The arrangement of furniture and fittings can also have an impact on lighting levels and thus on energy use. It is logical, for instance, to use natural light where it is available and put people, not storage areas, near natural light, and to ensure that furniture is not causing shadowing from either natural or artificial light. It also important that as many people as possible benefit from natural light; in many offices it has been customary to place managers and managing directors along perimeter walls in large offices, a strategy that prohibits natural light reaching the majority of the staff.

22.7.2 HVAC systems

There are small adjustments that can be made to HVAC systems that provide major energy savings and do not entail the replacement of whole systems. These include the following (Harding *et al.*, 1997):

- set thermostats to comfortable levels of between 20 and 24°C – this will vary depending on local climatic conditions, but generally these are comfortable temperatures in cool and warm climates – this can reduce the heating and cooling demands of the HVAC system
- arrange air-conditioning zones so that rooms are not air conditioned when not in use
- organize after hours switching for air conditioning in conjunction with lighting systems
- use fans in summer to reduce the load on air conditioning
- have the air-conditioning system calibrated so that correct airflow rates are achieved
- ensure windows are properly insulated/protected against heat gain and loss
- investigate the potential for waste heat recovery from other areas.

22.7.3 Energy consumption of equipment

Equipment in buildings may not only use energy directly but also output energy in the form of heat, increasing the demand on HVAC systems, particularly in warmer climates. According to the US Environmental Protection Agency, computers and their components use 3–5% of all commercial power. This costs US businesses approximately US$2.1 billion per annum, while the extra energy required to cool interior environments due to the extra heat from this equipment is 36 billion kilowatt-hours (Townsend, 1997).

Research in the US (Townsend, 1997) has shown that the cost savings in the order of US$50 000 were possible in an average office – these cost savings were achieved by using Energy Star computers, combining printer/fax machine/scanner and networking with computers, and reducing the number of photocopiers to a minimum number of high-volume Energy Star copiers.

David Jones, a major retail store in Australia, is expected to save AU$49 000 per annum for 15 years on the energy being saved by enabling the Energy Star option on their office equipment (including computers, monitors, fax machines and photocopiers), and by phasing out old equipment and replacing with equipment with the Energy Star function (SEDA, 1999).

Electronic equipment, such as copiers and printers, also consumes power when it is not actually doing anything useful. This is called 'standby power' which is the power that is used when the equipment is idle or even when it is apparently switched off. The average home in the US uses about 50 W of standby power which is equivalent to around 5% of total electricity use in US houses. Given that there are over 100 million homes in the US, standby power accounts for roughly 8 GW of generation – this is equal to the output of eight large power stations. It is estimated that total power consumption from standby power use is close to twice that amount, as commercial and industrial buildings also have many electronic items with a standby mode (Meier, 1999).

Selecting equipment that is energy efficient in use and in standby mode, reducing the amount of equipment needed or using equipment that performs more than one task can significantly reduce energy use in an organization.

22.7.4 Transport

If an organization is not fortunate enough to be situated close to public transport or bicycle paths, there are other strategies that may reduce the use of private automobiles for travel to and from work. These can include the following:

- organize a company minibus to pick staff up within the local area
- organize carpooling within the company – with the availability of computers this can be easily and efficiently organized in large businesses
- set up and encourage telecommuting for staff who are not needed personally within the work environment every day
- encourage staff to use alternative transport systems if available by having regular 'car free' days (say, once a month or once a fortnight) and impose fines on those who do drive.

22.7.5 Educational programmes

It is not until a building is occupied and in operation that the energy-efficient systems and designs that have been put in place can really be tested, and the energy targets of the organization will seldom be met without the co-operation of the occupants in using and maintaining the systems.

It is important that all staff understand the strategies that have been established and how they should be operated. Simple things such as placing signs near light switches to remind people to switch off when leaving a room unoccupied remind individuals of how they can continue to save energy and play a part. In buildings that rely on occupants to physically change ventilation and shading systems to suit different climatic conditions it may be necessary to have a more formal educational process. Neglecting education in these circumstances can have serious consequences. This was demonstrated in the case of a faculty building at the University of New South Wales in Australia: when occupants moved into their new passively controlled building, staff were not properly trained in the use of the systems. As a result extra heaters were used in winter, fans were needed in summer and external personnel were required to come in each day to adjust the ventilation systems to suit the weather conditions of the day. This situation has been corrected but at a cost to the university as well as the environment (Máté, pers. research, 2001).

22.8 Conclusion

Saving resources is not only good environmental practice, but it also makes good business sense. Putting systems and strategies in place that reduce resources is not difficult and when included as part of an environmental policy and EMS, can set high quality standards within a company and a philosophy for staff and stakeholders to adopt for now and the future.

References and bibliography

Andrews, S. (1998) *Waste Wise Construction Program – Review, A Report to ANZECC*. Natural Heritage Trust (Canberra: Commonwealth Department of Environment).

AS/NZS (1996) *AS/NZS ISO 14004:1996 Environmental Management Systems – General Guidelines to Principles, Systems and Supporting Techniques* (Standards Australia and Standards New Zealand)

BP Amoco Australia (2000) *Environmental Improvement Programme*, internal document.

BP Australia (1997) *BP's Melbourne Central Environmental Improvement Project*, internal document.

BRBA (2002) Buy Recycled Business Alliance home page. www.brba.com.au.

BRE (2002) *Environmental Assessment of Buildings – BREEAM*. www.bre.co.uk/sustainable/breeam.html

Edwards, B. (ed.) (1998) *Green Buildings Pay* (London: E. & F.N. Spon).

Energy Design Resources (1999) Unexpected benefits. *The Newsletter*, **1** (2) winter, 1–2.

Energy Star (2001) *Lighting For Your Home*. http://yosemite1.epa.gov/estar/consumers.nsf/content/lighting.htm

Fuji Xerox Australia (undated) *Eco Manufacturing Centre – Sustaining our environment through better technology*. Company brochure. www.fujixerox.com.au/environment/

Gottfried, D. (ed.) (1996) *Sustainable Building Technical Manual – Green Building Design, Construction & Operations*, US Green Building Council, Public Technology Inc.

Harding, R., Sharp, L. and Hewitt, C. (1997) Down-to-Earth Officecare – A Practical Guide to Environmental Action in the Office. Gray, C. (ed.), Fuji Xerox, Australia.

ISO (1996) *ISO 14001:1996, Environmental Management Systems – Specification with guidance for use* (Geneva: International Organization for Standardization).

Jackson, S. (1997) *The ISO 14001 Implementation Guide – Creating an integrated management system* (New York: John Wiley)

Kincaid, J., Walker, C. and Flynn, G. (1995) *Model Specifications of Construction Waste Reduction, Reuse and Recycling* (Research Triangle Park, NC: Triangle J Council of Governments).

Lawson, B. (1996) *Building materials, Energy & the Environment – Towards ecologically sustainable development* (Canberra: Royal Australian Institute of Architects).

Máté, K. (1997) *Report for Step 1 of Greening UTS*, University of Technology Sydney.

Meier, A. (1999) Standby power – a quiet use of energy. *CADDET Newsletter*, **4**, November, 9–11.

NSW Government (2001) *New South Wales Sustainable Solutions for the Future*. www.cabinet.nsw.gov.au/pdfs/sustain.pdf

Ontera (undated) Company brochure. Ontera Pty Ltd. www.ontera.com.au

Pilatowicz, G. (1995) *Eco-Interiors – A Guide to Environmentally Conscious Interior Design* (New York: John Wiley).

QUT (1999) *Integrating Ecologically Sustainable Development and Information Technology (ESDIT) for more efficient building design, Stage 1 Final Report*, Queensland University of Technology (CSIRO).

RMI (2001) *Buildings and Land* (Snowmass, CO: Rocky Mountains Institute). www.rmi.org/sitepages/pid13.asp

Schepers, H. (2001) The Impacts of Different Features in Delivering 5 Star Buildings. *Building Greenhouse Rating Scheme Seminar*, Sydney, June (Sydney: Sustainable Energy Development Authority).

SEAV (1999) *Australian Commercial Building Sector – Greenhouse Gas Emissions 1990–2010, Executive Summary Report 1999*, Australian Greenhouse Office (Canberra: DISR).

SEAV (2000) *Model Technical Specifications for Commercial & Public Buildings Version 1*, Sustainable Energy Authority of Victoria (Melbourne: SEAV).

SEAV (2001) Sustainable Energy Authority of Victoria.
www.seav.vic.gov.au/building/ESCB/index.html

SEDA (1999) *David Jones – The Quality Option that Saves you Money* (Sydney: Sustainable Energy Development Authority). www.seda.nsw.gov.au/pdf/david_jones.pdf

SEDA (2001) *Delivering a New 5-Star Office Cost Effectively – Building Greenhouse Rating Scheme Seminar*, Australian National Maritime Museum, Sydney, June 2001, Sustainable Energy Development Authority, NSW.

Townsend A. K. (1997) *The Smart Office – Turning your Company on its Head* (Olney MD: GILA Press).

USGBC (2002) *LEED Green Building Rating System*. US Green Building Council.
www.usgbc.org

Wastebusters Ltd (1997) *The Green Office Manual* (London: Earthscan).

23

Outsourcing

Constantine J. Katsanis*

Editorial comment

There is a worldwide trend towards outsourcing specialist non-core services to consultants and suppliers. Outsourcing is now a hot topic for facility management and is changing the way organizations restructure and procure support services. By concentrating on what is generally described as 'core business', organizations can take advantage of greater expertise through engagement of specialists that have this non-core activity as their main interest. Better service is expected through the introduction of performance-based contracts set against agreed benchmarks, but where this is not realized it provides an opportunity for using a competitor and therefore some incentive exists to deliver a high quality service.

But outsourcing is not always the right decision nor in the interests of an organization in the long term. For example, it is generally acknowledged that outsourcing leads to a transfer of expertise from the organization to others from where it is then purchased back at a higher rate. If the quality of the service is not improved then there is no gain, as managerial supervision must still occur. Response rate and level of care may also be lower due to typical commercial realities. There are some support services that are routinely outsourced, like cleaning, security, landscaping and car fleet management, and other support services that are seldom outsourced, like strategic planning and day-to-day maintenance.

An interesting aspect to outsourcing arises with the formation of alliances and partnerships. These are hybrid outsourcing solutions, where the support service is handled by a consortium of organizations that share in the provision and delivery of the service. Not unlike traditional co-operatives, the costs of supply are reduced while keeping the expertise and control largely in-house.

Outsourcing offers the potential for value enhancement and is worth careful consideration. It enables workplace strategies to concentrate on core business with the

* Ryerson University, Canada